Science and Engineering of One- and Zero-Dimensional Semiconductors

NATO ASI Series

Advanced Science Institutes Series

A series presenting the results of activities sponsored by the NATO Science Committee, which aims at the dissemination of advanced scientific and technological knowledge, with a view to strengthening links between scientific communities.

The series is published by an international board of publishers in conjunction with the NATO Scientific Affairs Division

A	**Life Sciences**	Plenum Publishing Corporation
B	**Physics**	New York and London
C	**Mathematical and Physical Sciences**	Kluwer Academic Publishers Dordrecht, Boston, and London
D	**Behavioral and Social Sciences**	
E	**Applied Sciences**	
F	**Computer and Systems Sciences**	Springer-Verlag
G	**Ecological Sciences**	Berlin, Heidelberg, New York, London,
H	**Cell Biology**	Paris, and Tokyo

Recent Volumes in this Series

Volume 208—Measures of Complexity and Chaos
edited by Neal B. Abraham, Alfonso M. Albano, Anthony Passamante, and Paul E. Rapp

Volume 209—New Aspects of Nuclear Dynamics
edited by J. H. Koch and P. K. A. de Witt Huberts

Volume 210—Crystal Growth in Science and Technology
edited by H. Arend and J. Hulliger

Volume 211—Phase Transitions in Soft Condensed Matter
edited by Tormod Riste and David Sherrington

Volume 212—Atoms in Strong Fields
edited by Cleanthes A. Nicolaides, Charles W. Clark, and Munir H. Nayfeh

Volume 213—Interacting Electrons in Reduced Dimensions
edited by Dionys Baeriswyl and David K. Campbell

Volume 214—Science and Engineering of One- and Zero-Dimensional Semiconductors
edited by Steven P. Beaumont and Clivia M. Sotomayor Torres

Volume 215—Soil Colloids and their Associations in Aggregates
edited by Marcel F. De Boodt, Michael H. B. Hayes, and Adrien Herbillon

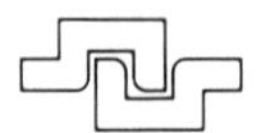

Series B: Physics

Science and Engineering of One- and Zero-Dimensional Semiconductors

Edited by

Steven P. Beaumont and

Clivia M. Sotomayor Torres

University of Glasgow
Glasgow, Scotland, United Kingdom

Plenum Press
New York and London
Published in cooperation with NATO Scientific Affairs Division

Proceedings of a NATO Advanced Research Workshop on
Science and Engineering of One- and Zero-Dimensional Semiconductors,
held March 29-April 1, 1989,
in Cadiz, Spain

Library of Congress Cataloging-in-Publication Data

NATO Advanced Research Workshop on Science and Engineering of One- and
Zero-Dimensional Semiconductors (1989 : Cádiz, Spain)
Science and engineering of one- and zero-dimensional
semiconductors / edited by Steven P. Beaumont and Clivia M.
Sotomayor Torres.
p. cm. -- (NATO ASI series. Series B, Physics ; vol. 214)
"Proceedings of a NATO Advanced Workshop on Science and
Engineering of One- and Zero-Dimensional Semiconductors, held March
29-April 1, 1989 in Cadiz, Spain"--T.p. verso.
"Published in cooperation with NATO Scientific Affairs Division."
Includes bibliographical references.
ISBN 0-306-43417-2
1. Semiconductors--Congresses. 2. Semiconductors--Design and
construction--Congresses. I. Beaumont, Steven P. II. Sotomayor
Torres, C. M. III. North Atlantic Treaty Organization. Scientific
Affairs Division. IV. Title. V. Series: NATO ASI series. Series
B, Physics ; v. 214.
QC610.9.N364 1989
537.6'22--dc20 89-48932
CIP

A Division of Plenum Publishing Corporation
233 Spring Street, New York, N.Y. 10013

Printed in the United States of America

SPECIAL PROGRAM ON CONDENSED SYSTEMS OF LOW DIMENSIONALITY

This book contains the proceedings of a NATO Advanced Research Workshop held within the program of activities of the NATO Special Program on Condensed Systems of Low Dimensionality, running from 1983 to 1988 as part of the activities of the NATO Science Committee.

Other books previously published as a result of the activities of the Special Program are:

Volume 148 INTERCALATION IN LAYERED MATERIALS
edited by M. S. Dresselhaus

Volume 152 OPTICAL PROPERTIES OF NARROW-GAP LOW-DIMENSIONAL STRUCTURES
edited by C. M. Sotomayor Torres, J. C. Portal, J. C. Maan, and R. A. Stradling

Volume 163 THIN FILM GROWTH TECHNIQUES FOR LOW-DIMENSIONAL STRUCTURES
edited by R. F. C. Farrow, S. S. P. Parkin, P. J. Dobson, J. H. Neave, and A. S. Arrott

Volume 168 ORGANIC AND INORGANIC LOW-DIMENSIONAL CRYSTALLINE MATERIALS
edited by Pierre Delhaes and Marc Drillon

Volume 172 CHEMICAL PHYSICS OF INTERCALATION
edited by A. P. Legrand and S. Flandrois

Volume 182 PHYSICS, FABRICATION, AND APPLICATIONS OF MULTILAYERED STRUCTURES
edited by P. Dhez and C. Weisbuch

Volume 183 PROPERTIES OF IMPURITY STATES IN SUPERLATTICE SEMICONDUCTORS
edited by C. Y. Fong, Inder P. Batra, and S. Ciraci

Volume 188 REFLECTION HIGH-ENERGY ELECTRON DIFFRACTION AND REFLECTION ELECTRON IMAGING OF SURFACES
edited by P. K. Larsen and P. J. Dobson

Volume 189 BAND STRUCTURE ENGINEERING IN SEMICONDUCTOR MICROSTRUCTURES
edited by R. A. Abram and M. Jaros

Volume 194 OPTICAL SWITCHING IN LOW-DIMENSIONAL SYSTEMS
edited by H. Haug and L. Banyai

Volume 195 METALLIZATION AND METAL–SEMICONDUCTOR INTERFACES
edited by Inder P. Batra

Volume 198 MECHANISMS OF REACTIONS OF ORGANOMETALLIC COMPOUNDS WITH SURFACES
edited by D. J. Cole-Hamilton and J. O. Williams

Volume 199 SCIENCE AND TECHNOLOGY OF FAST ION CONDUCTORS
edited by Harry L. Tuller and Minko Balkanski

SPECIAL PROGRAM ON CONDENSED SYSTEMS OF LOW DIMENSIONALITY

PREFACE

This volume comprises the proceedings of the NATO Advanced Research Workshop on the Science and Engineering of 1- and 0-dimensional semiconductors held at the University of Cadiz from 29th March to 1st April 1989, under the auspices of the NATO International Scientific Exchange Program.

There is a wealth of scientific activity on the properties of two-dimensional semiconductors arising largely from the ease with which such structures can now be grown by precision epitaxy techniques or created by inversion at the silicon-silicon dioxide interface. Only recently, however, has there burgeoned an interest in the properties of structures in which carriers are further confined with only one or, in the extreme, zero degrees of freedom. This workshop was one of the first meetings to concentrate almost exclusively on this subject: that the attendance of some forty researchers only *represented* the community of researchers in the field testifies to its rapid expansion, which has arisen from the increasing availability of technologies for fabricating structures with small enough (sub - $0.1\mu m$) dimensions.

Part I of this volume is a short section on important topics in nanofabrication. It should not be assumed from the brevity of this section that there is little new to be said on this issue: rather that to have done justice to it would have diverted attention from the main purpose of the meeting which was to highlight experimental and theoretical research on the structures themselves. In Part II will be found papers reporting results on transport of electrons through laterally-confined devices, either one-dimensional waveguides or zero-dimensional quantum dots. The waveguide model of 1-D transport and the role of quantum interference in determining the terminal properties of these devices emerge clearly from the contributions. Moreover, recent research investigating the properties of quasi-zero dimensional structures shows how close we have come to fabricating artificial molecules.

Laterally periodic systems are discussed in Part III. This work shows that artificial lattices can be fabricated by lithographic means rather than by growth, and that the electronic levels this created can be controlled by external voltages, offering the possibility of a tunable bandstructure. Part IV is concerned with the optical properties of 1- and 0-D structures, an important area of investigation from which might arise optical sources modulators and detectors of enhanced performance. Part V is a short section reporting new results in two-dimensional systems.

The organising committee extends its sincerest thanks to all those representatives of the University and City of Cadiz who contributed to the smooth running of the workshop. Professor Rafael Garcia Roja, the local organiser, deserves our especial gratitude for his unstinting efforts before, during and after the meeting. We are also greatly appreciative of financial support from the Science and Engineering Research Council (UK), US Army Research Office (London), US Air Force European Office of Aerospace Research, US Office of Naval Research, the University and City of Cadiz.

S P Beaumont
C M Sotomayor Torres

Summer 1989

CONTENTS

PART 1 : FABRICATION OF 1- AND 0-D STRUCTURES

PART 2 : TRANSPORT IN 1- AND 0-d STRUCTURES

PART 3 : PERIODIC 1- AND 0-D STRUCTURES

PART 4 : OPTICAL PROPERTIES OF 1- AND 1-D STRUCTURES

PART 5 : 2-DIMENSIONAL SYSTEMS

FABRICATION OF BURIED GaInAs/InP QUANTUM WIRES BY ONE-STEP MOVPE GROWTH

Y.D. Galeuchet, P. Roentgen, S. Nilsson and V. Graf

IBM Research Division, Zurich Research Laboratory
CH-8803 Rüschlikon, Switzerland

I. INTRODUCTION

One- or zero-dimensional structures such as quantum well wires (QWWs) or quantum dots (QDs) have received considerable attention in recent years because of their physical properties and their potential device application. Nanostructures with two or three dimensions smaller than the de Broglie wavelength of the electrons and holes have shown interesting phenomena such as for example quantum interference effects in small rings, quenching of the Hall effect in quantum wires or quantization of the ballistic resistance in narrow conducting channels.[1] New optical properties such as increased exciton binding energy [2] or strong nonlinear effects [3] are also expected in such structures. Some theoretical calculations have shown that semiconductor lasers with QWWs or QDs as active region should result in ultimate device performance.[4]

In this paper, we will present a novel approach for the fabrication of buried low-dimensional GaInAs/InP structures using a one-step metalorganic vapor phase epitaxy (MOVPE) technique. By using patterned or SiO_2 masked InP substrates and plane selective growth, buried GaInAs layers with simultaneously vertical and lateral quantum size dimensions are obtained without any need for submicron lithography and processing techniques. The main advantage of our one-step growth technique over conventionally fabricated QWWs is that buried QWWs can be obtained without interface contamination or surface defects due to etching, air exposure or regrowth step. Spatially resolved low-temperature cathodoluminescence (CL) of selectively grown structures shows strong luminescence and almost lattice matching of the as-grown buried GaInAs wires.

II. CONCEPTS FOR ONE-STEP GROWN BURIED QWWs

To date, the most common method to make QWWs or QDs is to grow GaAs/AlGaAs quantum wells and to define lateral nanostructures by high-resolution lithography and dry etching. This process has shown its ability to

Science and Engineering of One- and Zero-Dimensional Semiconductors
Edited by S.P. Beaumont and C.M. Sotomajor Torres
Plenum Press, New York, 1990

produce good structures for transport measurements [5] but not for the study of optical properties because of the inevitable surface damages caused by processing.[6] Because of a very large surface-to-volume ratio, QWWs and QDs are very sensitive to surface defects. Since the GaInAs/InP system has a surface recombination velocity ~100 times smaller than the GaAs/AlGaAs system,[7] this should be a more forgiving material system for studying the optical properties of III-V semiconductor nanostructures. Nevertheless, even this system shows surface recombination effects when the feature sizes are reduced below 100 nm by standard techniques.[6] Buried structures which are inherently free of exposed interfaces should then eliminate the surface problems in nanostructures.

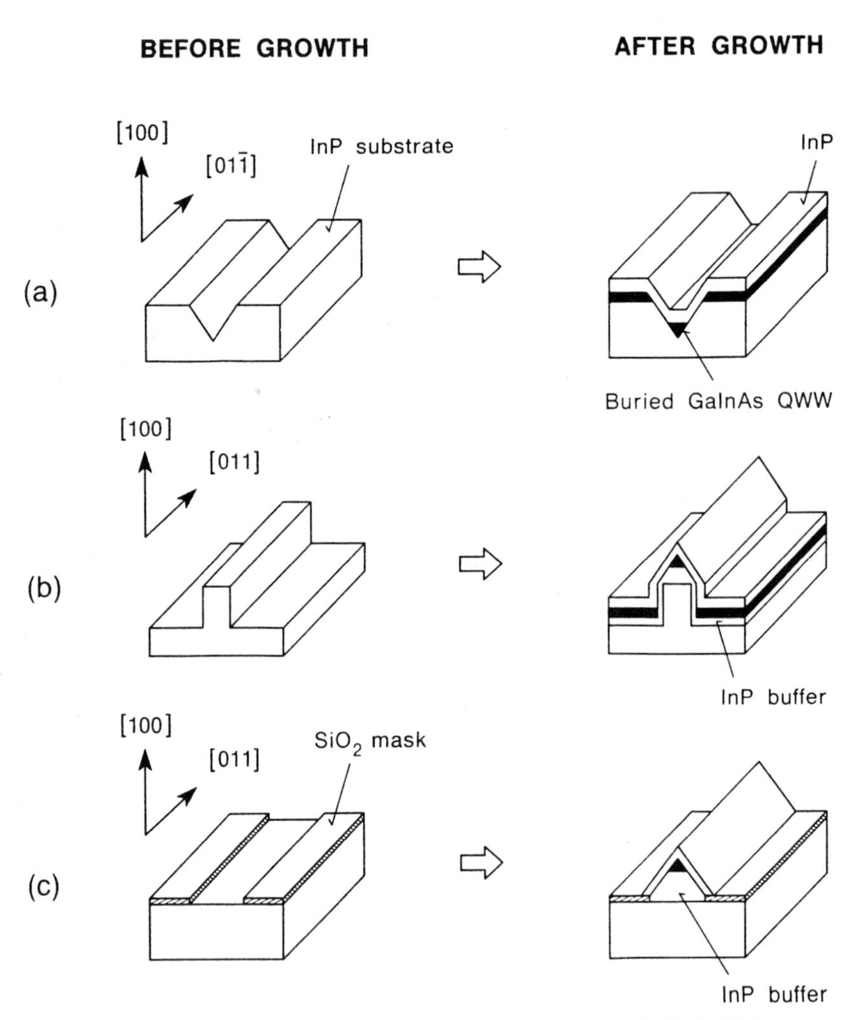

Fig. 1. Examples for the fabrication of buried GaInAs/InP QWWs by one-step MOVPE. Buried GaInAs QWW (black) (a) grown on [01$\bar{1}$] oriented V-groove etched in (100) InP substrate, (b) grown on [011] oriented rectangular mesa and (c) selectively grown within a stripe open in a dielectric mask deposited on the InP substrate.

Impurity-induced disordering offers one route to define buried QWWs or QDs.[8] However, structures with lateral dimensions smaller than 500 Å and large spatial filling factors are difficult to achieve with this method.[9] Recently, a new approach has been proposed to fabricate as-grown buried nanostructures by controlling nucleation and growth kinetics during molecular beam epitaxy on vicinal substrates.[10] The vicinal growth epitaxy, however, has only been shown for the GaAs/AlGaAs system. Therefore, our technique of combining the GaInAs/InP material system and the creation of in-situ buried quantum structures by MOVPE growth on structured substrates promises the fabrication of ideal low-dimensional structures for optical applications.

It has been demonstrated that buried GaInAs/InP ridge lasers or waveguides can be grown on nonplanar substrates using a one-step MOVPE process.[11] The fabrication of these devices is based on the effect that GaInAs can be selectively deposited onto (100) surfaces of nonplanar substrates, resulting in plane selective growth. In a previous paper,[12] we have shown that the {111} InP crystal facets are no-growth planes for GaInAs whereas InP grows on both (100) and {111}. This effect can be used to fabricate buried QWWs if the substrates have appropriate geometries.

In Fig. 1, we propose some design examples of GaInAs QWWs buried in InP during a single growth step. The left-hand side of the figure presents the geometry of the substrate before growth. The right-hand side shows the result of an alternative deposition of GaInAs/InP layers on these substrates. Fig. 1a shows the plane selective growth of a GaInAs QWW in a $[01\bar{1}]$ oriented V-groove limited by {111}A side facets. In Fig. 1b, a buried QWW is obtained by growing GaInAs/InP layers on a rectangular mesa oriented along [011]. The same method can be applied with selective area epitaxy as shown in Fig. 1c, when GaInAs/InP layers are grown through a patterned SiO_2 mask deposited on the InP substrate.

III. GROWTH STUDIES ON NONPLANAR SUBSTRATES

We have studied the MOVPE growth behavior of GaInAs/InP layers on nonplanar substrates. For all experiments, (100) InP substrates, misaligned 2° off towards nearest [011], were patterned with grooves or mesas by wet chemical etching through a SiO_2 etch mask.[12] After the removal of this mask in buffered HF, the samples were slightly etched in Br_2:CH_3OH and immediately loaded in the growth chamber. The growth evolution on the etched structures was evaluated by growing GaInAs/InP layer sequences by MOVPE at 20 mbar or 100 mbar with trimethylgallium, trimethylindium, AsH_3 and PH_3 gas sources. For most layers, a V/III ratio of 170 and a growth temperature of 630°C were chosen.

The buried QWW design in Fig. 1a has been fabricated and has resulted in a 80×100 nm buried GaInAs QWW.[12] We have shown and discussed in Ref. 12 that under the above growth conditions, GaInAs does not grow on the {111}A InP facets, in contrast to InP. However, although almost perfect V-grooves limited by these facets can be obtained, we found that this design is not well suited because irregular nucleation occurs along the groove.[12]

We have also realized the design proposed in Fig. 1b. For that purpose, a rectangular mesa limited by {110} and (100) facets was wet etched along [011] in the InP substrate and a layer sequence of nominally 3000 Å GaInAs buffer, 2000 Å InP and 20 periods of 140 Å GaInAs quantum wells and 240 Å InP barriers was grown. Fig. 2 shows scanning electron microscope (SEM) micrographs of (011) stained etched cross sections of the grown structure. In Fig. 2a, it is clearly observed that GaInAs (dark) has not grown on the vertical {110} side facets of the mesa nor on the {111}B facets which have developed on top of it during the growth. The GaInAs layers deposited on the mesa top are one-step buried quantum wells. It is noteworthy that the lateral size of the buried layers decreases when the growth proceeds. This unique feature allows the fabrication of a buried GaInAs QWW on top of the {111}B facetted structure without limiting lithography constraints, as shown in Fig. 2b. In this magnified view of a similar structure as in Fig. 2a, the top GaInAs layer has dimensions of 30×30 nm. In contrast to the design proposed in Fig. 1a, the {111}B facets which limit the structure are very smooth and regular along the mesa.[12] An interesting feature is the fact that although 20 GaInAs quantum wells have nominally been grown, only 9 are seen in Fig. 2b. The reason is that when the two {111}B facets bind together, no (100) facet is available any longer for the growth of GaInAs and no material is deposited. On the other hand, InP grows slowly on these {111}B facets and the InP barriers, which remain after the binding of the two facets, merge and form a thick burying InP layer.

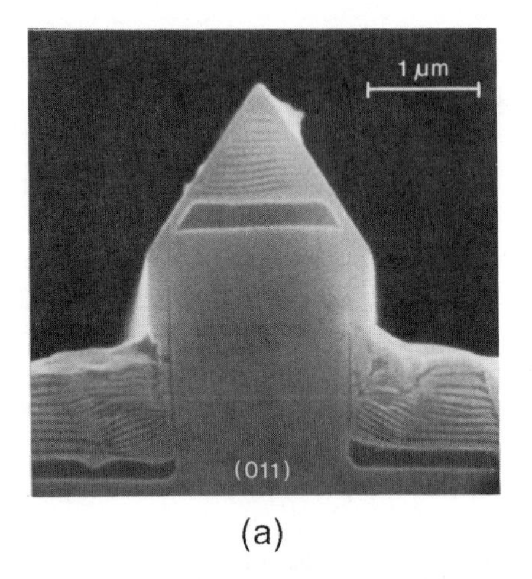

(a)

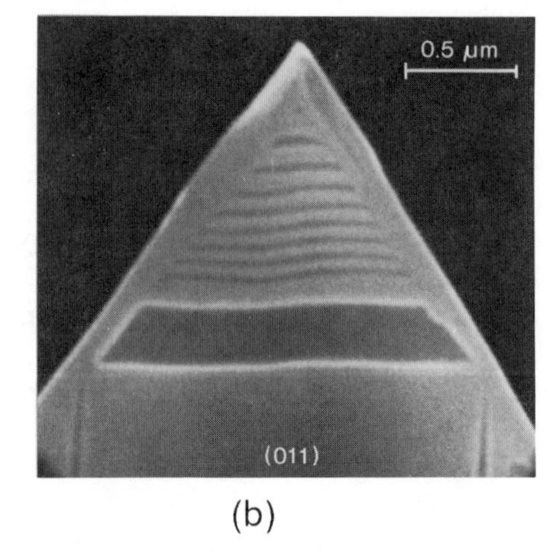

(b)

Fig. 2. (a) SEM picture of the (011) cross section of GaInAs (dark)/ InP (light) epitaxial layers grown on a rectangular InP mesa. Buried GaInAs quantum wells with submicron lateral dimensions, grown in one MOVPE step, are visible on top of the mesa. (b) Magnified view of a similar structure. The top GaInAs layer is a 30×30 nm buried QWW.

The growth on nonplanar substrates has shown that GaInAs QWWs can be as-grown buried in InP within a single growth step. However, GaInAs layers are simultaneously deposited on the (100) surfaces between the mesas or the grooves (see the black layers in Fig. 1a and 1b). These layers, much larger than the QWWs, prevent high-performance device application and disturb the characterization because of their intense contribution. As shown in Fig. 1c, this could be avoided by using selective area growth on SiO_2 masked substrates.

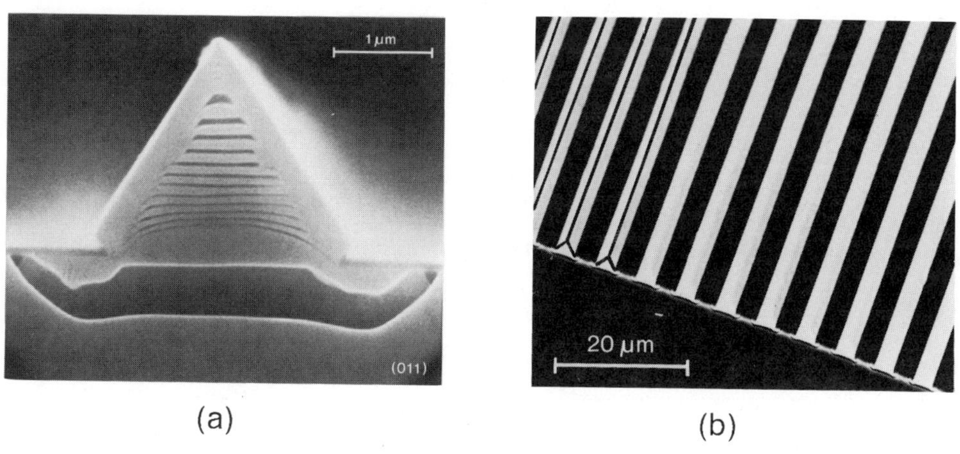

(a) (b)

Fig. 3. (a) SEM picture of the (011) cross section of buried GaInAs (dark)/ InP (light) layers grown by selective area MOVPE in a single growth step. The thick top GaInAs layer has average dimensions of 100×200 nm. (b) Tilted view of selectively grown structures shown in (a). The selectivity of the growth on the SiO_2 mask (black) is very good and smooth {111}B facets have developed along [011] over more than 75 μm.

IV. SELECTIVE AREA GROWTH STUDIES

For selective epitaxy experiments, growth was performed on (100) InP:S substrates coated with 100 nm of SiO_2. Stripes 2.5 μm wide and 4 μm apart were defined in the mask along [011] by conventional lithographic techniques and a nominal layer sequence of 3000 Å GaInAs buffer, 2000 Å InP and 20 periods of

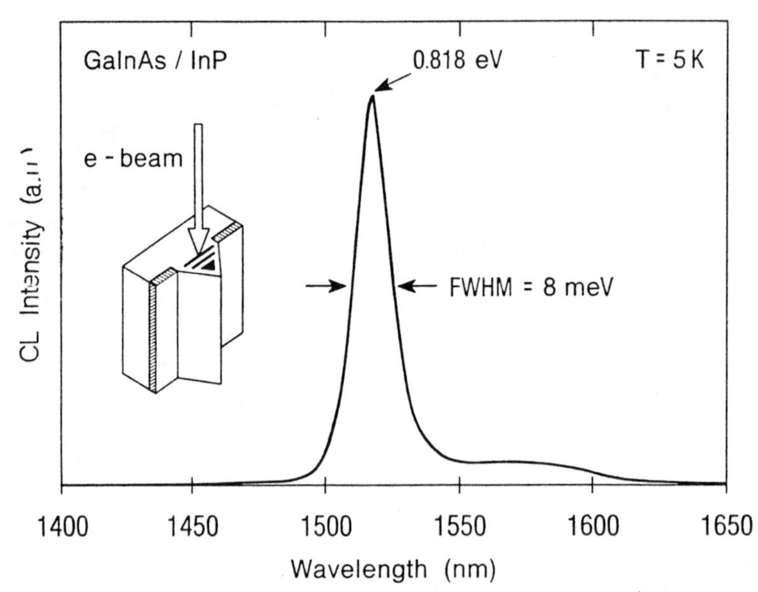

Fig. 4. CL spectrum (T=5K) of a single selectively grown GaInAs/InP structure shown in Fig. 3. The luminescence was excited with 7keV electrons perpendicular to the cleavage plane.

100 Å GaInAs quantum well and 200 Å InP barriers was grown. Fig. 3a presents an SEM picture of the (011) stained cross section of the sample after growth. As in growth experiments on nonplanar substrates, {111}B facets have developed and plane selective growth of GaInAs (dark) is obtained. Fig. 3b presents a tilted view of structures shown in Fig. 3a. We observe a very good selectivity of the growth, i.e. there is no material deposition on the SiO_2 mask (black), and very regular and smooth {111}B facets along [011] over more than 75 μm.

Narrowing of the lateral dimension of the epilayers as growth proceeds allows submicron buried GaInAs wires to be easily fabricated. The average dimensions of the thick top layer in Fig. 3a is 100×200 nm. Transmission electron microscopy (TEM) measurements on a planar unmasked substrate grown simultaneously shows that the nominal layer thickness is 100 Å for GaInAs and 200 Å for InP. However, a large increase of the layer thicknesses is observed when the width of the (100) top surface decreases (see Fig. 3a). This effect is due to a strong contribution of molecules diffusing onto the (100) growth plane from the SiO_2 and the no-growth neighboring facets. More details of this phenomenon will be presented elsewhere.[13]

V. CATHODOLUMINESCENCE MEASUREMENTS

The luminescence properties of the selectively grown buried layers (see Fig. 3) were investigated by CL in a SEM modified with a CL collector and a continuous flow helium cold stage. The CL was dispersed in a 0.22 m double spectrometer and detected by a Ge photodetector cooled to 77 K. In order to increase the spatial resolution, the SEM was operated at 7 keV, reducing the generation volume of electron-hole pairs to a Bethe range of $\sim$ 0.5 μm. Fig. 4 shows a typical 5 K CL spectrum of a single selectively grown structure. The spectrum was measured with the beam perpendicular to the cleavage plane, as shown in the insert of Fig. 4. The near band edge emission which peaks at 1516 nm indicates an acceptable lattice matching ($\Delta a/a \sim$ -1500ppm) of the buried GaInAs layers. By comparison, a lattice matched unmasked planar sample grown simultaneously gives a near band edge emission at 1542 nm. The 28 nm difference comes from small stochiometry variation ($\sim$ 2% Ga rich) in the buried layers and can be readjusted by fine tuning the growth parameters. The strong luminescence and the FWHM of 8 meV indicates high-quality material. The low energy tail of the spectrum can be attributed to defects arising from the heavily doped substrate into the GaInAs buffer layer.

VI. CONCLUSION

We have presented some new concepts for the fabrication of as-grown buried GaInAs/InP QWW and shown the growth conditions where lower bandgap GaInAs layers are fully embedded in InP layers within one MOVPE growth step. By combining the use of patterned or SiO_2 masked InP substrates and plane selective growth, we have realized buried layers with simultaneously vertical and lateral quantum size dimensions, without the need for submicron lithography and processing techniques. In comparison to nonplanar growth designs, the fabrication of buried QWW by selective area MOVPE appears to be more promising since it combines interesting features such as non-parasitic (100) surfaces and smooth, well-controlled, closely spaced and high-quality buried layers, as confirmed by CL measurements. This new technique can be further extended to the fabrication of as-grown buried QWW arrays where strong two-dimensional optical confinements effects should be observable.

The authors would like to acknowledge the technical assistance of D.J. Webb and W. Heuberger during sample preparation, R. Spycher of the EPF Lausanne for TEM measurements and Prof. F.K. Reinhart for his advice.

REFERENCES

1. See e.g. Springer Serie in Solid State Physics Vol.83: " Physics and Technology of Submicron Structures", H. Heinrich, G. Bauer and F. Kuchar, eds., Berlin Heidelberg, 1988, and references therein.
2. I. Suemune and L.A. Coldren, Band-mixing effects and excitonic optical properties in GaAs quantum wire structures - Comparison with the quantum wells, IEEE J. Quantum Electr. QE-24: 1778 (1988)
3. L. Banyai, I. Galbraith and H. Haug, Biexcitonic nonlinearity in GaAs/AlGaAs quantum wells and wires, J. Phys. 49: C2-233 (1988)
4. See e.g. M. Asada, Y. Miyamoto and Y. Suematsu, Gain and the threshold of the three-dimensional quantum-box lasers, IEEE J. Quantum Electr. QE-22: 1915 (1986)
5. S.P. Beaumont, C.D.W. Wilkinson, S. Thoms, R. Cheung, I. McIntyre, R.P. Taylor, M.L. Leadbeater, P.C. Main and L. Eaves, Electron beam lithography and dry etching techniques for the fabrication of quantum wires in GaAs and AlGaAs epilayers systems, 1988, in Ref. 1
6. A. Forchel, B.E. Maile, H. Leier and R. Germann, Fabrication and optical characterization of semiconductor quantum wires, 1988, in Ref. 1
7. K. Tai, R. Hayes, S.L. McCall and W.T. Tsang, Optical measurement of the surface recombination in InGaAs quantum well mesa structures, Appl. Phys. Lett. 53: 302 (1988)
8. J. Cibert, P.M. Petroff, G. J. Dolan, S. J. Pearton, A.C. Gossard and J.H. English, Optically detected carrier confinement to one and zero dimension in GaAs quantum well wires and boxes, Appl. Phys. Lett. 49: 1275 (1986)
9. Y. Hirayama, S. Tarucha, Y. Suzuki and H. Okamoto, Fabrication of a GaAs quantum-well-wire structure by Ga focused-ion-beam implantation and its optical properties, Phys. Rev.B 37: 2774 (1988)
10. J.M. Gaines, P.M. Petroff, H. Kroemer, R.J. Simes, R.S. Geels and J H. English, Molecular-beam epitaxy growth of tilted GaAs/AlAs superlattices by deposition of fractional monolayers on vicinal (001) substrates, J. Vac. Sci. Technol. B6: 1378 (1988)

11. M.D. Scott, J.R. Riffat, I. Griffith, J.I. Davies and A.C. Marshall , CODE: A novel MOVPE technique for the single stage growth of buried ridge double heterostructure lasers and waveguides, J. Cryst. Growth 83: 820 (1988)
12. Y.D. Galeuchet, P. Roentgen and V. Graf, Buried GaInAs/InP layers g rown on nonplanar substrates by one-step low-pressure metalorganic vapor phase epitaxy, Appl. Phys. Lett. 53: 2638 (1988)
13. Y.D. Galeuchet, P. Roentgen, S. Nilsson and V. Graf, One-step grown buried GaInAs/InP layers by selective area MOVPE, to be submitted to Appl. Phys. Lett.

DRY-ETCHING DAMAGE IN NANO-STRUCTURES

C.D.W.Wilkinson and S.P.Beaumont

Department of Electronics and Electrical Engineering
The University
Glasgow G12 8QQ,Scotland,U.K.

In the past few years a great deal of new physics has emerged from the examination of structures made in semi-conducting materials with dimensions on a nano-metric scale -some examples are the observation of Aranhov-Bohm oscillations in a ring, the quantised resistance observed with point contacts and in wires with restrictions and the continuing saga of the photoluminescence from quantum dots and the allied hope of more efficient laser performance.

The most used approach to pattern formation in a resist at nanometric sizes is electron beam lithography; and lines with widths down to 10nm on thin substrates and 15nm on solid substrates have now been made by a number of groups. The translation of this pattern in an ion-resistant material (resists, dielectrics and metals have been used in this role) into a relief pattern in the semiconductor is done by etching. Most devices of interest require structures which have high aspect ratios,and as wet chemical etching is in general isotropic in its action, it is unsuitable for the production of such structures.

However dry etching using directed ions does lead to anisotropic profiles and has been widely applied in the Silicon semiconductor industry to make structures with vertical walls. While dramatic cliff-like structures (see Figure 1) can be made using reactive ion etching of III-V semiconductors an important question is to evaluate the change in the underlying semi conducting material caused by the influx of etching ions.
This paper is concerned with the damage which is inflicted on a III-V semiconductor when it is bombarded with energetic ions and in particular with the damage caused by reactive ion etching of GaAs in $SiCl_4$ and in mixtures of methane and hydrogen. The term 'damage' needs to be defined: for the present purposes, damage is taken as 'any change in the physical characteristics of the material particularly as manifest in its electrical or optical behaviour which results from the etching'. It is important to note that, from this definition, the effect of dry etching could (and in fact, at times, does) - manifest itself in only optical and not electrical effects or vice versa.

Science and Engineering of One- and Zero-Dimensional Semiconductors
Edited by S.P. Beaumont and C.M. Sotomajor Torres
Plenum Press, New York, 1990

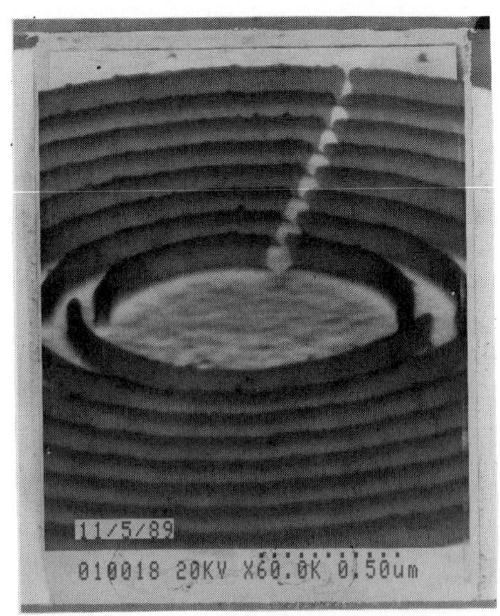

Figure 1. 40 nm wide curved structures etched into Gallium Arsenide using Silicon Tetrachloride

Often in dry etching the aim is to produce a ridge or groove and the effect of the bombardment of ions may well be different in the side walls than on the flat surface. Given an adequate removal rate of material, one might expect that the surface damage is independent of the etch depth while the side wall damage would increase with the ion exposure time.

Reactive ion etching

A schematic diagram of an reactive ion etching machine is shown in figure 2. Gas at a pressure of 10 to 50mT is admitted to a vacuum chamber containing two electrodes excited by rf (typically 13.6 MHz) power. The earthed electrode is larger than the powered electrode on which sits the sample to be etched. Partial rectification of the rf power takes place (the ions move more slower than the electrons) and a d.c. voltage builds up across the electrodes. The behavior of the system depends on the purity of the gases (trace gases with lower ionisation potential can give rise to an disproportionate number of ions), the nature of the walls and the degree of confinement of the plasma away from the walls, and the preconditioning of the chamber (for example the use of oxygen to burn off any polymeric impurities). While the state of the system can usefully be analysed by residual gas analysis, Langmuir ion probes and optical spectroscopic means, a certain element of art also enters into the process.

Evaluation of Schottky junctions

When a rectifying contact is made between a semiconducting surface and a metal, the resulting forward current I at a bias voltage V is given by

$$I = I_0 \{ e^{\frac{eV}{nkT}} - 1 \}$$

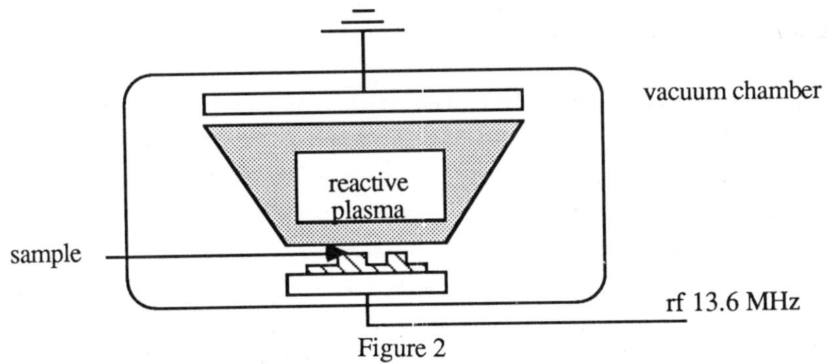

Figure 2

Methods for the evaluation of dry etching damage

where for a perfect diode the ideality factor n is 1. For Titanium-Gold contact on lightly doped GaAs ideality factors of the order of 1.04 are obtained. If such a Schottky metal semiconductor contact is made on a surface following dry etching of the surface, it is often found that the ideality factor has increased. However more information can be extracted from I-V measurements of Schottky junctions on dry etched semiconducting surfaces. In a study of the use of methane-hydrogen for the reactive ion etching of GaAs,[1] monitoring of the forward current at a fixed bias level, the reverse leakage current and the barrier height revealed more about the effects of the etching and annealling procedures.

Photoluminescence

The intensity of the peak near infra-red photoluminescence light emitted by GaAs and InP has been found to decrease after reactive ion etching in Silicon tetrachloride; and the photoluminescent intensity is found to decrease monotonically with an increase in ideality factor of diodes made on the surface damaged surface.[2] Monitoring of the photoluminescent intensity is a particularly convenient technique with semiconductors such as n- doped InP which do not form good Schottky contacts.

Raman Spectroscopy

Raman scattering from semiconductors is sensitive to the condition of the surface layers. In particularly, for light scattered in the near backward direction from a (001) surface of GaAs, for a perfect sample while scattering from the longitudinal optical (LO) phonons is expected, scattering from the transverse optical (TO) phonons is forbidden on symmetry grounds. However when a sample which has been reactive ion etched in Silicon Tetra-chloride is examined[3] a line in the spectrum corresponding to the forbidden TO phonons is observed. Moreover the intensity of this line increases with the power used in the etching process. This technique of observing dry etch damage has the advantage of being passive and in principle by making observations as a function of wavelength it should be possible to deduce the total depth of material affected by the damage.

Observation of the cut-off width of narrow epitaxial wires

A schematic of a wire made in heavily doped epitaxial GaAs on a semi-insulating substrate is shown below in Figure 3(a). The surface states formed on the free surface of GaAs pin the Fermi level to the middle of the energy gap(see Figure 3(b)); thus a depleted layer is formed around the surface of the wire. If the conductance of a series of wires of different widths is plotted ; the width at zero conductance (at cut-off) would be expected to be twice the depletion width[4]. While this is found to be true for wet etched wires, for wires made by reactive ion etching, the cut-off width is considerably wider than twice the depletion width. The excess can be regarded as a measure of (twice) the extent of sidewall damage.

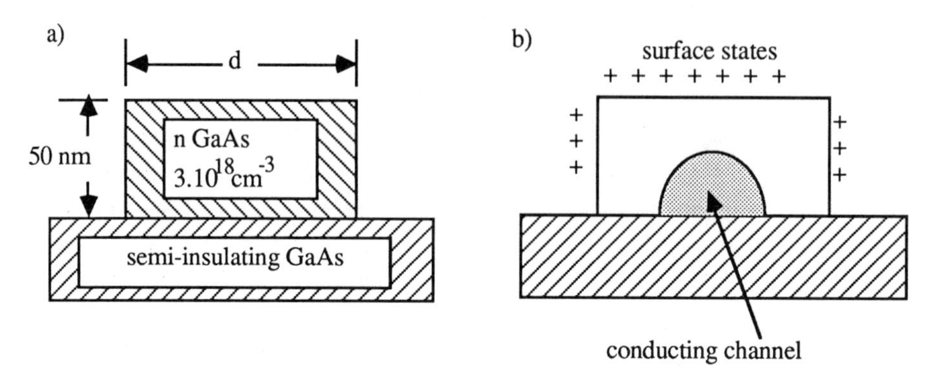

Figure 3

This technique has the advantage of revealing the sidewall damage rather than the surface damage and it allows a direct measurement of the width of damage. A very recent application of the method was to determining the sidewall damage caused by reactive ion etching in a plasma formed by a microwave source with an external magnet giving the conditions for electron cyclotron resonance[5] . It was found that using a mixture of Freon 12 and Helium, very low damage could be obtained by etching for a very short time. For longer etching some increase in the cut-off width was found - which shows in this case that the sidewall damage does depend on the total doseage.

TEM observation of damage in narrow wires

All the techniques described so far do not give direct visualisation of the damage. However a tall ridge etched into Gallium Arsenide can be used as a specimen in a transmission electron microscope and the images formed by the beams diffracted by the crystalline lattice reveal the position and nature of the damage[6].

It is convenient to make a ridge by dry etching, at least 0.5μm high, and as narrow as possible to avoid multiple scattering of electrons, and then to leave this ridge on a wet etched mesa some few microns high. the specimen is then placed in a transmission electron microscope,with the direction of the examining beam as shown in Figure 4.

As the specimen is crystalline, a number of diffracted beams are formed. The (000) beam is the bright field image, of greater interest is the (200) beam. For a perfect crystal this should present an dark image, as the intensity is directly related to (n_{Ga} - n_{As}) and the number density of Gallium atoms n_{Ga} and arsenic atoms n_{As} are equal in a perfect crystal. Thus any variation in contrast in the (200) image of a specimen with parallel walls will be due to a local variation of the stoichiometry of the GaAs. With a sample of GaAs etched in silicon tetrachloride, a bright edge to the wall appeared decaying away from the edge and being observable over a distance of 150nm. This implies that there is a change in the stoichiometry (probably a loss of Arsenic) caused by the dry etching.

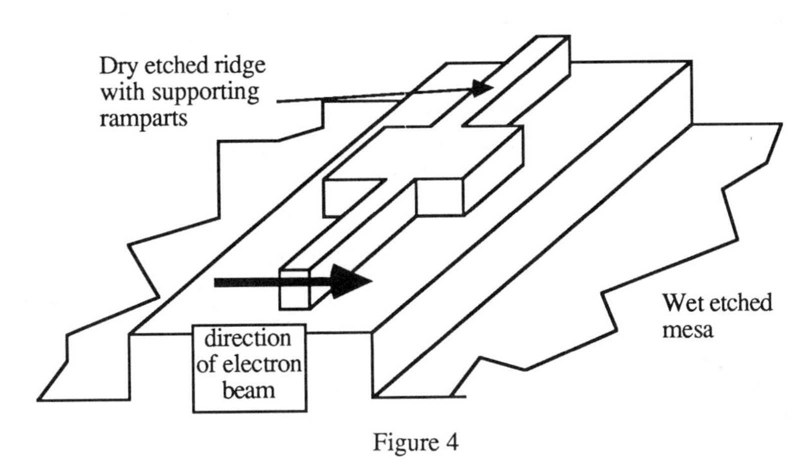

Figure 4

Conclusions

A variety of methods for unravelling the nature of the damage caused by dry etching have been discussed. The surface science techniques are available to ascertain changes in surface composition, such as XPS and Auger spectroscopy also have a strong role to play. In seems likely that in the near future a microscopic model of dry etch damage in III-V semiconductors may be possible.

Acknowledgements

Many people have contributed to the work described in this paper. On the dry etching work Rebecca Cheung, Steven Thoms, Gordon Doughty, Nigel Cameron,Hazel Arnott, Ray Darkin and Jimmy Young, for MBE growth and characterisation, Colin Stanley, Nigel Johnson,Alistair Kean,Martin Holland and John Cochrane, concerned with high resolution lithography and quantum wires Steven Thoms, Mathusa Rahman,Ian McIntyre,Jennifer Thompson, Douglas McIntyre, Dave Gourlay, and Helen Wallace, with the Raman spectroscopy Clivia Sotomayor Torres and Morag Watts , and with the TEM examination of the samples John Chapman of the Department of Physics and Astromony and A. Birnie. The work has been funded in part by the Science and Engineering Research Council.

References

1) R.Cheung, S.Thoms, I.McIntyre, C.D.W.Wilkinson and S.P.Beaumont,'Passivation of

donors in electron beam defined nanostructures after methane/hydrogen reactive ion etching', *J.Vac.Sci.Technol.***6**,1911,1988

2)G.F.Doughty, S.Thoms, R.Cheung and C.D.W.Wilkinson,'Dry etching damage to Gallium Arsenide and Indium Phosphide', *IPAT 87 Proceedings*,284,1987

3) M.Watt,C.M.Sotomayor-Torres,R.Cheung,C.D.W.Wilkinson, H.E.G.Arnot and S.P.Beaumont 'Raman scattering of reactive ion etched GaAs',*J.Mod.Optics*,**35**,365,1988

4) S.Thoms,S.P.Beaumont,C.D.W.Wilkinson,J.Frost and C.R.Stanley,' Ultrasmall Device Fabrication using dry etching of GaAs' , *Microelectronic Engineering*,**5**,249.1986

5) R.Cheung,Y.H.Lee,C.M.Knoedler,K.Y.Lee,T.P.Smith III and D.P.Kern, 'Sidewall damage in n^+ GaAs quantum wires from reactive ion etching', *Appl. Phys. Lettrs.*,***54***,2130,1989

6) R.Cheung,A.Birnie,J.N.Chapman and C.D.W.Wilkinson, 'Evaluation of dry etch damage by direct TEM observation' to be published in *Proc.Microcircuit Engineering 1989 (Cambridge).*

FABRICATION AND QUANTUM PROPERTIES OF

1D AND 0D NANOSTRUCTURES IN III-V SEMICONDUCTORS

H.Launois, D.Mailly, Y.Jin, F.Pardo
Laboratoire de Microstructures et de Microélectronique- CNRS
196 Av.Henri Ravera -92220 Bagneux- France

A.Izrael, J.Y.Marzin and B.Sermage
Laboratoire de Bagneux- Centre Paris B - CNET
196 Av.Henri Ravera -92220 Bagneux- France

INTRODUCTION

Nanofabrication at the "Groupement Scientifique CNET-CNRS" in Bagneux has been performed using either a JEOL 5D2U e-beam pattern generator, or a modified Philips 515 scanning electron microscope. In this paper we present some of the results we obtained on the transport and optical properties of 1D an 0D structures in III-V semiconductors.

-I- VARIANCE OF UNIVERSAL CONDUCTANCE FLUCTUATIONS

In mesoscopic systems quantum interference during the elastic travelling of the electron yields unusual behavior of the transport properties. For instance, a sufficient modification of the interference pattern give rise to a change of the conductance of the order of $\delta g= e^2/h$ independent of the material, of the precise shape of the mesoscopic sample , and of the average value of the conductance. This effect is known as the Universal Conductance Fluctuations [1]. First proposed by Imry the random matrix theory[2] has given a very fundamental explanation of this universal variance. Following Landauer the conductance of a system can be expressed through the transfer matrix. One can then define a random matrix which represents a given impurities configuration, the conductance is then a linear statistics of the eigenvalues of the matrix[3]. Dyson and Wigner in nuclear physics have already developped a Random Matrix Theory (RMT) which explains the universal value of the variance of the energy level excitation of the complex nucleus. The variance of the conductance depends only on the three possible universality classes[4] characterized by three values of a parameter β. For systems without spin-orbit coupling $\beta=1$ if time reversal symmetry is present and $\beta=2$ without time reversal symmetry .For systems with spin-orbit coupling $\beta=4$. One can then show that the fluctuations have a Gaussian distribution:

$$P(\delta g)=\exp-\beta(\delta g/(2e^2/h))^2$$

An immediate consequence of this theory is the reduction by a factor of two of the fluctuations by applying a magnetic field, which corresponds in RMT

Science and Engineering of One- and Zero-Dimensional Semiconductors
Edited by S.P. Beaumont and C.M. Sotomajor Torres
Plenum Press, New York, 1990

language to a transition between two ensembles of statistics. Experimentally the observation of such transition requires a method for changing the disorder configuration under a fixed applied magnetic field. This can be done by moving a scatterer over a distance larger than the Fermi wavelength [5].

We have used a GaAlAs/GaAs heterojunction with the following geometry: 50Å undoped GaAs caplayer / 400Å of Si doped GaAlAs / 50Å of undoped GaAlAs. At T=4K the two dimensional electron gas has a mobility μ=130 000 cm^2/Vs and an electron density n=8.2 $10^{11} cm^{-2}$. Using Argon etching we have defined a 1μm wide and 37μm long wire. The etching was performed down to the GaAs buffer layer. The width of the 2DEG is then smaller than that of the etched structure due to depletion effects (W=0.7μm).

At T=50 mK we have applied a voltage pulse into the sample: 0.4V amplitude and 15ms duration. After such a voltage pulse has been applied (see figure 1) the resistance of the sample increases abruptly to about 50Ω and then decreases back to around its initial value within 15 minutes. This Fermi relaxation is then followed by resistance jumps. We claim these jumps are due to the modification of the impurity potential which changes the interference pattern of the electrons. Indeed after etching the surface of the heterojunction is left with defects, which can trap electrons. By applying a voltage pulse, we can populate or depopulate these traps. Since the impurity potential is locally modified when a trap releases or captures an electron, we have a system where the impurity configuration is changed with time. The frequency of the jumps decreases with time, but their typical amplitude does not. Accordingly, the resistance of the sample when cooled down to 50 mK without voltage pulses is very stable and does not show any jump.

If one records the conductance while sweeping the magnetic field, one gets a so called "magnetofingerprint" (MFP). The fluctuation of the conductance is due to the phase shift of the electron wavefunction as a function of the magnetic field. The exact shape of this fluctuation depends on the exact position of the impurities in the sample (thus the origin of the name). The magnetofingerprint recorded before and after a voltage pulse are shown on figure 2. Some of the features are identical, showing that the two MFP are not completely decorrelated.

We have recorded the histogram of the conductance jumps . The obtained points have been fitted using a Gaussian law . We have first compared the conductance distribution of the jumps at H=0.2T and the conductance distribution of the MFP. The two Gaussian curves are comparable showing that

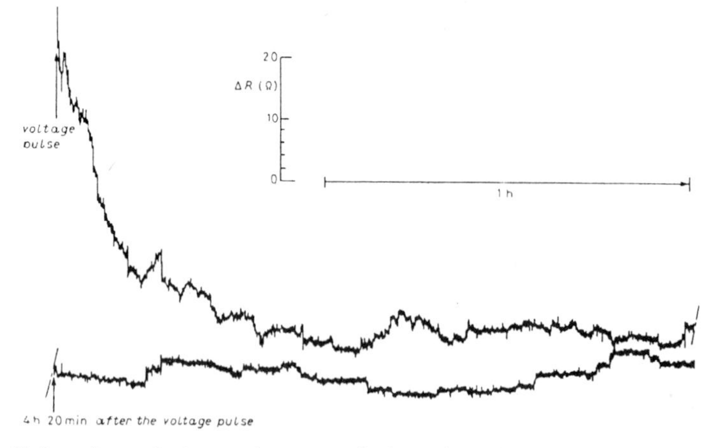

Fig.1 Relaxation of the resistance of the wire after a voltage pulse and appearance of the jumps

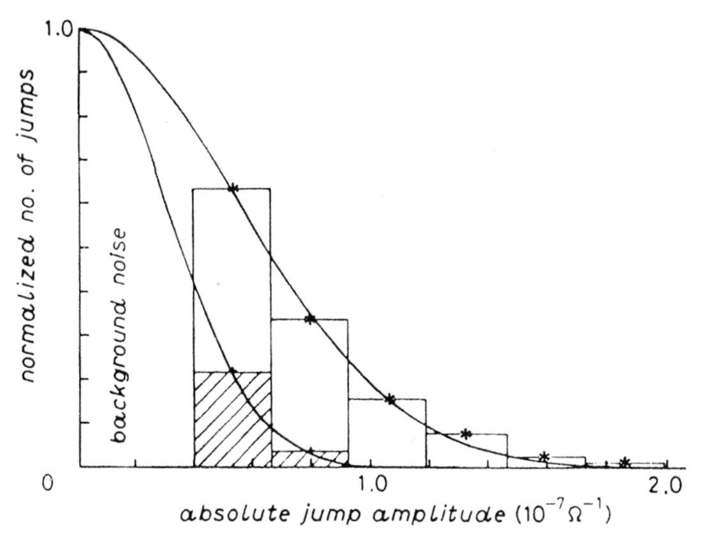

Fig.3 Normalized number of jumps versus absolute amplitude. Stars correspond to H=0, and crosses to H=0.2T

our pulse method indeed provides a good sample averaging technique or that the ergodic hypothesis usually admitted for the MFP is valid.

We show on figure 3 histograms and gaussian fits from measurements in zero magnetic field and in a 0.2T magnetic field. The jump distribution without magnetic field is characterized by a larger variance. This shows us that time reversal symmetry yields larger fluctuations. We obtained a ratio of 3.2 for the two variances. This higher value compared to the theoretical value of 2 can be attributed to the effect of Zeeman splitting [6]. Indeed the Zeemann energy $E_g = g\mu_B H$ is not negligeable at H=0.2T compared to the correlation energy Ec. Therefore spin degeneracy is almost broken and one has to add uncoherently the contribution of the spin up and spin down giving a factor 4 instead of 2 for the ratio of the two variances.

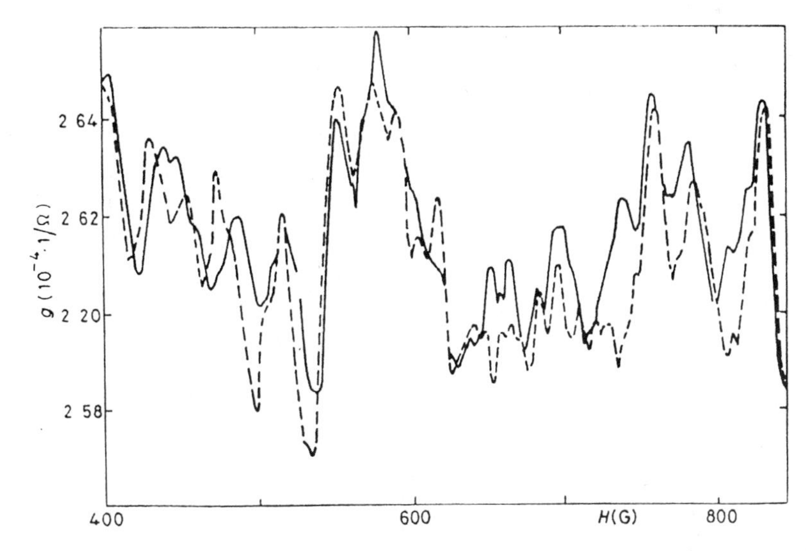

Fig.2 Detail of two magnetofingerprints. The solid and dashed lines are respectively the MFP before and ten hours after a voltage pulse.

-II- PERPENDICULAR TRANSPORT THROUGH 1D STRUCTURES AT THE AlGaAs/GaAS INTERFACE

Transport properties in low dimensionality structures has during many years been studied mainly in directions parallel to the low dimensionality, in spite of the fact that the initial proposal of semiconductor superlattices was directed towards negative differential resistance (NDR) effects in coherent transport perpendicular to superlattices[7]. This was due to the difficulty of producing superlattices of quality high enough to obtain long inelastic mean free path. During the last years a lot of interest arised in coherent transport perpendicular to short barriers produced by heteroepitaxy, and some studies

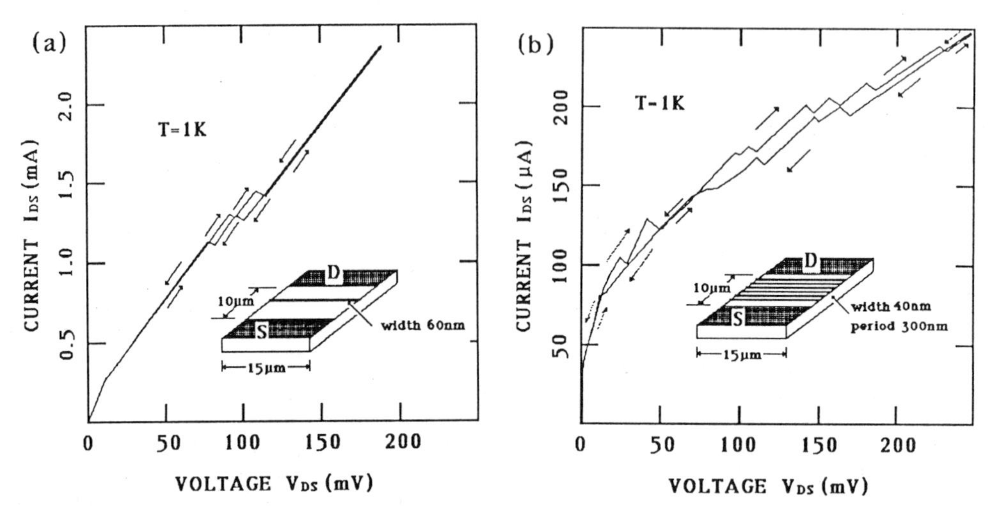

Fig.4 I(V_{DS}) characteristics at 1K for (a)the short gate TEGFET and (b) the 20 lines gate TEGFET

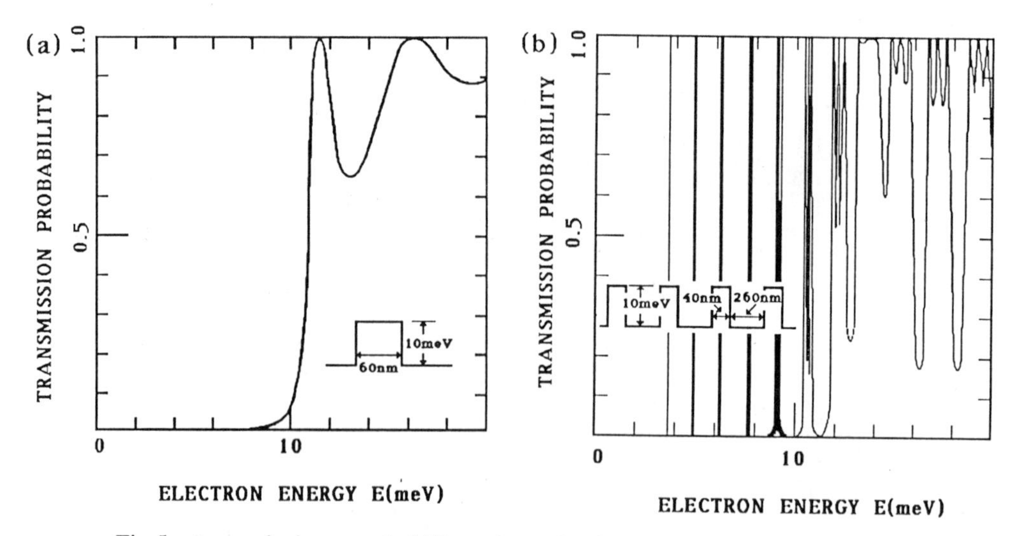

Fig.5 transmission probability through (a)a single barrier and (b) a succession of four barriers

also concerned field induced lateral barriers at the Si/SiO_2 interface[8], and at the AlGaAs/GaAs hetrerojunction interface[9-12]. In these last cases the additional confinement of the electrons in the 2D gas allowed the clear observation of the predicted superlattice minibands. The observation, or the absence of strong NDR effects has been reported for lateral 2D superlattices.

In order to study the properties of coherent transport perpendicular to field induced barrier in a system with long inelastic mean free path, we have fabricated TEGFET-like structures, including either an ultra-short gate (one barrier) or a succession of barriers at a good AlGaAs/GaAs interface. The epilayer, grown by MBE on semi-insulating GaAs, consisted of a 2 µm undoped GaAs, a 20nm $Al_{.3}Ga_{.7}As$ spacer layer, planar doping by Si, a 10nm undoped $Al_{.3}Ga_{.7}As$ layer, and finally an undoped GaAs cap layer. The mobility and carrier concentration obtained were the following:

$\mu = 7.2\ 10^3\ cm^2/Vs$ and $n = 5.9\ 10^{11} cm^{-2}$ at room temperature

$\mu = 7.1\ 10^5\ cm^2/Vs$ and $n = 4.5\ 10^{11}\ cm^{-2}$ at 4K.

The device fabrication was entirely done using e-beam lithography. MESA insulation was done by chemical etching with an $H_2O_2:H_3PO_4:H_2O$ solution. Ohmic contacts for source and drain were obtained by rapid thermal annealing for one minute at 450°C of Ni/Ge/Au/Ni/Pt/Au. The source-drain distance was equal to 10µm. The nanostructure gates were obtained by deposition of a 40nm thick gold layer, after e-beam patterning of 150nm thick PMMA . Two kinds of devices were fabricated: single line gate , 60nm wide, and 20parallel lines, 40nm wide, with a 300 nm period.

We observed small NDR and hysteresis effects at 1K in both kinds of devices, as shown in figure 4 while no anomalous phenomena was obtained from 4K to 300K. The hysteresis in the NDR region can be accounted for by a serial resistance (~ 100Ω) effect.

These data have to be considered as extremely preliminary. The reproducibility as well as the effect of gate polarisation have to be checked on different wafers and technological runs. We can nevertheless point out that they could be qualitatively in agreement with a description by tunneling through the large barriers. We have estimated the tunneling probabilities, either through a single barrier (figure 5a) or through a multiple barrier (4 barriers on figure 5b), using a model square barrier at the interface 40nm wide and 10meV large. NDR effects occur in both cases. They would be in agreement with the experimental results if the electron energy corresponds only to 1/10 of V_{DS}: with the large source drain distance of 10µm, an inelastic mean free path of the order of 1 µm would be needed. Further work is clearly needed before quantitative estimations can be made.

-III-LOW TEMPERATURE PHOTOLUMINESCENCE OF ETCHED QUANTUM WIRES

Theoretical estimates of the optical properties of quantum wires and dots predict attractive properties for optoelectronic devices. With the reduction of the dimensionality, one expects for example the narrowing of the gain spectrum of quantum wire lasers as well as an increase of the excitonic oscillator strength[13]. It is thus important to study experimentally the optical quality of such nanostructures and to evaluate with this respect the impact of their fabrication processes. Although many results concerning the transport properties of 1D and 0D nanostructures have been reported, very few publications were devoted to their optical properties. Clear evidence of lateral quantum confinement was only obtained using fabrication techniques for which the lateral dimensions are difficult to control[14-16]. On the other hand, recent progresses in e-beam lithography and reactive ion etching (RIE) techniques allow to control precisely and with a unique flexibility the lateral quantum confinement was only obtained using fabrication techniques for which the lateral dimensions are difficult to control[14-16]. On the other hand, recent progresses in e-beam lithography and reactive ion etching (RIE) techniques allow to control precisely and with a unique flexibility the lateral dimensions of the wires and dots. Some luminescence results[17] were recently reported on this type of nanostructures, either on GaAs/AlGaAs or InP/InGaAs(P) systems, without epitaxial regrowth.

Wafers used for the fabrication of wires and dots were made in two different material systems: (1) A GaAs/AlGaAs wafer was grown by molecular beam epitaxy with a 8 nm GaAs well between two 150nm and 90nm $Al_{.3}Ga_{.7}As$ layers. (2) A InP/InGaAsP layer was grown by metalorganic vapor deposition, with a 9 nm quaternary quantum well covered with a 600nm InP layer. Careful mapping of the luminescence wavelength showed the very good lateral homogeneity (differences lower than a few meV/cm) of both samples.

The geometry of our samples is as follows: on 1x1 cm^2 samples we dispatched some 2D reference areas of $500x500\mu m^2$. Between them we patterned $80x80\mu m^2$ areas , with a 500 µm pitch in both directions, consisting of arrays of wires and dots od various sizes and pitches: the width of the wires varied from 1µm to 30nm while the diameter of the dots varied from 60nm to 30nm.

Both samples were patterned with conventional e-beam naolithography, using a JEOL 5DIIU machine, on a 150nm thick PMMA resist. After a lift-off process, we obtained the metallic mask suitable to each RIE process used to define wires and dots in GaAs/AlGaAs or InP/InGaAsP system.After a chemical removal of the metallic mask, the nanostructures were observed by SEM(figure 6). The picture shows that the technological processes (lithography and dry etching) leading to lateral dimensions as small as 30 nm are now well controlled.

C.W. and time resolved photoluminescence (P.L.) were investigated at 10 and 77K respectively. The C.W. photoluminescence was excited using the 514.5nm line of an Ar^+ laser, and detected with either a cooled GaAs photomultiplier or a cooled Germanium detector. In the time resolved experiments, the source consisted in a R6G synchronously pumped dye laser, providing 3ps pulses at 0.6 µm. The P.L. was dispersed through a .32m focal

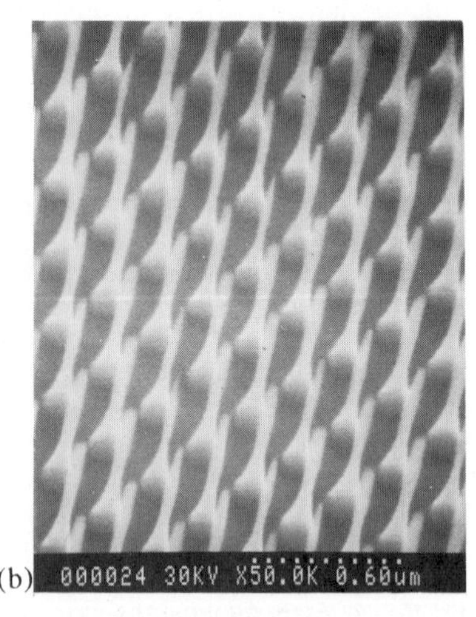

Fig.6 SEM micrographs of (a) quantum wires and (b) quantum dots etched in GaAs/AlGaAs system

length monochromator and resolved in time by a synchroscan Hamamatsu streak camera. The overall time resolution was around 15 ps.

We will discuss here the optical data obtained , in the two systems studied, on the wire structures (width w). The results obtained for GaAs/AlGaAs are similar to those already reported by Mayer and al.[17]: a steep decrease with the wire width was observed in the 8nm quantum well emission around 1.56eV. For quantum size wires (w≤40nm) P.L. is hardly observed. No clear-cut high energy shift associated with lateral confinement could be evidenced, and a high energy broadening of the P.L. spectrum is observed instead. The quenching of the P.L. intensity is correlated with the decrease of the exciton life-time inside

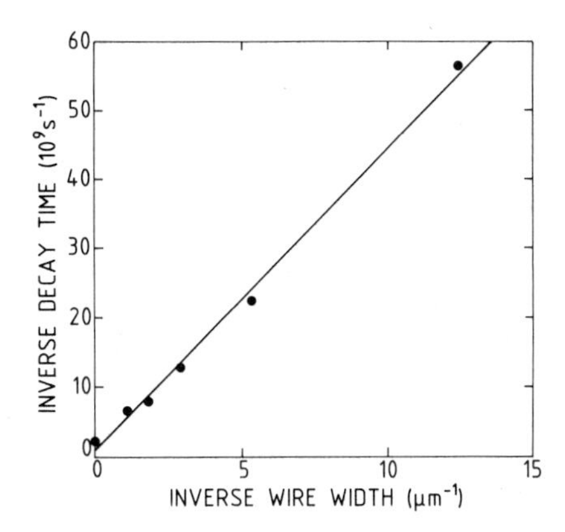

Fig. 7 Photoluminescence decay time as a function of the wire widths in GaAs/AlGaAs system. The straight line corresponds to $1/\tau = S/d$ with $S = 4 \times 10^5$ cm/s

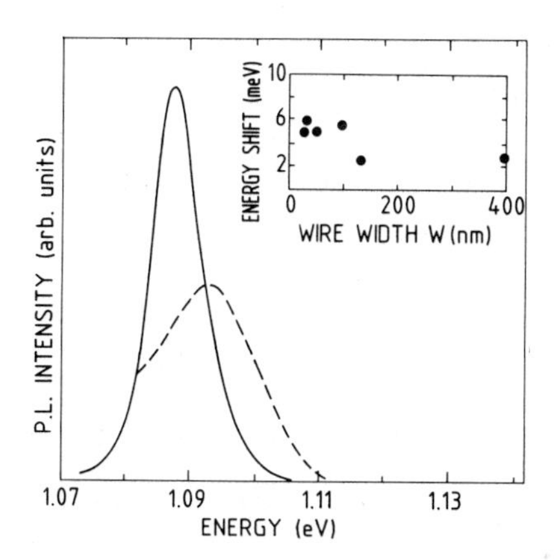

Fig.8 :Luminescence spectra of wires , with widths <30nm, (dashed line) and of 2D reference area (full line). Insert: High energy shift of the luminescence exciton peak related to the wire width

the GaAs well. Figure 7 shows these life-times, obtained from the P.L. decay times, as a function of w. These data can be accounted for considering a surface recombination velocity $S=4\times10^5$ cm/s at the well lateral edges. One should note that the P.L. efficiency is further reduced by surface recombination at the AlGaAs surfaces of the carriers photocreated in this barrier material, in particular when w becomes smaller than the AlGaAs thicknesses. No increase of S could be attributed to the etching process, in agreement with reference 17.

In the InP/InGaAsP system these effects are expected to be less pronounced, due to the lower bulk material S values. This is indeed the case, and C.W. P.L. of the wires was observed easily down to quantum sizes. The P.L. decay times correspond now to a value $S=2\times10^5$ cm/s. Figure 8 shows the luminescence spectrum obtained for w≤30nm together with the 2D quantum well reference spectrum. The wires P.L. peak is indeed shifted by 6meV towards higher energy. Unfortunately, as shown in the inset, a significant high energy shift is already observed for much broader wires. Though part of the shift may be due to lateral quantum confinement, we suggest that the InGaAsP well material is mismatched with respect to InP and that some elastic strain occurs, when the wire width becomes of the order of its height(750nm)

These results point out the interest of the study of the emission as a function of w to evidence meaningfully any lateral confinement effect on the P.L. . Additionally, the starting quantum well properties have to be precisely determined, in particular in InP/InGaAsP where it is built from an eventually lattice mismatched alloy material. Finally, while the progresses in the patterning processes allow us to design vertical edges wires with widths down to less than 30nm, their properties are still dominated by surface related phenomena.

ACKNOWLEDGMENTS

The authors wish to thank B.Etienne and V.Thierry-Mieg (MBE growth), A.Mircea (MOCVD process), J.Etrillard and L.Henry(RIE process), F.R.Ladan and C.Mayeux(contact metal deposition), without whom this work could not have been done.

REFERENCES

1- For a review see S. Wasburn and R. Webb , Ad. Phys.35,375(1986)
2- Y. Imry Europhys. Lett. 1, 249, (1986)
3- K.Muttalib, J-L. Pichard and A.D. Stone, Phys. Rev. Lett. 59, 2475,(1987)
4- F. J. Dyson, J. Math. Phys., 3, 140, (1962)
5- S. Feng, P. A. Lee and A.D. Stone Phys. Rev. Lett., 56, 1960, (1986).
6- D. Mailly, M. Sanquer, J-L Pichard and P. Pari , Europhys. Lett. 8,471(1989)
7- L.Esaki and R.Tsu, J.Res.Dev. 14,61(1970)
8- A.C.Warren, D.A.Antonidis,H.I.Smith and J.Melngailis, IEEE Electron Device Lett.6,294(1985)
9- G.Bernstein and D.K.Ferry, J.Vac.Sci.Tecnol.B5,964(1987)
10- Y.Jin, D.Mailly, F.Carcenac, B.Etienne and H.Launois, Microcircuit Engineering 6,195 (1987)
11- T.Oshima, M.Okada, M.Matsuda,N.Yokoyama and A.Shibatomi, Superlattices and Microstructures 5,247(1989)
12- K.Ismail, W.Chu,D.A.Antoniadis and H.I.Smith, Appl.Phys.Lett.52,1071(1988)
13- M.Asada, Y.Miyamoto and Y.Svetmatsu, Japanese Journal of Applied Physics 24 (1985) L95, and references therin
14- J.Cibert, P.M.Petroff, G.J.Dolan,S.J.Pearton, A.C.Gossard and J.M.English, Appl.Phys.Lett.49(1986)1275
15- Y.Hirayama, S.Tarucha, Y.Suzuki and H.Okamoto, Phys.Rev.B37(1988)2744
16- A.Forchel, G.Trankle,U.Cebiella, H.Leier,B.E.Maile to be published
17- G.Mayer,B.E.Maile, R.Germann, A.Forchel, H.P.Meier to be published.

BALLISTIC ELECTRON TRANSPORT IN A GATED CONSTRICTION

T J Thornton, M L Roukes, A Scherer, B van der Gaag

Bellcore
331 Newman Springs Road
Red Bank
NJ 07701
USA

INTRODUCTION

It is now clear that the conductance of a short, narrow constriction in a 2DEG is quantised in units of $2e^2/h$ [1 to 8]. This quantisation has so far only been demonstrated in short, split-gate FETs [9] in which the width of the constriction can be continuously varied thereby changing the number of occupied quantum channels (subbands). Squeezing the constriction reduces the number of i occupied subbands and the resistance increases in a step like fashion with plateaus at values of $R=h/2ie^2$. Split gates have been used in various configurations to demonstrate the magnetic depopulation of 1D subbands [10], electron focussing [11] and the non-additivity of ballistic resistors [12]. In this paper we show that the same quantisation exists in devices of constant width but variable carrier concentration (Fermi energy) and we discuss preliminary results from two devices of different aspect ratio.

GATED CONSTRICTIONS BY ION DAMAGE PATTERNING

Recent theoretical modelling has highlighted the importance of resonance effects when standing waves can be established in a ballistic constriction [5,6,7,8]. These resonances are similar to organ pipe modes with open ended boundary conditions and are expected to be strong at low temperatures in constrictions which are longer than they are wide (aspect ratio greater than unity). The fact that resonances have not been experimentally observed is probably due to the soft confining potential formed between the split gates. Indeed, adiabatic funelling of electrons into and out of the constriction has been evoked to explain the accuracy of the quantisation [4]. To study the effect of geometry on the quantisation of the ballistic resistance we have used ion damage patterning [13] to make short, sharply defined quantum wires which have a fixed width but a variable carrier concentration. This is in contrast to the constrictions formed between split gates where the channel width is the principal variable. A schematic diagram of the device is shown in Figure 1. Optical lithography was used to define ohmic contacts and mesas in a GaAs:AlGaAs heterojunction with $\mu(4.2K) = 700{,}000 cm^2V^{-1}s^{-1}$ and a corresponding carrier concentration of $2.5 \times 10^{15} m^{-2}$. Large pads of SrF_2 were then patterned by electron beam lithography and lift-off with spacings of 40, 200 and 600nm. A further lift-off stage was used to deposit a fine metal gate perpendicular to the gaps in the SrF_2 as shown schematically in Figures 1 and 2. Figure 3a is an electron micrograph of the device and Figure 3b is a close up of the gated region showing the step coverage into and out of the gaps in the SrF_2. The gate was deposited on the surface of the SrF_2 so that the carrier concentration would only be

Science and Engineering of One- and Zero-Dimensional Semiconductors
Edited by S.P. Beaumont and C.M. Sotomajor Torres
Plenum Press, New York, 1990

modulated in the narrow region. Having the gate under the SrF_2 would lead to an inhomogeneous electron density near the mouth of the constriction.

The final stage in the fabrication is to define the constriction by ion damage patterning [13] (Figure 2). The devices used in this work were exposed to 100 eV Ne^+ ions for times short enough that less than 50Å of material was removed from the surface of the heterojunction. However the damage induced in the material extends far below the surface and can be used to shut off conduction in regions of the 2DEG which are not masked by the SrF_2 or the gate metal. To produce optimally defined devices the ion dose has to be precisely controlled. Too little dose would result in wires which are not properly defined at low temperatures whereas over exposure results in a large lateral depletion length and an irregular geometry. Measuring the device resistance in-situ during the exposure ensures that each device receives the optimum definition dose. By this method it is possible to make narrow, ballistic channels with a well defined aspect ratio.

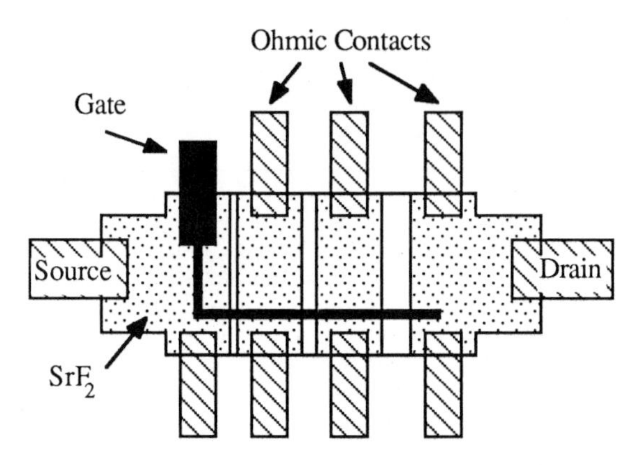

Fig. 1 A plan view of the devices used to make gated constrictions of variable carrier concentration. The mesa is 40μm long with 8μm between voltage probes. The gate width was 400nm.

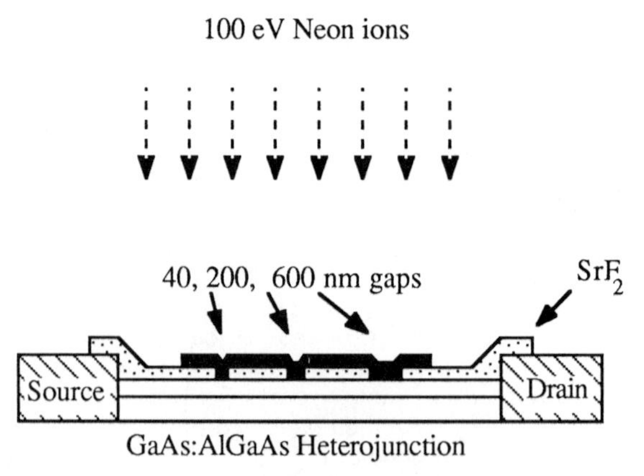

Fig. 2 A cross section through the device shown in Fig. 1. The 2DEG beneath the gate and the SrF_2 is masked from the ion beam and is not damaged during the exposure to the low energy ion beam

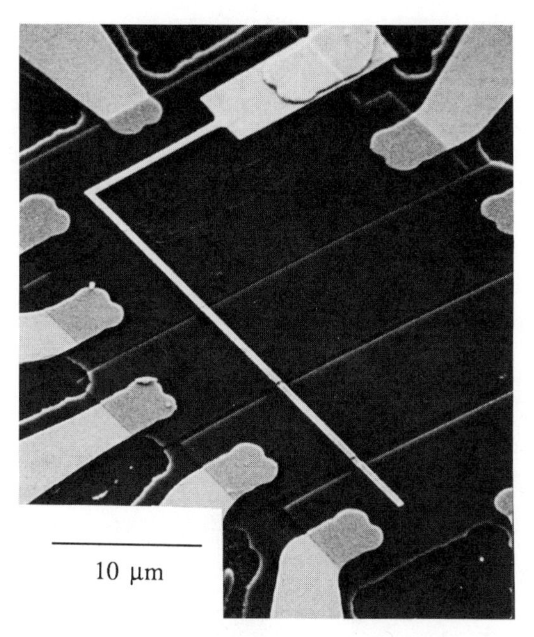

Fig. 3a Electron micrograph of the gated constrictions showing the ohmic contacts, SrF_2 pads and the gate wire.

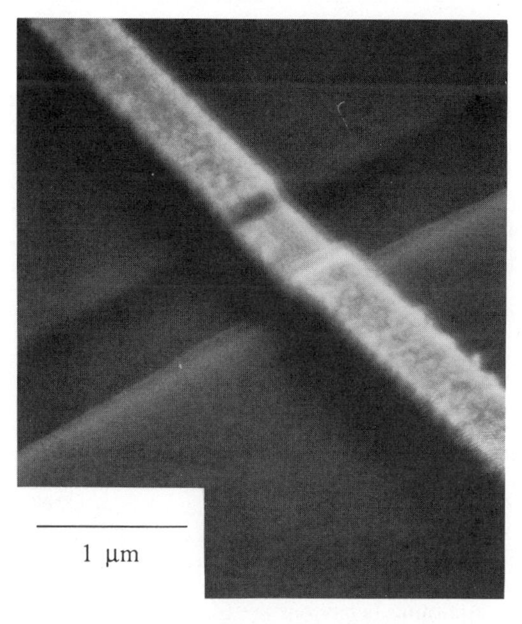

Fig. 3b Close up of the gated region showing the step coverage into and out of a 600nm gap between the SrF_2 pads

To confirm that the width of our gated constrictions did not vary significantly with gate voltage we studied the magnetic depopulation of one dimensional subbands in long gated wires defined by ion damage patterning. The results for a channel of length 8μm are shown in Figure 4 in the form of a fan diagram showing the Landau level index, N_L, plotted against inverse magnetic field for a number of different gate voltages. The curvature is due to the presence of one dimensional subbands at low magnetic fields [10,14]. The subbands evolve into Landau levels at high fields and the deviation from a straight line in

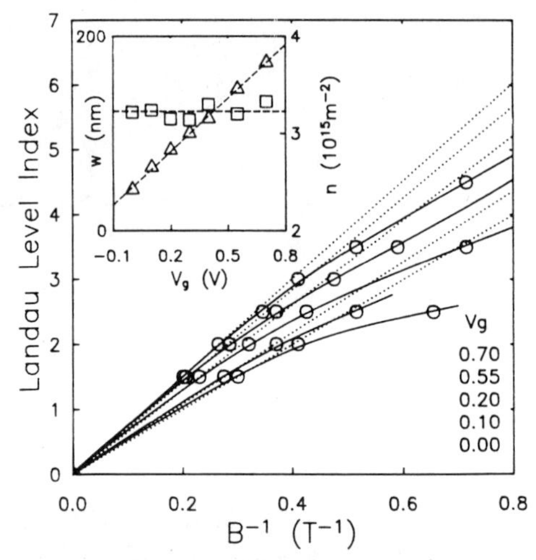

Fig. 4 A fan diagram showing the Landau Level index plotted against inverse magnetic filed for a number of different gate voltages (circles). The dotted lines are linear fits to the high field data (B>3T). The inset shows the wire width (squares) and carrier concentration (triangles) calculated at each gate voltage as discussed in the text.

Figure 4 occurs when a cyclotron orbit can fit into the channel width ie when $W \geq 2\sqrt{(2N_L+1)\hbar/eB}$ (at least to within a numerical factor which will depend on the confining potential). A lower bound to the width determined in this way is plotted against gate voltage in the inset of Figure 4 along with the corresponding carrier concentration calculated from the high field gradient. Over a wide range of gate voltage for which the carrier concentration varies approximately linearly the channel width is almost constant with a mean value of 120nm. A similar behaviour is expected in shorter wires and confirms the assumption that in the gated constrictions the principal effect of the gate voltage is to vary the carrier concentration while having little effect on the width.

Figure 5 shows the the pinch-off characteristics of a gated region 600nm long and 400nm wide at a temperature of 2K. The resistance is measured in a four terminal configuration before inverting to obtain the conductance and the series resistance of the reservoirs either side of the constriction is expected to be negligible ($R \leq 50\Omega$). The conductance plotted against gate voltage shows steps which lie on or close to the quantised values but the steps themselves are not ideally flat presumably because of the comparatively

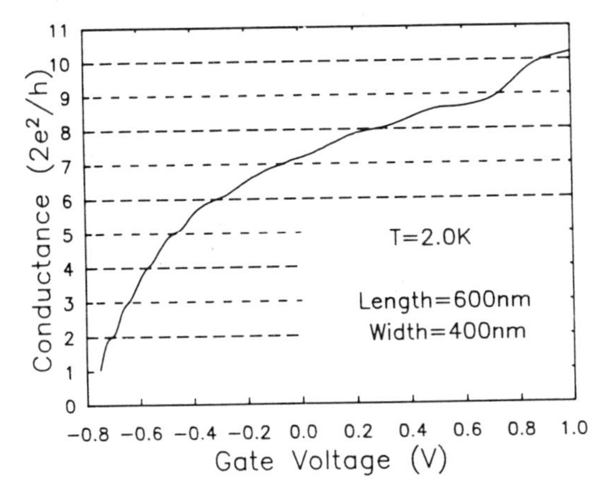

Fig. 5 The conductance of a gated constriction with an aspect ratio greater than unity. The dashed lines indicate the values at which quantised plateaus are expected to occur.

high temperatures at which these results were taken. At zero gate voltage there are seven occupied subbands whereas a simple calculation assuming a parabolic confining potential suggests that up to 11 subbands would be occupied at a carrier density of 2.5×10^{15} m^{-2}. A lower bound estimate to the electrical width would therefore be 300nm suggesting a depletion width of 50nm either side of the gate mask. A depletion width of 50nm is close to the resolution limit of ion damage patterning and means that these constrictions are more sharply defined than those formed between split gates.

Numerical simulations of split gate constrictions [5,6,7,8] have calculated the device conductance in terms of the transmission of electron waves through the constriction. From a Landauer type formula $g=2e^2/h\ Tr(tt^{\dagger})$ [15] it is expected that the conductance will be "exactly" quantised when $Tr(tt^{\dagger}) = N$ the number of occupied quantum channels (here t is the matrix which gives the probability of transmission through the device). There are however a number of effects which can reduce this inequality to $Tr(tt^{\dagger}) < N$. In particular an impurity in the vicinity of the constriction has a dramatic effect on the quantisation as shown by the numerical simulations of van der Marel and Haanappel [7] and Tekman and Ciraci [8].In general they have shown that the conductance is lowered by the presence of an impurity and the plateaus deviate considerably from the quantised values. Such an effect is consistent with the data shown in Figure 6 taken from a much shorter constriction (nominal length 200nm). This shorter constriction was measured on the same device as the longer one discussed above and was exposed to the same ion dose so that the electrical width is expected to be much the same in each case. Although there are well defined steps they do not in general have quantised values and the conductance at a given gate voltage is less than that of the longer device (cf Figure 5).

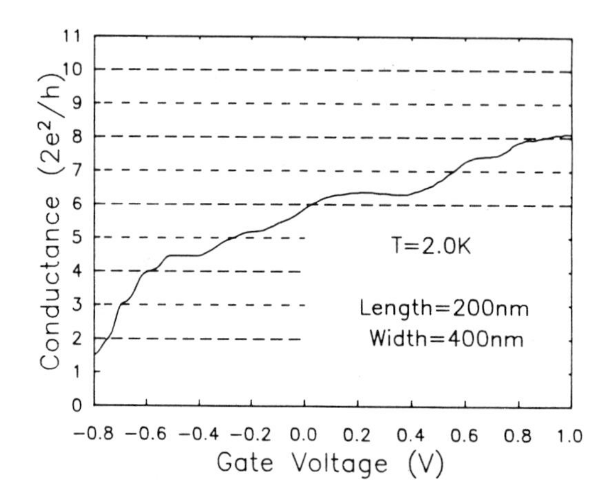

Fig. 6 The conductance of a constriction of aspect ratio less than unity. Again the dashed lines indicate the quantised values of the conductance.

Backscattering from the potential step at the open ends of the constriction is expected to have a similar effect as impurities in the channel. The potential step arises because of the difference in chemical potential between the wide regions and the gated constriction as pointed out by Kirczenow [16]. If the potential step varies slowly over many Fermi wavelengths it can be considered adiabatic in the sense discussed in reference 4 and the back scattering would not be significant. This is presumably the case in the longer constriction where the conductance is approximately quantised. For the shorter constriction the potential step is more abrupt and the loss of quantisation is consistent with the expected increase in back scattering.

In conclusion we have measured the conductance of gated constrictions of variable carrier concentration and well defined aspect ratio and have seen quantised steps similar to those observed in split gate devices.

REFERENCES

(1) B J van Wees, H van Houten, C W J Beenakker, J G Williamson, L P Kouwenhoven, D van der Marel, C T Foxon Phys Rev Lett, 60, 848 (1988)
(2) D A Wharam, T J Thornton, R Newbury, M Pepper, H Ahmed, J E F Frost, D G Hasco, D C Peacock, D A Ritchie, G A C Jones J Phys C:Solid State Phys 21, L209 (1988)
(3) Y Imry in "Directions in Condensed Matter Physics" G Grinstein and Gmazenko eds, World Scientific Press, Singapore p101 (1986).
(4) L I Glazman, G B Lesovich, D E Khmelnitskii, R I Shlukher Pys'ma vZhETF 48, (1988)
(5) G Kirczenow Solid State Comm 68, 715 (1988)
(6) A Szafer, A D Stone Phys Rev Lett 62, 300 (1989)
(7) D van der Marel, E G Haanappel Phys Rev B 39, 5484 (1989)
(8) E Tekman and S Ciraci submitted to Phys Rev B
(9) T J Thornton, M Pepper, H Ahmed, D Andrews, G J Davies Phys Rev Lett
(10) K-F Berggren, T J Thornton, D J Newson M Pepper Phys Rev Lett 57, 1769 (1986)
(11) H van Houten, B J van Wees, J E Mooij, C W J Beenakker, J G Williamson, C T Foxon Europhys Lett 5, 712 (1988)
(12) D A Wharam, M Pepper, H Ahmed, J E F Frost, D G Hasco, D C Peacock, D A Ritchie, G A C Jones J Phys C:Solid State Phys 21, L887 (1988)
(13) A Scherer, M L Roukes submitted to Appl Phys Lett
(14) K-F Berggren, G Roos, H van Houten Phys Rev B 37, 10118 (1988)
(15) See for example Szafer and Stone IBM J Res Dev 32, 384 (1988)
(16) G Kirczenow to be published

QUANTUM-STATE SPECTROSCOPY IN QUANTUM WIRES AND QUANTUM DOTS

T. P. Smith, III, H. Arnot, J. A. Brum, L. L. Chang, L. Esaki,

A. B. Fowler, W. Hansen, J. M. Hong, D. P. Kern,

C. M. Knoedler, S. E. Laux, and K. Y. Lee

IBM T.J.Watson Research Center
Yorktown Heights, NY 10598

ABSTRACT

We review the results obtained from spectroscopy of low-dimensional systems using capacitance measurements. This technique has provided prima facie evidence for spatial quantization in quantum wires and quantum dots. The energy level spacings scale appropriately with the size of the confining structure reflecting simple particle-in-a-box quantization. In addition we have obtained unprecedented agreement between theoretical calculations and our experimental results.

INTRODUCTION

In the past decade, (the past two years in particular) prodigious work has been done on electronic systems with less than two degrees of freedom. In many cases (the present work included) two-dimensional electronic systems have been further confined by various lithographic, etching, gating and related techniques. High-mobility GaAs-AlGaAs heterostructures have become the system of choice for these experiments because of the lower effective mass and the larger inelastic mean free path vis a vis Si devices, for example. The reduction of scattering and the preservation of phase over long distances have yielded a myriad of information on the fundamental aspects of electronic transport and confinement in artificial structures.

In this paper we will discuss several of the discoveries and observations we have made by studying low-dimensional electronic systems. While low-dimensional systems can be studied in a variety of ways we have focused on determining the density of states of these systems. This is because the density of states determines to a large degree the properties of these systems and changes dramatically with the degree of confinement.

Science and Engineering of One- and Zero-Dimensional Semiconductors
Edited by S.P. Beaumont and C.M. Sotomajor Torres
Plenum Press, New York, 1990

EXPERIMENTAL

In order to measure the density of states in mesoscopic structures we have devised a novel spectroscopic technique. As pointed out by Stern[2] and later by Luryi[3] the capacitance of a quantum structure is directly related to the density of states. If a two-dimensional model is used for the heterojunction confinement then the measured capacitance is approximately given by:

$$\frac{A}{C_{meas}} = \frac{A}{C_{ins}} + \frac{\gamma z_\theta}{\varepsilon_0 \kappa_c} + \frac{1}{e^2 dn/d\mu} \qquad 1$$

where A is the area of the wire or dot, C_{ins} is the capacitance of the AlGaAs

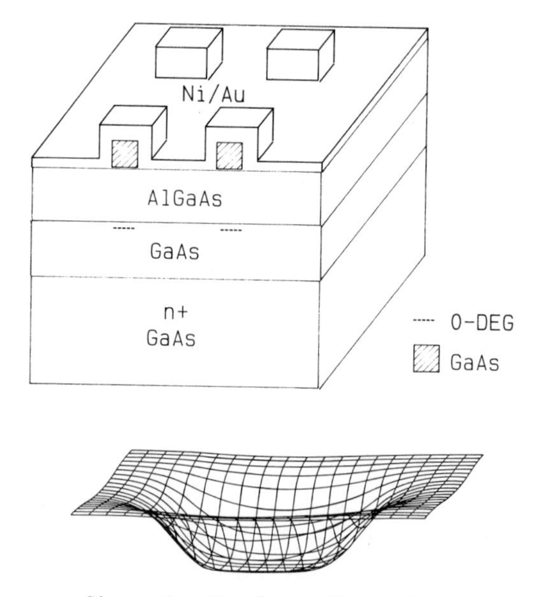

Figure 1. Sample configuration
The top portion shows a schematic diagram of the samples used to study spatial quantization effects. We have shown the configuration for a sample containing quantum dots but the configuration for quantum wires is very similar. The lower portion of the figure shows the lateral confining potential for a quantum dot.

layer, γ is a numerical constant of order unity, z_θ is the average position of the electrons in the wire or dot, κ_c is the relative dielectric constant of GaAs, and $dn/d\mu$ is the thermodynamic density of states. However, capacitance measurements can be complicated by series resistance effects and other parasitics. In

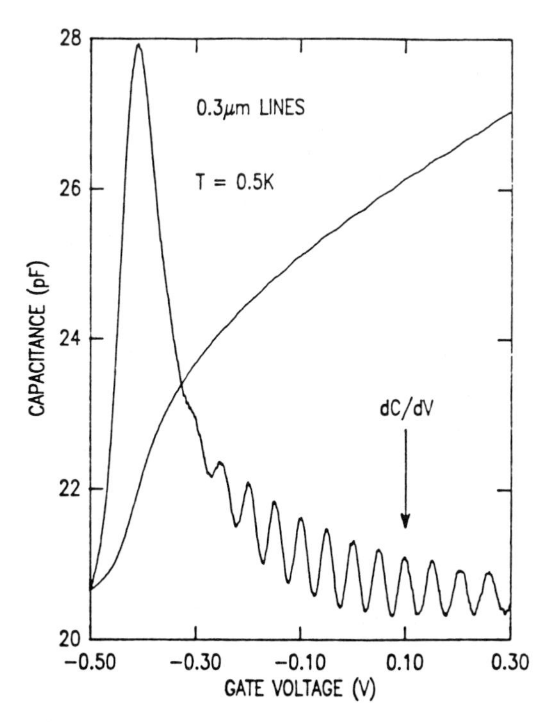

Figure 2. The differential capacitance and its derivative for a sample with 300nm wires. The wires are fabricated in the same fashion as the dots shown in Fig. 1.

addition, it is non-trivial to inject and remove charge from structures such as quantum dots. To avoid these problems we have employed a novel sample configuration[1].

A schematic diagram of our samples is shown in Fig. 1. Using molecular beam epitaxy, we grow a series of layers on a heavily doped n-type GaAs substrate. The confined electron systems reside at the interface between an undoped GaAs layer, and an undoped AlGaAs spacer layer. The separation between the n+ GaAs contact layer and the heterojunction is small enough that electrons can tunnel easily from one region to the other but large enough that the quantum confinement of the carriers at the interface is not significantly affected by the n+ layer. The AlGaAs layer is modulation doped with Si to provide a low-disorder two-dimensional electron system. Finally a relatively thick cap layer is placed on top of the structure. The sample is designed such that etching of the GaAs cap layer depletes the carriers at the heterojunction beneath the etched regions. We use selective reactive-ion etching to remove portions of the GaAs cap and define the wires and dots. The etch mask is either electron-beam resist which is exposed with an ultra-high resolution IBM VS-6 electron-beam lithography tool, or Ni-Au patterned using a lift-off technique. Each device contains a large array of dots or wires, and a metal gate is deposited over these grids. The metal gate on the top of the sample forms one electrode of the quantum capacitor and the n+ GaAs substrate the other.

In order to measure the density of states of quantum wires and dots we measure the differential capacitance of these structures. In our measurements we apply a small ac bias to the top metal gate and measure the differential capacitance as a function of dc gate bias. As mentioned above, charge is injected directly into the wires and dots from the n+ GaAs contact layer.

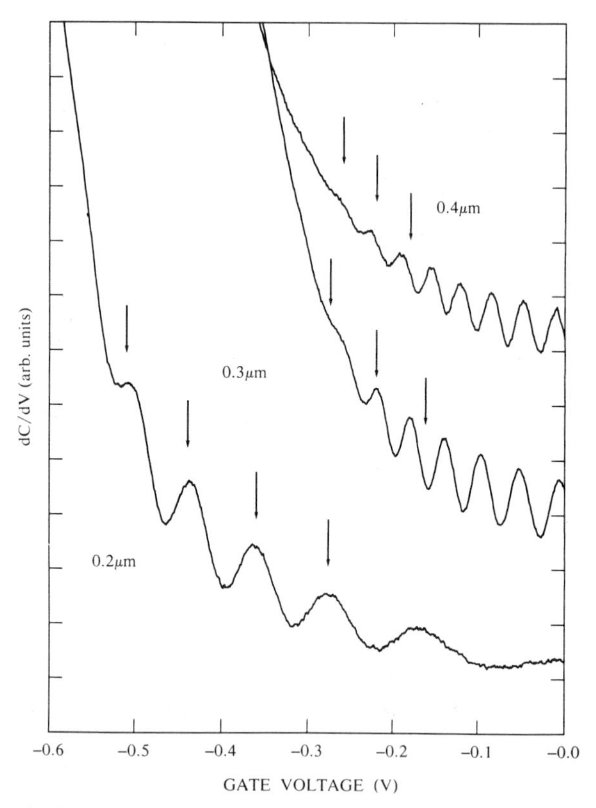

Figure 3. The derivative of the capacitance for samples with 200, 300, and 400nm wires. The arrow indicate the theoretical spacings for the oscillations in the capacitance. (after Lee et al.[1])

RESULTS

Fig. 2 shows the differential capacitance and its derivative for a sample with 300nm wires[4]. In general, the oscillations in the capacitance are fairly weak and we differentiate our signal to enhance the oscillatory component of the data. We see a large number of oscillations in the differential capacitance indicating that the confining potential is very uniform along the wires and that the wire-to-wire variations are very small. If this were not the case the oscillations at higher gate biases which arise from the higher-lying quasi-one-dimensional states in the wires would be smeared out. Another remarkable feature is that the oscillations are very regular. This is due to the fact that our confinement is electrostatic. Although the energy level spacings

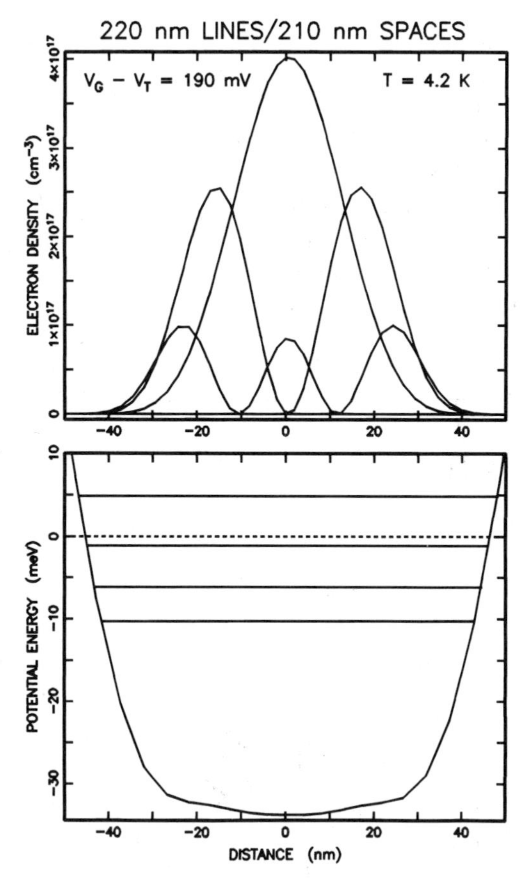

Figure 4. The potential, energy levels and charge distribution for a 220nm quantum wire with 3 quasi-one-dimensional subbands occupied.

are not uniform, the changes in the potential tend to reduce the differences in V_G separation.

As mentioned above, the period of the oscillations scales inversely with the width of the wire. This can be seen in Fig. 3. The period of the oscillations for the 200nm wires is 60-75 mV and the period for the 400nm wire is about half that value. This is what one expects: as the confinement is increased the energy-level spacings increase. However, one cannot extract the actual energy level spacings from our measurements directly. The energy level spacings can be estimated from the temperature, magnetic field, and width dependence but such estimates are not particularly accurate. To get a precise estimate fully self-consistent calculations for the electrostatic potential and the quantum states is required. The results of such a calculation[4] are shown in Fig. 3. The energy level spacings obtained from the calculation are shown as arrows. As can be seen, the agreement between the calculations and the experimental data is good for the 200nm and 400nm lines but not as good for the 300nm lines. The corresponding energy level spacings and charge distributions for the 200nm wires are shown in Fig. 4.

The results shown in Fig. 4 are for the gate bias corresponding to the leftmost arrow above the data from the 200nm wires. At this gate bias three quantum-wire subbands are occupied. At this voltage the energy level spacings are about 5meV for all the levels near the Fermi energy which is shown as a dashed line. The effective width of the channel at the Fermi energy is about 90 nm. The results for the other wire widths are similar: when only a few quasi-one-dimensional levels are occupied the effective width of the wires is less than half the lithographic width. This is because our confinement is produced by depletion of carriers in the heterostructure.

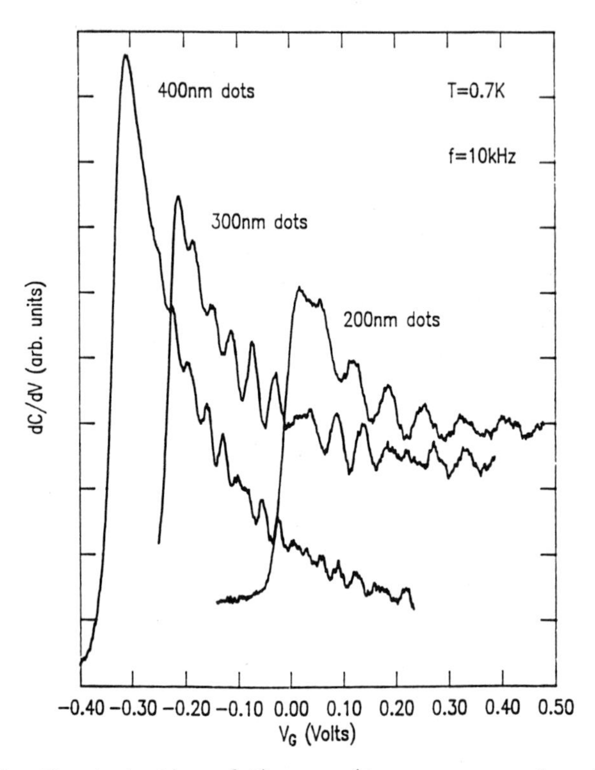

Figure 5. The derivative of the capacitance versus gate voltage for samples with three different size quantum dots.

While the fabrication of quantum wires is by no means trivial, the confinement of electrons in all three spatial dimensions is considerably more challenging. Nonetheless, we have observed quantum confinement in these systems. Fig. 5 shows the capacitance spectra for samples with three different sizes of quantum dots. The confinement scheme in these samples results in the confining potential shown at the bottom of Fig. 1. While the dots etched into the surface of the heterostructure are square with rounded corners, the potential is even smoother because of screening. The potential shown in Fig. 1 is a three-dimensional representation of the lateral potential versus energy. The same type of figure could be drawn for the vertical (z-direction) confining potential, with this direction showing the strongest confinement. Although the figure is not drawn to scale the confining potential for the quantum dots is also about twice as small as the lithographic size of the dots when only a few quantum levels are occupied.

Returning to Fig. 5, we see that the period of the oscillations in the capacitance scale appropriately with dot size. Although a fully self-consistent solution of this three-dimensional problem is still not complete, preliminary calculations[5] of the spacings for the oscillations in the capacitance for 200nm dots give good agreement with the measured spacings.

CONCLUSIONS

The results shown demonstrate convincingly that spatial confinement can be studied in semiconductor nanostructures. Having established this, we have also discovered a number of novel effects. We have seen a novel magnetic anisotropy in quantum wires[6] and we have found a striking difference between the magnetic response of quantum wires and quantum dots.[7] Capacitance spectroscopy has emerged as a powerful probe of few-electron systems and should continue to provide us with new insight into mesoscopic phyiscs

ACKNOWLEDGEMENTS

We would like to thank F. Stern and A. Kumar for sharing the results of thier calculations with us. The potential profile at the bottom of Fig. 1 is a result of their calculations. We are grateful to A. R. Williams for his support and interest in our work. We would also like to acknowledge the technical assistance of L. F. Alexander, M. S. Christie, and H. Luhn. This work was supported in part by the Office of Naval Research.

REFERENCES

1 K. Y. Lee, T. P. Smith, III, H. Arnot, C. M. Knoedler, J. M. Hong, and D. Kern, J. Vac. Sci. Tech. **B6**, 1856 (1988).

2 F. Stern, (IBM internal research report, 1972).

3 S. Luryi, Appl. Phys. Lett. **52**, 501 (1988).

4 T. P. Smith, III, H. Arnot, J. M. Hong, C. M. Knoedler, S. E. Laux, and H. Schmid, Phy. Rev. Lett. **59**, 2802 (1987).

5 A. Kumar, F. Stern, and S. E. Laux, (unpublished).

6 T. P. Smith, III, J. A. Brum, J. M. Hong, C. M. Knoedler, H. Arnot, and L. Esaki, Phys. Rev. Lett. **61**, 585 (1988).

7 W. Hansen, T. P. Smith, III, K. Y. Lee, J. A. Brum, C. M. Knoedler, J. M. Hong, and D. P. Kern, Phys. Rev. Lett. **62**, 2168 (1989).

ELECTRONIC PROPERTIES OF QUANTUM DOTS AND MODULATED QUANTUM WIRES

J. A. Brum
IBM T.J.Watson Research Center
Yorktown Heights, NY 10598 - USA

G. Bastard
Laboratoire de Physique de l'Ecole Normale Supérieure
F-75005 - Paris - France

ABSTRACT

We report the results of theoretical studies of the electronic properties of laterally confined electron systems in semiconductor heterostructures. The finite barrier effect is considered under the quasi-decoupled approximation. The side lengths are characterized by $L_x(L_y) >> L_z$. The effects of applied external electric and magnetic fields are considered. The evolution from the electrical confinement to magnetic confinement is discussed in detail for the quantum dots. Finally, we study the electronic structure of a quantum wire presenting a lateral constriction.

INTRODUCTION

In the last few years much attention has been devoted to the physics of low dimensional semiconductor heterostructures. Many interesting results have been obtained thanks to the high quality of the techniques necessary for their fabrication: epitaxial growth and lithographic patterning. The density of states of quasi-1 dimensional and quasi-0 dimensional heterostructures has been observed by capacitance measurements[1,2]. The influence of a magnetic field on those systems was also studied[3,4]. The quasi-1 dimensional system exhibited a strong anisotropy with respect to an external magnetic field[3] . The quantum dots showed the transition from the electrical confinement to the Landau confinement, with a complex energy level pattern for intermediate fields[4]. The tunneling through quantum dots has also been observed[5].

Very interesting results have been achieved in the physics of point contacts. The quantization of the conductance has been observed[6,7] and explained in terms of the one-dimensional carrier dispersion. More recently, the observation of optical anisotropy in a quantum wire has been reported[8,9].

On the theoretical side, several calculations have been reported aiming at predicting and explaining the experimental findings. Brum et al[10] calculated the charge transfer for a spike doped quantum wire. Laux and Stern[11] solved the self-consistent problem for a narrow gate Si-SiO_2 MOSFET and Lai and Das Sarma[12] proposed analytical variational calculations in narrow inversion MOSFET layers.

Science and Engineering of One- and Zero-Dimensional Semiconductors
Edited by S.P. Beaumont and C.M. Sotomajor Torres
Plenum Press, New York, 1990

The effects associated with electric and magnetic fields were studied by Brum and Bastard[13] and Berggren and Newson[14] in quantum wires. Bryant included the correlation effects in his study of the electronic structure of quantum dots[15].

In this paper we calculate the electronic structure of quantum dots for the electrons and holes. The finite barrier effect is considered by applying the decoupled approximation introduced for the quantum wires[13] . The effect of the magnetic field is discussed. In the second part, we introduce a simple method to study lateral constrictions in quantum wires. We apply it for the case of a symmetric and a non-symmetric constriction. Possible applications are discussed.

ELECTRONIC STRUCTURE IN QUANTUM DOTS

In the refs. 10 and 13 the effective-mass approximation was employed with the decoupling approximation to calculate the electronic structure of quantum wires including the presence of a magnetic field. The effects of the magnetic field depend dramatically on the direction of the applied field, as verified experimentally[3] . This anisotropy originates from the differences the carrier experiences between the strong epitaxial confinement and the less confined lateral direction as well as the free wire direction. For a quantum dot, the carrier motion is totally confined. The energy quantization in the presence of magnetic field has been studied theoretically mainly to understand the contribution of the surface states to the magnetization[16-18]. The evolution of the electrical confinement to the magnetic confinement depends on the symmetry of the initial confinement potential. Here we consider the simple case of a square shaped lateral confinement in the (x,y) plane of the order of a thousand Angstroms, and the z-direction confined by a small quantum well (of the order of tens of Angstroms). The strong difference between both confinements practically decouples the z and the x,y motion. Each z-related subband generates a family of x,y levels. Since we are interested in the effects of a magnetic field applied along the z-direction, we can neglect the coupling between levels from different z-related subbands. The zero of energy is then taken at the edge of the first z-related subband. In the effective mass approximation, considering a parabolic mass, the Hamiltonian for the electrons is written as

$$\{\frac{-\hbar^2(\nabla+\frac{e}{\hbar c}\vec{A})^2}{2m^*}+V(x,y,z)\}\Psi(\vec{r})=\varepsilon\Psi(\vec{r}) \qquad [1]$$

where m* is the carrier effective mass. The confinement potential can be written as

$$V(x,y,z)=V_b\{1-Y(\frac{L_x^2}{4}-x^2)Y(\frac{L_y^2}{4}-y^2)Y(\frac{L_z^2}{4}-z^2)\}=$$

$$=Y(z^2)-\frac{L_z^2}{4}+Y(\frac{L_z^2}{4}-z^2)\{1-Y(\frac{L_x^2}{4}-x^2)Y(\frac{L_y^2}{4}-y^2)\}=Y(z^2-\frac{L_z^2}{4})+$$

$$+Y(\frac{L_z^2}{4}-z^2)\{Y(x^2-\frac{L_x^2}{4})+Y(y^2-\frac{L_y^2}{4})-Y(x^2-\frac{L_x^2}{4})Y(y^2-\frac{L_y^2}{4})\} \qquad [2]$$

where Y(x) is the step function (Y(x) = 1 if x > 0, Y(x) = 0 if x < 0) and V_b is the barrier height. We obtain the energy levels by projecting the full Hamiltonian onto the basis formed by the decoupled solutions for the x,y and z motion. Only the first z-related family of levels is considered, and we can write

$$\Psi(\vec{r}) = \chi_1(z)\sum_{n,m} c_{n,m}\alpha_n(x)\beta_m(y) \qquad [3]$$

where $\chi_1(z)$ is the first level of the z quantum well and $\alpha_n(x)$ and $\beta_m(y)$ are the solutions of the x and y quantum wells, respectively. $c_{n,m}$ are obtained by the numerical diagonalization. In the low magnetic field regime considered here we may neglect the spin. The solutions we obtain are then doubly degenerate.

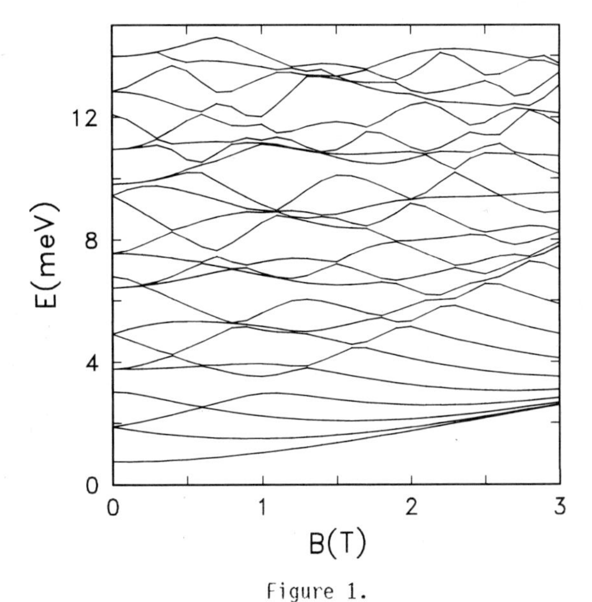

Figure 1.
Energy levels as a function of the magnetic field for a quantum dot with $L_x = L_y = 1200$Å, $m_e = 0.067m_0$ and $V_b = 600$meV.

In Fig. 1 we plot the energy levels as a function of the magnetic field applied along the z direction. The zero of energy is taken at the first z-related energy level, $L_x = L_y = 1200$ Å, $m^* = 0.067\ m_0$, where m_0 is the free electron mass. The barrier confinement is taken as 600 meV, which simulates the confinement for some of the actual structures studied[3,4]. Initially, the states present the typical degeneracy of the square confinement. As soon as the magnetic field is applied, they interact with each other, with some of the levels anticrossing and others crossing reflecting the coupling and lack of coupling between the levels. This complicated pattern evolves to Landau and surface levels and at high magnetic field to the highly degenerate levels. The transition between the three different regimes (electrical confinement, crossing and anticrossing levels and magnetic confinement) depends on the quantum level index. A clear convergence to Landau level energies is observed at high magnetic field. In Fig. 1 condensation of states into the fundamental Landau level is clearly observed and also the second and third Landau levels can be identified. The characteristics of the mixing between the levels depend strongly on the

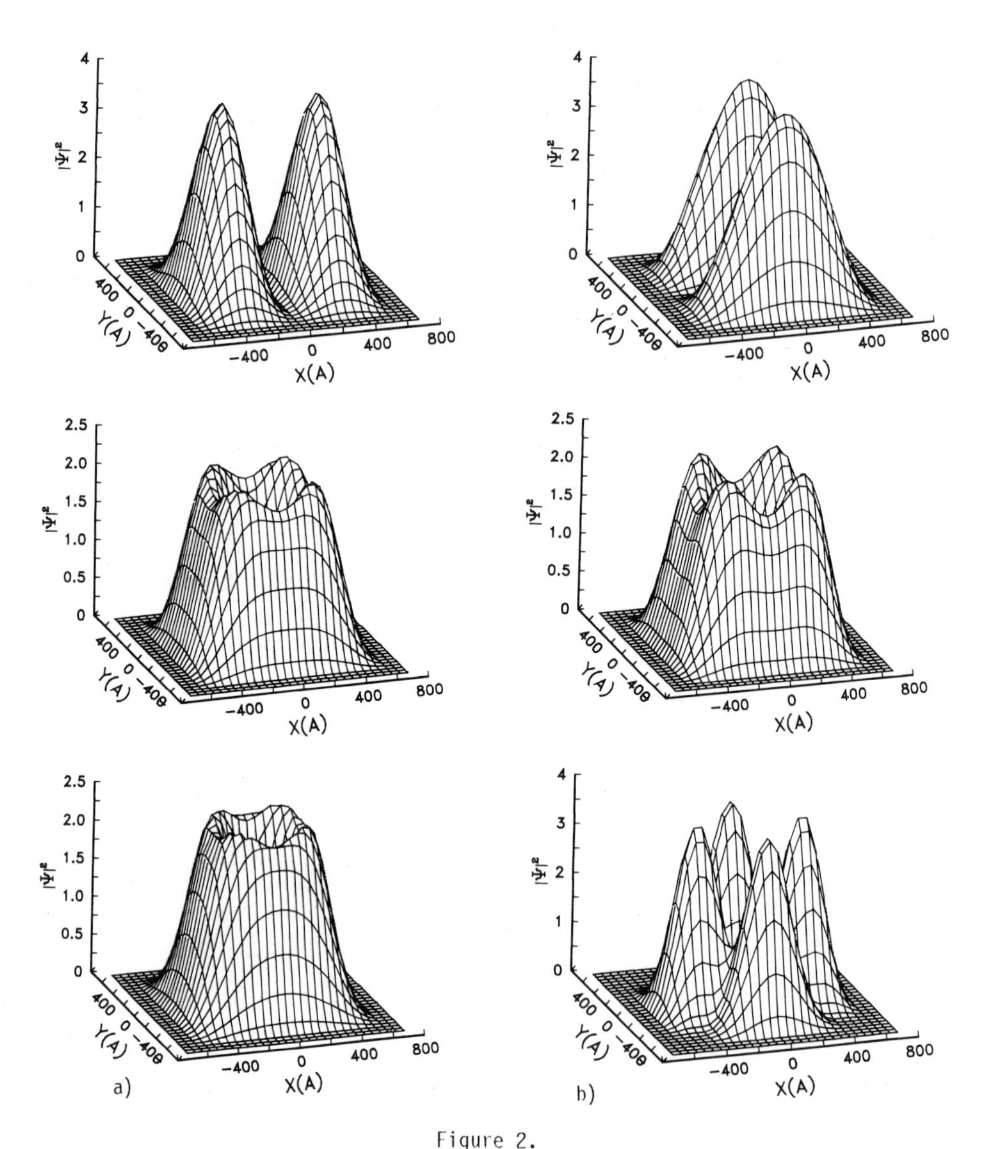

Figure 2.
Wave functions for three different values of the magnetic field. From the top to the bottom: 0, 0.2 T and 1.2 T. a) first surface level (second level in the energy diagram of Fig. 1) and b) second Landau level (second, third and fifth level, respectively, in the energy diagram of Fig. 1).

geometry of the lateral confinement. In Fig. 2 we plot the wave functions of some of the levels as they evolve with the magnetic field. The Fig. 2(a) shows the first surface level and the Fig. 2(b) shows the evolution of the levels that converge into the second Landau level. We can also observe the mixing between the wave- functions by calculating the expected value of the z orbital angular momentum of the energy eigensolutions as a function of the magnetic field (Fig. 3). Again, we observe the evolution of the quantum well levels (which, for the square configuration considered, have $<M_z> = 0$, in the absence of magnetic field) to the bulk Landau levels ($<M_z>$ = zero and integer negative values) and the surface Landau levels ($<M_z>$ = integer positive values). For example, the fundamental level interacts very weakly with others levels, keeping its average angular momentum close to zero. The two first excited quantum well levels have their degeneracy broken at very weak magnetic fields (not resolved in Fig. 3).

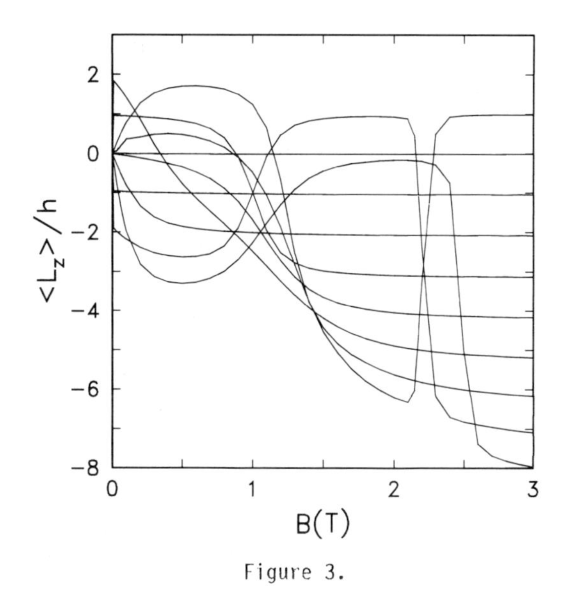

Figure 3.
$<M_z>$ as a function of the magnetic field for the quantum dot of Figure 1.

Their wave-functions are rearranged to reproduce the first excited Landau level ($<M_z> = 1$) and the first surface level ($<M_z> = -1$). Meanwhile the latter keeps this average angular momentum and converges to the first surface level, the former interacts with higher levels to finally converge to the third surface level. As observed in Fig. 1, once the levels rearrange themselves to follow the magnetic confinement, the quantum well levels originate the Landau levels with surface levels. At higher magnetic fields (which depend on the index of the Landau level), the effect of the interface is less pronounced, and the surface states converge to the Landau level energy, increasing the degeneracy. The strong difference between the square confinement and the magnetic confinement are responsible for the strong anticrossing between levels in the square shaped dot. At the other extreme, a parabolic confinement, as calculated by Darwin[16] , has the same symmetry as the magnetic field potential, and thus M_z is a good quantum number so the no anticrossing between levels is present.

QUANTUM WIRES WITH LATERAL MODULATION

The ability to fabricate high-quality nanostructures has permitted workers to engineer an entirely new kind of structures either for device applications or for the study of fundamental properties. Among the recent results, the physics of point contacts[5,6] and Aharonov-Bohm rings[19] are foremost. The modulation of quantum wires, by introducing lateral 'defects', indentations or constrictions, introduces interesting interference phenomena[20]. We introduce a simple but

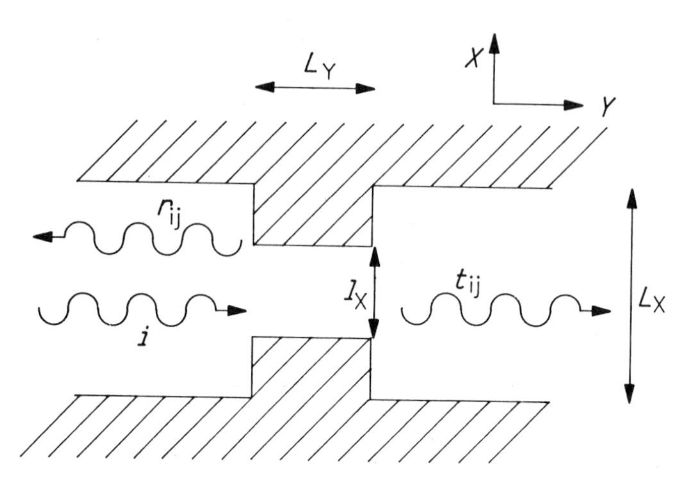

Figure 4.
Quantum wire showing a symmetric constriction. The origin of the x and y coordinates are taken at the bottom of the unperturbed wire and at the first interface unperturbed wire/defect, respectively.

powerful method to study the electronic properties of such structures. As an example of our method, we calculate the transmission coefficients for a quantum wire showing a lateral repulsive defect (see Fig. 4). The potential along the wire can be written as

$$V(x,y,z) = V_b Y(z^2 - \frac{L_z^2}{4}) Y(x^2 - \frac{L_x^2}{4}) + V_b Y(z^2 - \frac{L_z^2}{4}) Y(y(L_y^2 - y)) Y(\frac{l_x}{2} < |x| < \frac{L_x}{2}) \qquad [4]$$

Again, we apply the decoupled approximation. The z confinement being stronger, we consider only the first z-related subband. We neglect in the following its contribution and the zero of energy is taken as the edge of the fundamental z-related subband. We consider an incident mode at $y < 0$, associated to the nth x-related subband, in the wide region. The effect of the constriction is to scatter the incoming wave-function to different modes, being then reflected or transmitted. The general form of the income wave function is

$$f(x,y) = \alpha_n(x)e^{iq_ny} + \sum_m r_m(\varepsilon)\alpha_m(x)e^{-iq_my} \quad [5]$$

where $\alpha_m(x)$ is the wave-function of mth x-related subband and

$$\varepsilon = \varepsilon_m + \frac{\hbar^2 q_m^2}{2m^*} \quad [6]$$

where ε_m is the mth eigenvalue for the x motion. When $\varepsilon - \varepsilon_m < 0$ the solutions are

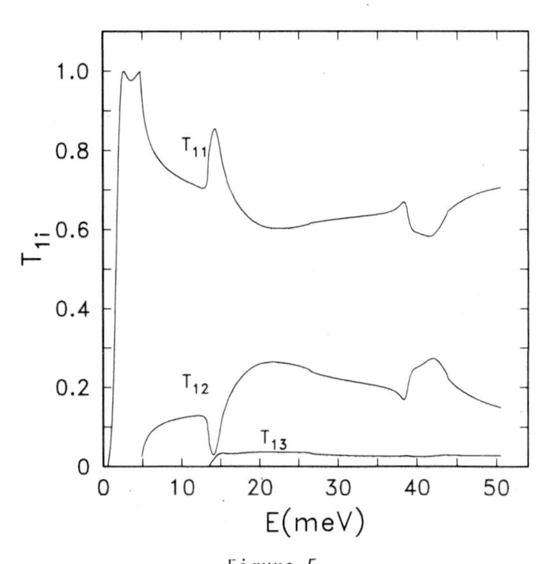

Figure 5.
Normalized transmissions for the first modes for an incident mode linked to the first x-related subband as a function of the energy.

evanescent. For the region $y > L_y$ we can write

$$f(x,y) = \sum_m \alpha_m(x)t_m(\varepsilon)e^{iq_m(y-L_y)} \quad [7]$$

The solutions of the constricted region are obtained by projecting the respective Hamiltonian onto the basis generated by the α_m's. The general expression for the wave-function is then

$$f(x,y) = \sum_j (d_je^{ik_jy} + e_je^{-ik_jy})\sum_m a_m(k_j^2)\alpha_m(x) \quad [8]$$

Once we determined a_m and k_j by the diagonalization of the Hamiltonian in the region $0 < y < L_y$, the coefficients r_m, t_m, d_j and e_j are determined by the boundary conditions at the interfaces. We then solve a linear system of 4Nx4N non-homogeneous equations which can be reduced to a 2Nx2N system (here N is the number of x-related eigenfunctions considered in our basis). In the example we are considering here, the modes will interact only with modes of same symmetry. We show in Fig. 5 the transmission coefficients as a function of the incident energy for the three first modes for an incident mode associated to the first

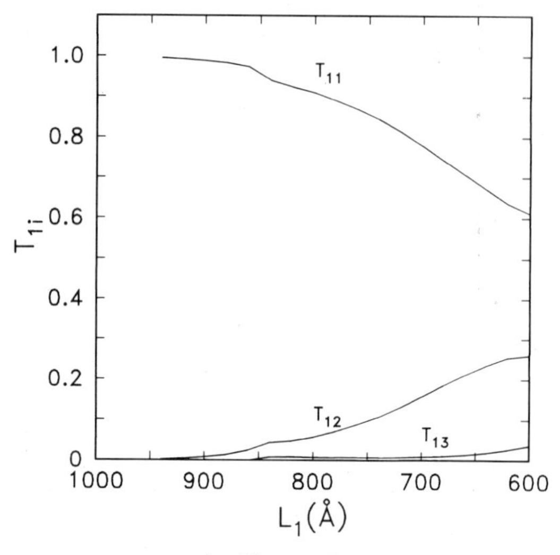

Figure 6.
Normalized transmissions for the first modes for an incident mode linked to the first x-related subband as a function of l_x for an energy of 20 meV.

x-related subband. The parameters we use are V_b = 600 meV, l_x = 600 Å, $L_y = 500$Å, $L_x = 1000$Å, m* = 0.067 m_0. Here, T_{1i} is the normalized transmission coefficient (= $\frac{q_i}{q_1} | t_{1i} |^2$), so to have an unitary flux[21]. We observe strong resonances in the transmission when a new mode becomes propagative, with a scattering between the different modes at higher energies. This can be understood in terms of the transversal kinetic energy, which is very weak at the edge of the x-related subband for each mode. In Fig. 6 we show the transmission coefficient as a function of l_x for an energy of 20 meV. All the others parameters remain the same. Although some weak structure is observed, for the parameters calculated, the main effect is the increase of the scattering between different modes when the defect size increases, leading to a decrease in the transmission by the first mode.

CONCLUSIONS

In conclusion, we have discussed the electronic structure of some low-dimensional heterostructures by combining the effective-mass approximation and the decoupled approximation. The electronic structure of quantum dots in the presence of a magnetic field was studied. A method to study modulated quantum wires was discussed. In our example, we analyzed the scattering by a narrow 1D region between two wider 1D regions. However, we can apply this method to the study of the transmission on point contacts[6,7] between 2D regions, by taking a very large value of L_x to simulate a quasi-continuous 2D gas. Further applications of the present method are envisioned.

Note added in proof- Upon completion of this work similar method was developed by Szafer and Stone[22] and G. Kirczenow[23] for the calculation of constricted quantum wires.

Acknowledgements- One of us (J.A.B.) is gratefully to L.L. Chang, W. Hansen, J. M. Hong and T.P. Smith, III for helpful discussions. This work was supported in part by the US Army Research Office.

REFERENCES

1. T.P. Smith, III, H. Arnot, J.M. Hong, C.M. Knoedler, S.E. Laux, and H. Schmid, Phys. Rev. Lett. **59**, 2802 (1987).

2. T.P. Smith, III, K.Y. Lee, C.M. Knoedler, J.M. Hong, and D.P. Kern, Phys. Rev. B **38**, 2172 (1988).

3. T.P. Smith, III, J.A. Brum, J.M. Hong, C.M. Knoedler, H. Arnot, and L. Esaki, Phys. Rev. Lett. **61**, 585 (1988).

4. W. Hansen, T.P. Smith, III, K.Y. Lee, J.A. Brum, C.M. Knoedler, D. Kern, and J.M. Hong, Phys. Rev. Lett. **62**, 2168 (1989) and T.P. Smith, III, et al, this volume.

5. M.A. Reed, J.N. Randall, R.J. Aggarwal, R.J. Matyi, T.M. Moore, and A. E. Wetsel, Phys. Rev. Lett. **60**, 535 (1988).

6. B.J. van Wees, H. van Houten, C.W.J. Beenakker, J.G. Williamson, L.P. Kouwenhoven, D. van der Marel, and C.T. Foxon, Phys. Rev. Lett. **60**, 848 (1988).

7. D.A. Wharam, T.J. Thornton, R. Newbury, M. Pepper, H. Ajmed, J.E.F. Frost, D.G. Hasko, D.C. Peacock, D.A. Ritchie, and G.A.C. Jones, J. Phys. C **21**, L209 (1988).

8. W. Hansen, M. Horst, J.P. Kotthaus, U. Merkt, and Ch. Sikorski, Phys. Rev. Lett. **58**, 2586 (1987).

9. M. Tsuchiya, J.M. Gaines, R.H. Yan, R.J. Simes, P.O. Holtz, L.A. Coldren, and P.M. Petroff, Phys. Rev. Lett. **62**, 466, (1989).

10. J.A. Brum, G. Bastard, L.L. Chang and L. Esaki, Proceedings of the 18th International Conference on the Physics of Semiconductors, Stockholm (1986) edited by O. Engstrom. World Scientific Singapore (1987), p. 505.

11. S.E. Laux and F. Stern, Appl. Phys. Lett. **49**, 91 (1986).

12. W.Y. Lai and S. Das Sarma, Phys. Rev. B**33**, 8874 (1986).

13. J.A. Brum and G. Bastard, Superlattices and Microstructures **4**, 443 (1988).

14. K.F. Berggren and K.J. Newson, Semicond. Sci. Technol. **1**, 327 (1986).

15. G. Bryant, Phys. Rev. Lett. **59**, 1140 (1987).

16. C.G. Darwin, Proc. Cambridge Phil. Soc. **27**, 86 (1930).

17. R.B. Dingle, Proc. Roy. Soc. London **A216**, 463 (1953).

18. M. Robnik, J. Phys. A **19**, 3619 (1984).

19. G. Timp, A.M. Chang, J.E. Cunningham, T.Y. Chang, P. Mankiewich, R. Behringer, and R. E. Howard, Phys. Rev. Lett. **58**, 2814 (1987).

20. F. Sols, M. Macucci, U. Ravaioli, and K. Hess, Appl. Phys. Lett. **54**, 350 (1989).

21. D. S. Fisher and P.A. Lee, Phys. Rev. B **23**, 6851 (1981).

22. A. Szafer and A.D. Stone, Phys. Rev. Lett. **62**, 300 (1989).

23. G. Kirczenow, J. Phys. Condens. Matter **1**, 305 (1989).

CONDUCTION IN n^+-GaAS WIRES

P.C. Main*, R.P. Taylor*, L. Eaves*, S. Thoms+,
S.P. Beaumont+, and C.D.W. Wilkinson+

*Department of Physics
University of Nottingham
Nottingham, UK.

+Department of Electronic and Electrical Engineering
University of Glasgow
Glasgow, U.K.

INTRODUCTION

For clear observation of interference between waves it is a necessary condition that the waves must maintain a constant phase relationship with respect to each other. This coherence is easily achieved with laser sources in optics. In metals and semiconductors coherence can be lost either by thermal smearing of the Fermi level or by the presence of inelastic scattering and so it is best achieved at low temperatures. Such interference phenomena have been observed in multiply-connected structures at low temperatures [1]. However, even in a simple wire the electrical resistance is modified by interference between scattered electron waves. Thus an ensemble of wires all prepared in identical macroscopic fashion would not all have the same resistance since the microscopic configuration of scatterers, which defines the interference paths, will be different for each wire. The application of a magnetic field changes the flux through each loop formed by the interfering waves and hence introduces a phase difference between the electron waves so that the magnetoresistance of each wire fluctuates as the field increases. Since the size and distribution of interference loops is determined by the microscopic scattering configuration, each wire has a unique magnetoresistance or "magnetofingerprint".

We have investigated the magnetoresistance of n^+ GaAs wires grown by MBE on semi-insulating substrates and fabricated using electron-beam lithography and a dry etch technique[2]. The wires have a height of 50nm, a length of 10 μm and have fabricated widths between 100 nm and 300 nm. The nominal Si doping was $5x10^{24}$ m^{-3}. In some of the wires, Shubnikov-de Haas oscillations are observable at high magnetic fields and analysis of these yields a free carrier density of $n = 4.5(\pm0.2)x10^{24}$ m^{-3}. The conductivity widths of the channels differ from the nominal widths due to sidewall depletion. This makes it difficult to estimate the elastic scattering length. Our best estimate is ~48 nm. The high degree of

Science and Engineering of One- and Zero-Dimensional Semiconductors
Edited by S.P. Beaumont and C.M. Sotomajor Torres
Plenum Press, New York, 1990

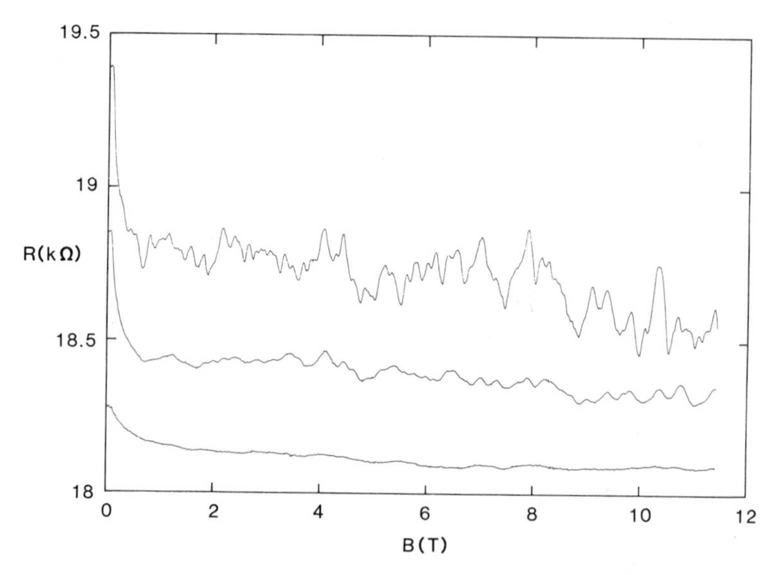

Figure 1 ACF in the magnetoresistance of a n^+ GaAs wire of nominal width 300 nm at 4.2K (top), 26.9K and 105K (bottom).

elastic scattering makes these wires ideal for studying the aperiodic conductance fluctuations associated with quantum interference.

The magnetoresistance of a wire of nominal width 100 nm is shown in Figure 1 for three temperatures where the magnetic field is applied perpendicular to the plane of the substrate. The resistance fluctuations, about 1% of the zero field resistance at 4.2K, persist to above 100K in sharp contrast to those seen in other systems[3]. As long as the temperature is maintained below ~100K then the magnetoresistance traces are entirely reproducible. Figure 1 shows that the zero field resistance increases as the temperature decreases, consistent with the expected corrections for weak localisation and electron interaction effects. As the temperature increases the amplitude of the aperiodic conductance fluctuations (ACF) falls but the peaks remain in the same field positions. The higher frequency fluctuations quench first. This can be understood qualitatively in terms of the interference loops. The larger area loops correspond to the fluctuations with smaller periods. To traverse a large loop requires very coherent electron waves and the degree of coherence decreases as the temperature rises.

On all the traces there is an initial negative magnetoresistance (NM). The magnitude of the NM appears to vary roughly in line with the amplitude of the ACF and also the field over which the NM occurs is roughly equivalent to the shortest ACF period. We identified the NM with weak localisation, or coherent back-scattering as described by Bergmann in his excellent review[4]. The ACF are thought to have a different origin, being due to direct interference loops[5], but the analysis described below casts some doubt on this interpretation for our wires.

It is worth emphasizing that the electronic motion is most certainly three-dimensional in these wires, unlike the ACF seen in GaAs/(Al)GaAs heterostructures[6]. In Figure 2 are three magnetoresistance traces, for a wire of nominal width 100 nm at 4.2K. The uppermost trace is with the magnetic field perpendicular to the plane of the substrate, the middle

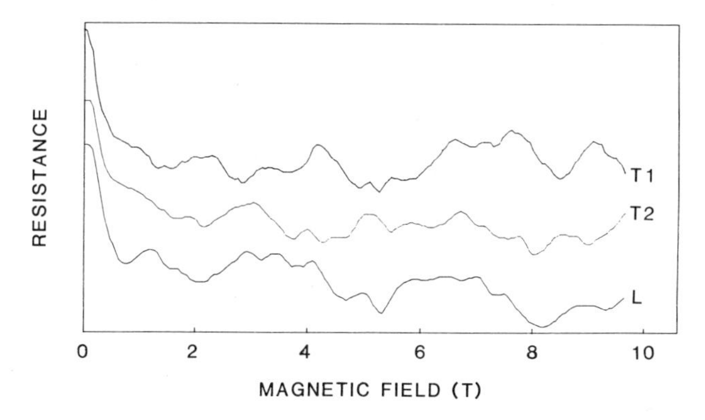

Figure 2 Magnetoresistance of a n^+ GaAs wire of nominal width 100 nm at 4.2K. The magnetic field is perpendicular to the substrate (T1), in the plane of the substrate but perpendicular to the current (T2) or parallel to the current (L).

trace with the field in the plane of the substrate but perpendicular to the current direction and the lowest trace is with the field parallel to the current direction. The results imply that the interference loops are oriented in three dimensions since it is the net flux through the loop which determines the phase shift between the interfering electron waves.

FOURIER TRANSFORM ANALYSIS

The conventional method of analysing the ACF is to use the correlation analysis of Lee and Stone [5,7]. This requires the subtraction of the NM from the raw data and it is clearly a somewhat arbitrary choice how one chooses to do this. However, using this approach we have been able to obtain some information about the coherence length, or phase breaking length, in this material [8]. In parallel with this procedure we also analysed our data using Fourier Transforms. At first we carried out the Fourier Transforms on magnetoresistance plots on which the NM had been subtracted. We found we were unable to do this consistently. To our surprise we could make consistent sense of the Fourier Transforms only if we included the NM as part of the data.

In Figure 3 we show the cosine Fourier Transforms of the magnetoresistance traces for a wire of nominal width 100 nm at different temperatures. The Transforms are for our full range of magnetic field -12T→12T. We can use a smaller range of field, say -8T→8T. In this case we obtain the same spectrum but with reduced resolution. Figure 3 shows clearly that as the temperature is increased the basic shape of the spectrum is unchanged. It is inconceivable that this could happen by chance. Therefore we are drawn empirically to the conclusion that the negative magnetoresistance and the ACF are manifestations of the same physical phenomenon. This conclusion is confirmed by comparing the total area under the Fourier Transform curve with the amplitude of the NM. This is shown in Figure 4. Each point corresponds to a different temperature and/or a different microscopic configuration of scatterers. The line drawn has slope one - the small intercept is probably due to problems in determining precisely the magnitude of the NM. The data shown cover a range of temperature from 2K to 100K.

The magnitudes of the various components in the Fourier spectrum (Figure 3) do not vary uniformly with temperature. In particular, the high frequency components decay more rapidly as the temperature is increased. This is because high frequency corresponds to large interference loops. Thus the amplitude of the peak at $1/B = 0.3T^{-1}$, which coresponds to an <u>effective</u> loop area of 700 nm^2, hardly varies at all when the temperature is changed from 4.2K to 28.8 K whereas the amplitude of the peak at $1.7T^{-1}$ changes by more than a factor of 3.

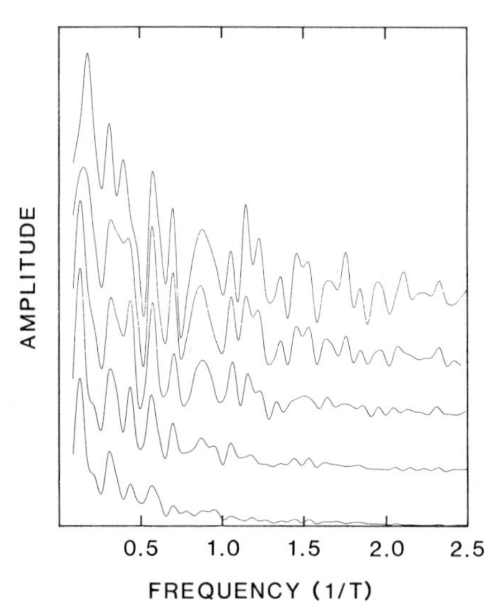

Figure 3 Fourier Transforms of the magnetoresistance for a wire of nominal width 300 nm at (from top to bottom) 4.2K, 6.8K, 16.4K, 28.8K and 51.5K. The zeros have been offset for clarity.

Our identification of the NM with the ACF leads us to consider a model in which both phenomena are due to the coherent backscattering loops used by Bergmann[4] to provide a physical picture for weak localisation. This model has enabled us to obtain values for the phase-breaking rate, τ_Φ^{-1} at various temperatures. The details are published elsewhere[6]. We find

$$\tau_\Phi^{-1} = A+BT$$

where $B = 3.8x10^{10}$ s^{-1} K^{-1} in good agreement with other workers [9,10], and A is a constant inversely proportional to the width of the wire. For example, $A= 3x10^{12}$ s^{-1} for a wire of nominal width of 300 nm. Below 10K, τ_Φ^{-1} is dominated by the temperature - independent term and this is reflected in the saturation of the amplitude of the ACF in this temperature range.

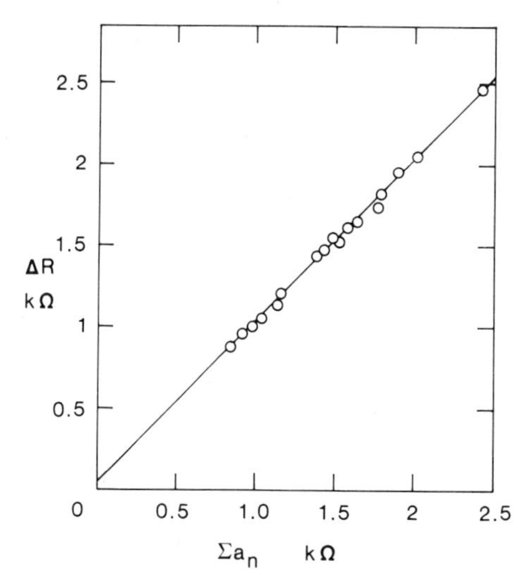

Figure 4 Negative magnetoresistance magnitude plotted against the area under the Fourier Transform spectrum. The line has slope 1.

Our interpretation of the ACF in terms of the same model as the NM, ie coherent backscattered loops, has provoked much criticism on the grounds that the two processes although similar in their reliance on interference effects are theoretically distinct. However, we would like to stress two points: (i) The equivalence of the NM and the ACF is an experimental observation from the data shown in Figures 2 and 3. This is not a matter of interpretation. (ii) It is possible that the n^+ wires behave in a different fashion from the other systems observed. There are clearly quantitative differences, for example in the temperature range over which the ACF are observed. Possibly there may also be important qualitative differences in distribution of scattering centres. There is a need for a consistent theory of the ACF which can assimilate our experimental observations.

ELECTRON HEATING EFFECTS

The ACF depend on the temperature of the electrons in the system. This is illustrated in Figure 5 where the magnetoresistance traces are taken at different measuring currents. In each case the lattice temperature is held at 4.2K.

Such electron heating effects are important both technologically and with regard to the physical processes leading to energy relaxation from the electron system. Note that the rate at which the electrons can lose energy is not necessarily the same as the phase-breaking scattering rate. For example, the latter may be due to electron-electron collisions which result in no overall loss of energy from the electron system.

We have investigated electron heating by measuring the resistance of the n^+ wires as a function of temperature using a range of rms excitation voltages. The results are shown in Figure 6. The resistances are plotted against $T_o^{-\frac{1}{2}}$, where T_o is the lattice temperature. To interpret the data we assume that the zero excitation resistance varies precisely as $T_o^{-\frac{1}{2}}$. By using this relationship we can relate the resistance as

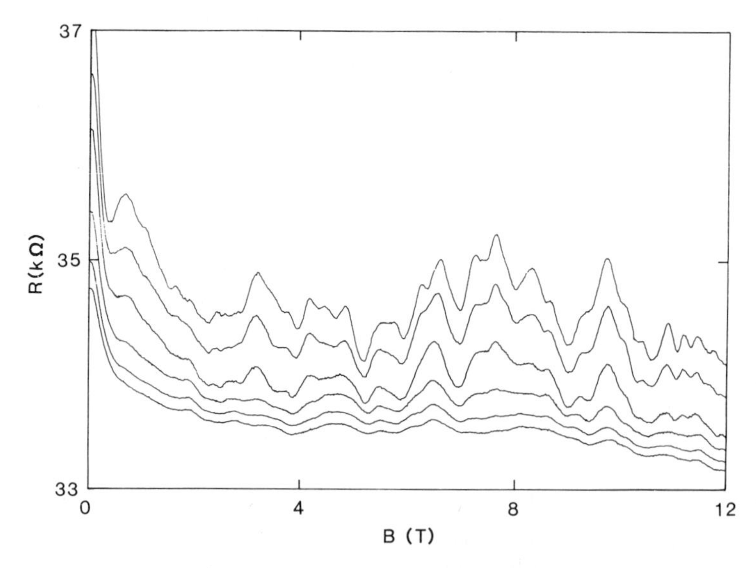

Figure 5 Magnetoresistance of a wire of nominal width 100 nm at 4.2K with rms excitation 1mV (top), 4mV, 8mV, 18mV, 28mV and 40mV (bottom).

measured at a given rms excitation to the electron temperature. Thus for any excitation voltage we can determine the electron temperature independent of lattice temperature.

We use a simple thermodynamic model to analyse the results, as shown in Figure 7. We assume that the electrons, the lattice and the helium all have well defined temperatures and that the excitations are coupled by thermal resistances. In our temperature range the Kapitza resistance, R_K, is negligible so we can assume the lattice temperature is held at the temperature of the helium bath [12].

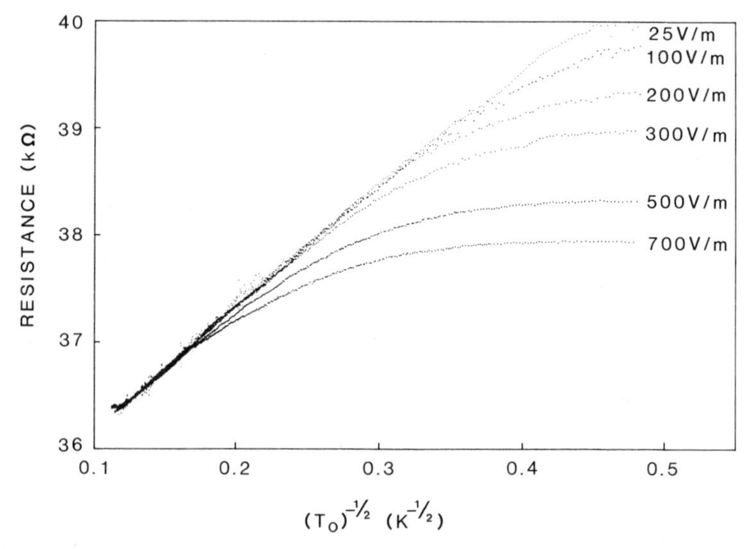

Figure 6 Resistance of a 100 nm wide wire plotted as a function of $(T_o)^{-\frac{1}{2}}$ at various rms electric fields.

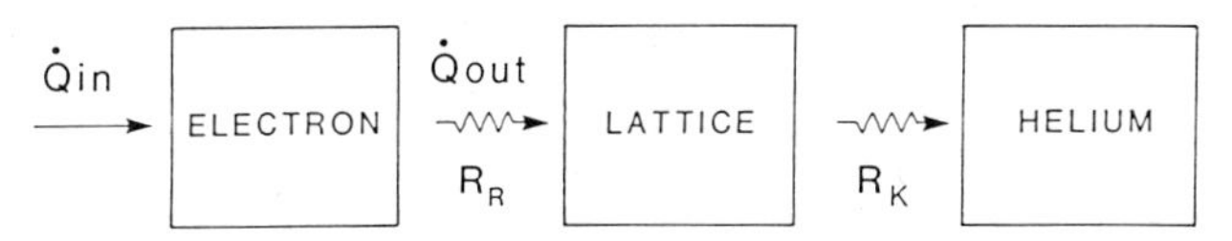

Figure 7 Thermodynamic model for electron heating.

By definition

$$d\dot{Q}_{out} = \frac{dT}{R_R} \tag{1}$$

where $\dot{Q}_{out}$ is the power lost by the electrons and T is the temperature difference between the electrons and the lattice. Also

$$d\dot{Q}_{IN} = 2VEdE/\rho \tag{2}$$

where V is the volume of the wire, ρ its resistivity and E is the electric field. We assume that the electrons have an energy relaxation rate

$$\tau_R^{-1} = \alpha T^p \tag{3}$$

where α and p are constants. By a simple thermodynamic arguement

$$\tau_R = R_R C_e = R_R \gamma T \tag{4}$$

where C_e is heat capacity of the electron system and γ is a constant. $C_e = \gamma T$ will be true as long as T is well below the Fermi temperature of the electrons which for this system is 1500K. Considering a steady state where $d\dot{Q}_{IN} = d\dot{Q}_{out}$ and integrating we obtain the expression

$$\frac{2}{2+p} \log \left[\frac{E}{T_0^{(2+p/2)}}\right] = \log \left[\left(\left(\frac{T}{T_o}\right)^{2+p} - 1\right)^{1/(2+p)}\right] + \frac{1}{2+p} \log \left[\frac{\alpha\gamma\rho}{V(2+p)}\right] \tag{5}$$

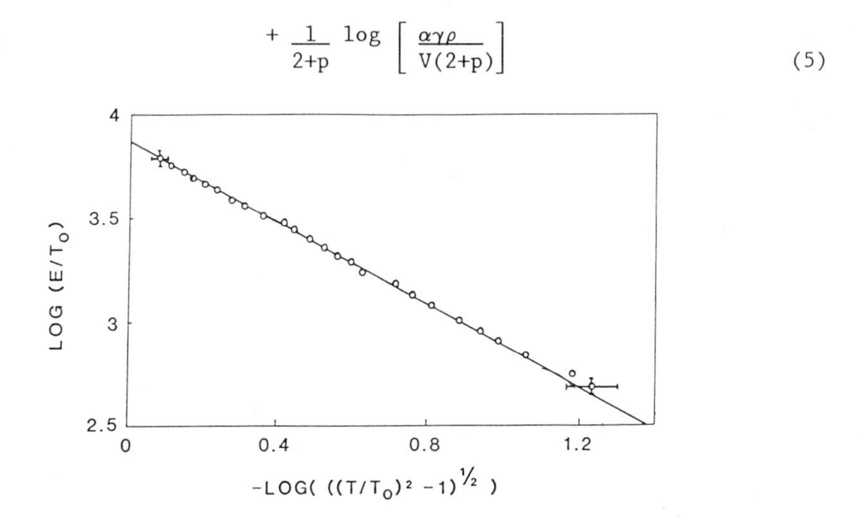

Figure 8 Equation 5 plotted using p = 0 for a lattice temperature of 22K.

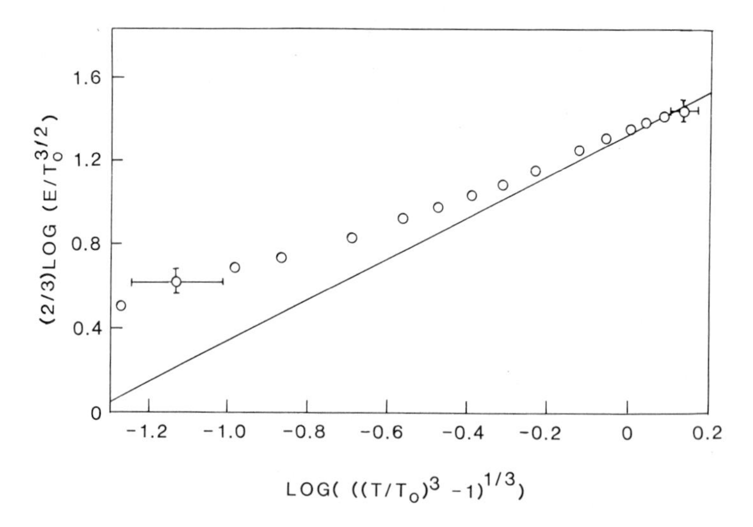

Figure 9 The same data as shown for Figure 8 plotted using p = ½.

If we ignore the variation of ρ with E, which is always less than 5%, we can determine p by plotting the left-hand side of the equation against the first term on the right-hand side. The second term on the right-hand side is then a constant. For the correct choice of p, the graph will have a slope of unity.

We find that such a plot will only lead to a slope of unity if we take p=0. This is illustrated in Figure 8 for T_o= 22K. The line drawn is of unit slope. Similar quality fits are obtained for all temperatures below 40K. The next best fits are obtained for p = ½, but these are very poor in comparison (see Figure 9). p = 0 implies a temperature independent energy relaxation rate. To obtain an absolute value for the scattering rate we take the intercept of all plots like Figure 8 and obtain α. We find α=1.4(±0.2)x10^9 s^{-1} at all values of T_o which is good confirmation of the consistency of the model. This scattering rate is much less than the phase-breaking rate. A suitable physical picture might be one in which strong scattering is taking place at the surface of the dry-etched wires. Each collision is sufficient to destroy the phase of a single electron, but several thousand collisions must take place before the energy of the electron is lost. It is straightforward to show that for an electron of energy ϵ losing energy $\delta\epsilon$ at each collision [13]

$$\tau_\Phi \sim \left(\frac{h^2 \tau}{(\delta\epsilon)^2} \right)^{1/3}$$

and

$$\tau_R \sim \tau \left(\frac{\epsilon}{\delta\epsilon} \right)^2$$

where τ is the time between collisions. If $\delta\epsilon \ll \epsilon$ then it is reasonable that $\tau_R \gg \tau_p$. Only if $\delta\epsilon \sim \epsilon$ will the times be comparable.

Other evidence that surface processes may be important is that the value of α can change drastically, by a factor of four on one occasion, after a wire has been left in air at room temperature for long periods.

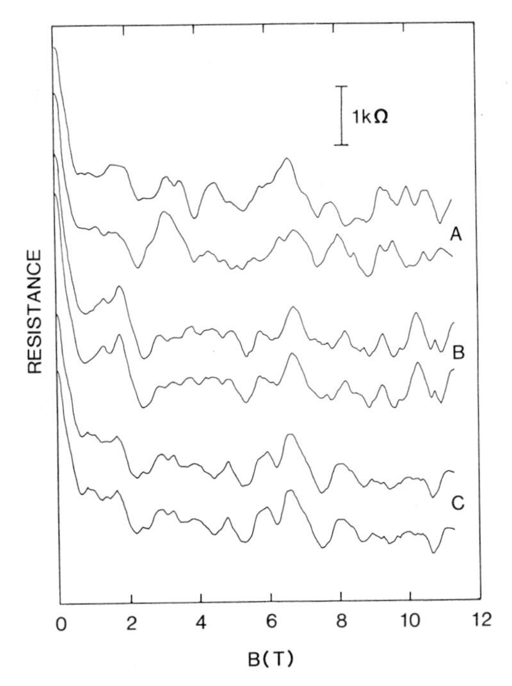

Figure 10 Three pairs of magnetoresistance traces taken at 4.2K for a wire of nominal width 100 nm. Pair C are two consecutive traces at 4.2K, pair B has an intermediate temperature of 30K and pair A an intermediate temperature of 130K.

STABILITY OF THE ACF

As long as the temperature is maintained below 100K, the magnetoresistance traces, such as those shown in Figure 1, are entirely stable over periods of many days. However, raising the temperature to some intermediate value, then cooling it again may result in a different ACF spectrum provided the intermediate temperature exceeds about 100K. This can be seen in Figure 10 for a wire of nominal width 100 nm. All traces were taken at T=4.2K and the bottom pair,C, are two consecutive traces taken to illustrate the stability. Pair B have an intermediate temperature of 30K, that is the wire was raised to 30K and held there for ~30 minutes in between the two traces. There is still essentially total correlation. However, for pair A, where the intermediate temperature is 130K, the two traces are appreciably different.

To investigate this effect in more detail we define a correlation function, F, to correlate the rth magnetoresistance trace with the $(r-1)^{th}$ trace.

$$F = \left\langle \frac{<(\Delta R(B))^2> - (Rr(B) - R_{r-1}(B))^2}{<(\Delta R(B))^2>} \right\rangle_B^{\frac{1}{2}} . \qquad (6)$$

Since the traces are taken digitally at 1000 different magnetic field values it is easy to determine $(R_r - R_{r-1})$ at any value of B. $<(\Delta R(B))^2>$ is the mean square deviation between 78 uncorrelated pairs averaged over all field values. It provides a normalisation parameter for determining the degree of correlation between two arbitrary traces. $< >_B$ means averaged over all field values.

For two identical traces $R_r = R_{r-1}$ and F = 1. For two completely uncorrelated traces $<(R_r(B)-R_{r-1}(B))^2> = <(\Delta R(B))^2>$ and F = 0. A graph of F plotted as a function of intermediate temperature is shown in Figure 11. The crosses indicate pairs of traces where the intermediate temperature was held for a time t <30 minutes and the circles for t >30 minutes.

Figure 11 shows that the microscopic scattering configuration is stable below ~ 80K. Above 80K charge is being redistributed in the wire. Comparing the form of Figure 11 with the behaviour of the carrier concentration in bulk GaAs of similar doping after illumination as seen by Theis[14] we provisionally identified the charge redistribution with changes in the population of the resonant DX level associated with the Si donor. However, at present we are not able to confirm this hypothesis and there may be some evidence for depopulation of surface traps[15]. One puzzling feature of our results is that the degree of de-correlation seems not to depend strongly on the time spent at the intermediate temperature. This may be a consequence of long experimental thermal time constants obscuring the true time-dependence.

Note that, in this experiment, the ACF act as a unique method of determining changes in the microscopic charge distribution. All bulk methods such as resistance, Hall effect etc. measure average properties and are not sensitive to detailed change. In fact we believe only about 50 scatterers out of a total of $\sim 10^5$ are changing their change state in the wire, approximately one in each phase-coherent sub-unit in the wire. This is sufficient to de-correlate strongly the ACF spectrum.

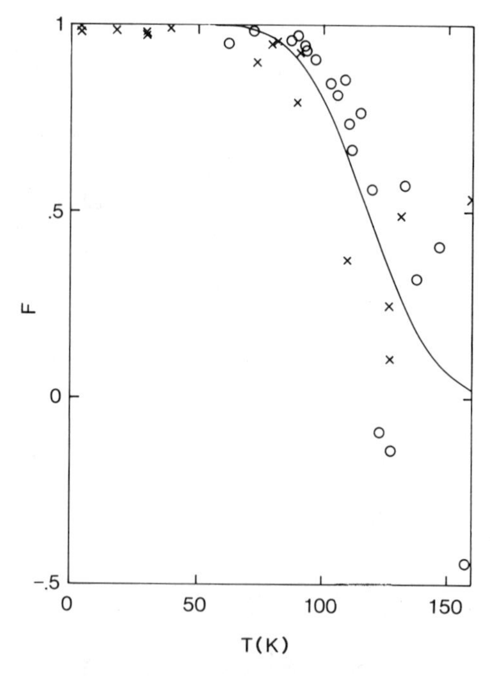

Figure 11 Correlation function F plotted as a function of intermediate temperature. See text for explanation of circles and crosses.

To provide a further tool to investigate the ACF we optically illuminated the wires at 4.2K using light of wavelength 1000 nm. Illumination does cause a de-correlation of the ACF spectrum. However, the situation is very complicated since following illumination, even if the wires are held at 4.2K, the ACF spectrum is time dependent, only stabilising after several hours in the dark.

CONCLUSION

We have studied the conductance properties of n^+ GaAs wires. We find conductance fluctuations in the low temperature magnetoresistance which appear to be intimately connected with the negative magneto-resistance usually associated with weak-localisation. The inelastic scattering processes appear to be dominated by surface scattering,both for the phase-breaking and energy relaxation rates, although the latter is much smaller than the former. The fabrication technique may be critical here in determining the degree of surface scattering. Finally, we have made an attempt to use the ACF as a unique probe of changes in the microscopic configuration of scatterers in the material.

This work is supported by the SERC.

REFERENCES

1. See for example R.A. Webb, S. Washburn, C.P. Umbach and R.B. Laibowitz, 1985, Phys.Rev. Lett. 54, 2696.
2. S. Thoms, S.P. Beaumont, C.D.W. Wilkinson, J. Frost and C.R. Stanley, 1986, "Ultrasmall Device Fabrication Using Dry Etching of GaAs" in Proceedings of Microelectronic Engineering 1986, H.W. Lehman and Ch. Bleicker, eds., Elsevier, Holland.
3. See for example S.B. Kaplan and A. Harstein, 1986, Phys. Rev. Lett. 56, 2403.
4. G. Bergmann, 1984, Physics Reports, 107, 1.
5. P.A. Lee, A.D. Stone and H. Fukuyama, 1987, Phys. Rev. B35, 1039.
6. R.P. Taylor, M.L. Leadbeater, G.P. Whittington, P.C. Main, L. Eaves, S.P. Beaumont, I. McIntyre, S. Thoms and C.D.W. Wilkinson, 1988, Surface Science, 196, 52.
7. W.J. Skocpol, 1987, Physica Scripta, T19, 95.
8. G.P. Whittington, 1989, Ph.D. Thesis, University of Nottingham.
9. S. Morita, N. Mikoshiba, Y. Koike, T. Fukose, M. Kitagawa and S. Ishida, 1984, J. Phys. Soc. Japan, 53, 2185.
10. E. Abrahams, P.W. Anderson, P.A. Lee and T.V. Ramakrishan, 1981, Phys. Rev. B24, 6783.
11. M.L. Roukes, M.R. Freeman, R.S. Germain, R.C. Richardson and M.B. Ketchen, 1985, Phys Rev. Lett., 55, 422
12. K. Gray, 1981, "Non Equilibrium Superconductivity, Phonons and Kapitza Boundaries", Plenum Press.
13. R.P. Taylor, 1988, Ph.D. Thesis, University of Nottingham.
14. T.N. Theis, 1987, Inst. Phys. Conf. Ser. 91, Chapter 1.
15. A. Long, Dept. of Physics, University of Glasgow, to be published.

ONE-DIMENSIONAL ELECTRON TRANSPORT IN EDGE-CHANNELS STUDIED WITH QUANTUM POINT CONTACTS

C.J.P.M. Harmans, B.J. van Wees, L.P. Kouwenhoven
Faculty of Applied Physics, Delft University of Technology
P.O.Box 5046, 2600 GA Delft, The Netherlands

J.G. Williamson
Philips Research Laboratories, 5600 JA Eindhoven, The Netherlands

1. INTRODUCTION

Electronic transport in a 2-dimensional electron gas (2DEG) has been studied widely, in particular following the discovery of the quantum Hall effect (QHE). Initially, the integer effect (IQHE) was explained as to emerge in the full width of the 2DEG area, i.e. as a bulk effect [1]. To explain the width of the quantized plateaux, the existence of localised states residing in the bulk was required.

More recently an alternative approach to the electronic transport in the integer regime has been proposed [2,3,4]. In this description, based on the Landauer formalism for the transport of electrons[5], magnetic field induced extended states at the boundaries of the sample carry the current. The resulting 1-dimensional magnetic edge-channels each contribute an equal amount of current, leading to a quantization of the conductance governed by the number of parallel channels.

In this paper we present experimental evidence for the existence and characteristics of magnetic edge states in high magnetic fields. To that purpose we have used quantum point contacts (QPCs), attached to a wide 2DEG area, as controllable current injector and voltage detector contacts, which can selectively probe specific Landau levels in the wide area. In section 2 we will briefly introduce electron transport in a high magnetic field in edge channels. The following section 3 describes a Hall experiment in the ballistic regime using QPCs as probes to the wide area. The resulting anomalous integer quantum Hall effect is attributed to the selective probing of the occupation of edge channels in the 2DEG bulk. Section 4 concentrates on experiments using the Shubnikov-de Haas resistance oscillations to study the scattering between adjacent edge channels at a single edge and between the edge channels at two opposite boundaries of the 2DEG. Finally, in section 5 the conclusions will be formulated.

2. ELECTRON TRANSPORT IN A HIGH MAGNETIC FIELD

In high magnetic fields the electron transport in a 2DEG cannot be described without accounting for the existence of the boundaries or edges of the sample. These boundaries result from the

Science and Engineering of One- and Zero-Dimensional Semiconductors
Edited by S.P. Beaumont and C.M. Sotomajor Torres
Plenum Press, New York, 1990

depletion of the 2DEG due to band bending, either from trapped charge at the edge surface or by external electrostatic means. This will lead to a rise of the local electrostatic potential. Also the Landau levels, into which the electrons are condensed in a high magnetic field, will be forced towards higher energy at the edges. At the intersection of the Landau levels with the Fermi energy 1-dimensional states will be formed extending along the edge. These states are called magnetic edge states or edge channels. Semi-classically these states can be seen to extend along the equipotential lines along which the orbiting electrons travels. As, with the Fermi energy residing between the Landau levels, these are the only available extended states, the net current will be transported through these states exclusively.

Using the Landauer formalism[4] the two-terminal conductance between two contacts (or reservoirs) attached to the 2DEG, which transmit and receive electrons through 1-dimensional quantum channels, is given by $G_{2t}=T(e^2/h)$, with T denoting the total transmission probability through the channels. In a wide 2DEG area in high magnetic fields the edge states associated with the N lower lying Landau levels located at one edge are well seperated from the states at the opposite edge. This makes the contribution to the transmission due to these channels equal to $T_i=1$ per channel. If in the bulk the highest filled Landau level (with index N+1) is close to the Fermi energy, the associated edge states at the two boundaries may couple partially, leading to back-scattering of the electrons and so the transmission T_u for this channel will be less than unity. This leads to a total transmission $T=N+T_u$. Quantization G_{2t} will occur if $T_u=0$, i.e. if no coupling between the opposite edge states exists. Using basically the same arguments for a four-terminal configuration leads to a quantization of the Hall conductance $G_H=N(e^2/h)$.

Now the question arises how the transport will be affected by scattering. As explained by Büttiker[4] a large magnetic field strongly affects the nature of the scattering proces, modifying it from isotropic to forward scattering. This implies that an electron which moves in an edge state and is scattered from a local potential disturbance at the edge, will continue to propagate along the same edge in the original direction, only acquiring an extra phase shift in this proces. Clearly this does not imply that the scattered electron conserves its edge channel quantum number, i.e. intra-edge scattering is still allowed. The effectiveness of this last scattering process will depend on the mixing of the different edge channel wavefunctions induced by the scattering potential. By experimentally studying the spatial distribution over macroscopic distances of the difference in population of two adjacent channels we will show that this scattering can be very small.

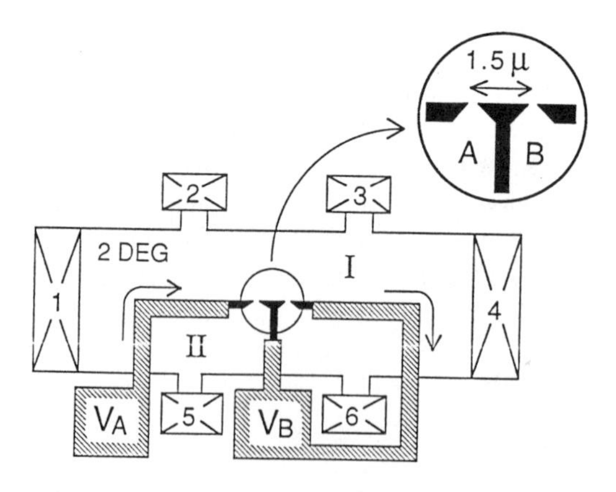

Figure1. Schematic layout of the double QPC device. The gates define two adjacent quantum point contacts (QPCs) A and B. The transmission of each QPC can be controlled individually by the gate voltages V_A and V_B. The arrows indicate the electron flow for the magnetic field orientation chosen in the experiment of section 3.

If the width of the 2DEG is reduced locally[6], ultimately the depletion potentials of the two edges will start to overlap, resulting in a rise of the electrostatic potential in the middle of the constriction. If the constriction is of a limited length, a saddle-shaped potential barrier will result. A structure of this type, called a quantum point contact (QPC), provides a powerful tool to investigate quantum transport through edge channels. In particular if the constriction is realised by controlled external electrostatic depletion using a metallic split-gate, the variable nature of this structure (by varying the gate potential) provides a controlled transmittance of the upper channel and of the number of channels being transmitted. It is this energy- or mode-selective nature of the QPC which we have employed to study edge channels in the 2DEG, either by using the QPC to selectively inject current into those edge channels which are

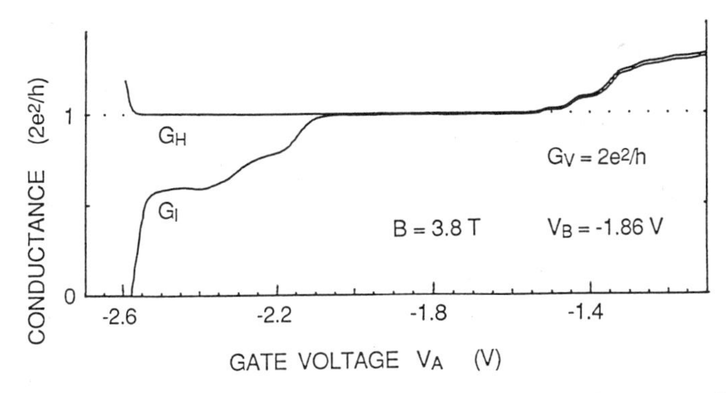

Figure 2. The Hall conductance G_H and the two-terminal conductance G_I of the current injector QPC A.The conductance G_V of the voltage probing QPC B is held constant at $2e^2/h$. It shows G_H to be equal to the largest of the two conductances G_I or G_V.

transmitted through the QPC, or to selectively measure the electrochemical potential of the carriers populating those edge channels which are transmitted by the QPC.

3. SELECTIVE PROBING OF EDGE CHANNELS: ANOMALOUS INTEGER QUANTUM HALL EFFECT

As indicated before, under IQHE conditions, the Hall or transverse conductance, being the ratio of the current carried by N edge channels and the electrochemical potential difference between the opposite edges, is quantized and given by $G_H=N(e^2/h)$. This expression is only valid if the contacts attached to the edges emit and absorb all edge channels equally. If any channel is only partially transmitted, G_H can differ from its exact, quantized value. We want to study the effect of partial filling of magnetic edge channels by employing the energy selective transmission of a QPC.

To that purpose (Fig.1) two split-gate QPCs are attached to the edge of a wide 2DEG area. The high mobility GaAs-AlGaAs heterostructure has an electron density of $n_e \approx 3.5\ 10^{15}$ /m^2 and a mobility $\mu \approx 80$ m^2/Vs, resulting in an elastic mean free path (at zero magnetic field) of $l_e \approx 8$ μm. The distance between the two QPCs of 1.5 μm ensures that the transport will be ballistic in the region between the contacts, provided the temperature is sufficiently low (inelastic length $l_i \gg l_e$).

Assuming the current I is injected and extracted through contacts 4 and 5, and the Hall voltage is measured between the contacts 1 and 6, we calculate $G_H=eI/(\mu_6-\mu_1)$. In each QPC the number of fully transmitted channels N will depend on the applied gate voltage, with $N \leq N_L$, the number of bulk 2DEG edge channels. First, take $N_A<N_B$ and take $\mu_1=0$ for convenience. For the given orientation of the magnetic field all electrons being emitted by contact 4 are absorbed by contact 1, so the electrochemical potentials will be identical, i.e. $\mu_1=\mu_4$. Clearly, the injected current through the N_A+1 channels in QPC A will be distributed unequally among the channels, as the lowest N_A channels will be fully populated up to the electrochemical potential μ_5 of the emiting contact, with the upper channel being populated only partially due to the transmission T_A ($T_A<1$). The total current equals $(e/h)(N_A+T_A)\mu_5$. As the voltage probe QPC B transmits all channels of QPC A, all electrons flow into contact 6. This will result in $\mu_6=\mu_5$, and so $G_H=(e^2/h)(N_A+T_A)$. For the cases $N_A>N_B$ and $N_A=N_B$, G_H can be calculated in a comparable way[7], leading to $G_H=(e^2/h)(N_B+T_B)$ for $N_A>N_B$ and $G_H=(e^2/h)(N+T_A)(N+T_B)/(N+T_AT_B)$ for $N=N_A=N_B$. In this way it is found that G_H is quantized, if the QPC with the largest conductance is quantized (for $N_A \neq N_B$) or if both QPCs are quantized ($N_A=N_B$). It has to be emphasised that this quantization is independent of the number of edge channels N_L in the bulk area, reason to denote this as the anomalous IQHE.
It has to be remarked that we implicitely have assumed that all current carried by the N_A+1 edge channels leaving QPC A and running in between QPC A and B, is retained in these channels .

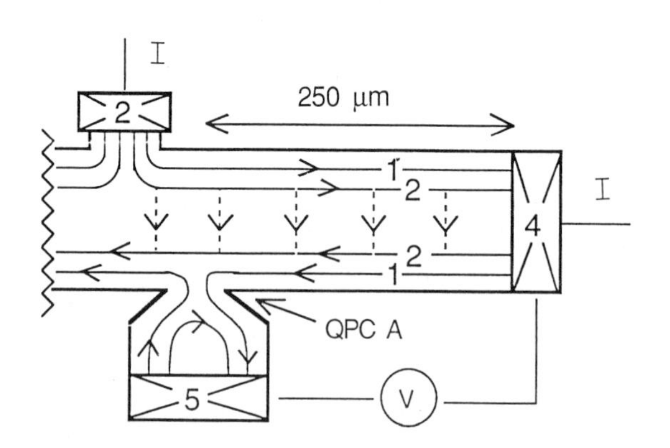

Figure 3. Edge channel pattern to demonstrate the suppression of the Shubnikov-de Haas resistance oscillations by selective detection of edge channel population. The orientation of the magnetic field and the resulting electron flow is opposite to that show in fig. 1.

If any scattering between adjacent channels would occur in this area, most notably to any of the "empty" bulk channels N_A+2, .., N_L, these electrons may not enter QPC B, and μ_6 would drop below μ_5, leading to a loss of quantization.
Fig. 2 shows the measurement of G_H (at a fixed magnetic field of 3.8 T) with the voltage detector QPC B kept at a constant transmission ($N_B=2$ and $T_B=0$) and the current injector QPC A varied by sweeping the gate voltage. Clearly G_H follows the two-terminal conductance of whichever QPC having the largest conductance, in accordance with the theory outlined above. From this and comparable experiments[7] it is concluded that virtually no scattering between edge channels located at a single edge occurs in the area between the two QPCs, i.e. over approximately 1.5 µm. As this distance is comparable to the mean free path (at zero magnetic field), it does not convincingly demonstrate the suppression of scattering at high B fields, as discussed in the introduction. So, it is of interest to study these scattering processes over considerably longer distances.

4. REDUCTION OF EDGE CHANNEL SCATTERING AND THE SUPPRESSION OF THE SHUBNIKOV-DE HAAS RESISTANCE

In section 3 experimental evidence has been presented showing that no appreciable inter-edge channel scattering (i.e. between different edge channels at a single boundary) occurs between the QPCs. To study these scattering processes over longer distances, we have used the longitudinal or Shubnikov-de Haas (SdH) resistance. Again a QPC is used to selectively probe the occupation of particular bulk edge channels.

As discussed earlier, inter-edge channel scattering in the IQHE regime will not result in a reduction of electrons transported from the injecting to the absorbing contact, and so this will not lead to resistance. However, in between the Hall plateaux where the Landau level associated with the upper edge channel passes through the Fermi level, electrons may find a path through the bulk region to traverse the sample from one edge to the other. This type of scattering will lead to a loss of forward moving electrons at one edge, resulting in a drop of the (average) electrochemical potential lengthwise this edge, i.e. longitudinal resistance.

It has to be noted that, if no scattering occurs between the adjacent channels at one edge, only the electrochemical potential of the electrons moving in the upper edge channel will be affected. This implies that under this condition selective probing of the edge channels by a QPC will yield either a normal SdH resistance peak in case of transmission of the upper channel through the QPC or a suppressed peak if the upper channel is not transmitted.The experiments were performed on the same sample as before (Fig 1). The contacts 2 and 4 were used for current injection. In a 3-terminal configuration the voltage difference between contact 4 and QPC A (leading to contact 5) was measured. Fig. 3 shows the electron flow for two bulk Landau levels, including the inter edge scattering of the upper edge channel. Fig. 4 displays the experimental

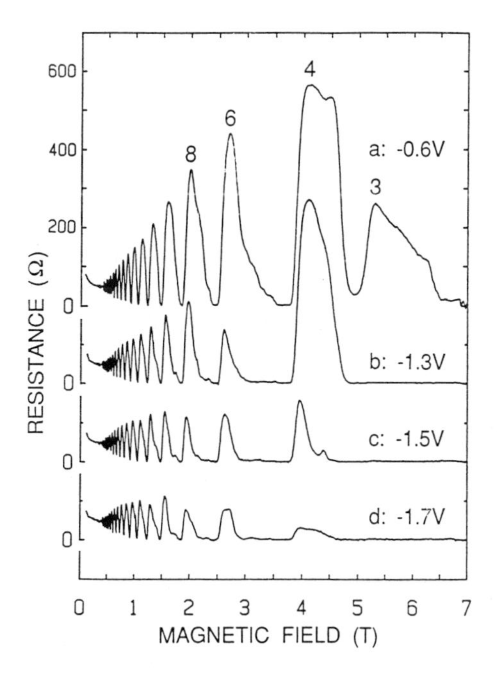

Figure 4. Experimental Shubnikov-deHaas resistance oscillations obtained for the configuration shown in fig.3. Curve (a) shows the regular SdH resitance pattern, curves (b), (c) and (d) show the suppression of the SdH signal.

results obtained for different gate voltages, i.e. different numbers of channels being transmitted through the QPC. Trace (a) shows the SdH resistance at V_g=-0.6 V. At this gate voltage the QPC is just formed and it transmits all channels residing in the bulk, so a regular SdH signal is found. Traces (b), (c) and (d) are obtained for a gradually more negative gate voltage, leading to a proportionally decreasing number of transmitted channels. From the figure it is clearly seen that the less channels there are transmitted through the QPC, the more SdH peaks are found to be suppressed or even disappeared. At V_g=-1.7 V (trace (d)) the N=3 peak has virtually disappeared, the N=4 peak is strongly suppressed and the other peaks (up to N=12, for B>1.5 T) are reduced substantially. A comparison of these results with the number of transmitted channels in the QPC as obtained from the measurement of the conductance of the QPC, confirms that only those SdH peaks which are associated with channels being transmitted by the QPC are fully developed, while all others are suppressed[8].
These experimental results lead to two conclusions. First it shows that the SdH resistance originates from the back-scattering from one edge of the sample to the other of electrons propagating in the edge channel associated with the highest filled Landau level, when this level crosses the Fermi energy. Secondly, it shows that a non-equilibrium population at two adjacent edge channels can persist over macroscopic distances of the order of the sample dimensions. In particular the last result indicates the strong reduction of scattering of electrons populating these channels in a strong magnetic field[4,9].This implies the transport to occur nearly fully adiabatically, i.e. under the conservation of quantum numbers.

5. CONCLUSIONS

The experiments presented here underline the interesting properties of the quantum point contact as a tool to investigate electron transport in a 2-dimensional electron gas on microscopic and macroscopic scale, employing their adjustable energy selective transmission characteristics. The results obtained can be explained fully using the concept of edge channels induced in a high magnetic field. This fact strongly suggests that the explanation of the integer quantum Hall effect has to be based on this edge channel model. Both the anomalous integer quantum Hall effect and the suppression of the Shubnikov-de Haas resistance clearly show the strong reduction of scattering in edge channel transport in high fields, leading to adiabatic transport over distances of typical sample sizes, i.e. hundreds of micrometers.

ACKNOWLEDGEMENT

We thank M.E.I. Broekaart and C.E. Timmering at the Philips Research Laboratories and the Delft Center for Submicron Technology (CST) for their contribution in the device fabrication, C.T. Foxon and J.J. Harris at the Philips Redhill Research Laboratories for the high mobility heterostructures, and the Stichting F.O.M for financial support for this project.

REFERENCES

1. "The Quantum Hall Effect", ed. R.E. Prange and S.M. Girvin, Springer Verlag, NewYork, 1987
2. B.I. Halperin, Phys. Rev. B25, 2185 (1982)
3. P.Streda, J. Kucera and A.H. MacDonnald, Phys. Rev. Lett. 59, 1973 (1987); J.K. Jain and S.A. Kivelson, Phys. Rev. Lett. 60, 1542, (1988)
4. M. Büttiker, Phys. Rev. B38, 9375 (1988)

5. R. Landauer, IBM J. Res. Dev. 1, 223 (1957)
6. B.J. van Wees, H. van Houten, C.W.J. Beenakker, J.G. Williamson, L.P. Kouwenhoven, D.van der Marel and C.T. Foxon, Phys. Rev. Lett. 60, 848 (1988); B.J. van Wees, L.P. Kouwenhoven, H. van Houten, C.W.J. Beenakker, J.E. Mooij, C.T. Foxon and J.J. Harris, Phys. Rev. B38, 3625 (1988); D.A. Wharam, T.J. Thornton, R. Newbury, M. Pepper, H. Ahmed, J.E.F. Frost, D.G. Hasko, D.C. Peacock, D.A. Richie and G.A.C. Jones, J. Phys. C21, L209 (1988)
7. B.J. van Wees, E.M.M Willems, C.J.P.M. Harmans, C.W.J. Beenakker, H. van Houten, J.G. Williamson, C.T. Foxon and J.J. Harris, Phys. Rev. Lett. 62, 1181 (1989)
8. B.J. van Wees, E.M.M. Willems, L.P. Kouwenhoven, C.J.P.M. Harmans, J.G. Williamson, C.T. Foxon and J.J. Harris, Phys. Rev. B39, 8066 (1989)
9. T. Martin and S. Feng, preprint (1989)

ELECTRON WAVEGUIDE JUNCTIONS:

SCATTERING FROM A MICROFABRICATION-IMPOSED POTENTIAL

M.L. Roukes, T.J. Thornton, A. Scherer,
J.A. Simmons†, B.P. Van der Gaag, and E.D. Beebe

Bellcore
Red Bank, New Jersey 07701

INTRODUCTION: JUNCTIONS AS SCATTERERS

In *quantum wires*, ultranarrow conduction channels having very low disorder, electrons can scatter from potentials which are radically different from those encountered in the bulk. Entirely new types of scattering potentials can be created by microfabrication, since ballistic devices having *overall* dimensions much smaller than the transport mean free path, and widths only slightly larger than one electron wavelength, can now be fabricated.[1]

In recent experiments on these quasi one-dimensional systems two important types of laterally-imposed potentials have emerged: those created by *constrictions*, and those created by *junctions*. Quantized resistance steps, observed in ballistic point contacts,[2] are a direct manifestation of a "contact resistance" between a broad region with closely-spaced modes and an adjacent narrow constriction having widely-spaced modes.[3] In multi-terminal devices, geometrical scattering at intersections (junctions) along a quantum wire, where connections to voltage probes are made, can dominate random impurity scattering. This has been demonstrated rather dramatically in recent experimental work. Quenching of the Hall resistance was one of the first graphic results of junction scattering observed.[4] Two closely-related discoveries followed: the resistance associated with a bend in a narrow conducting path was demonstrated[5] and, more recently, negative four-terminal resistance at a cross junction has been observed[6], induced by the ballistic transfer of electrons into the "wrong" probe. Through these and other recent experiments, and in conjunction with current theoretical work[e.g., 7-14], a cohesive picture of quantum transport at a junction is emerging.

These phenomena are all observed in samples patterned from high mobility two-dimensional electron gas heterojunctions (2DEGS). In these structures transport is ballistic for distances exceeding the physical separation of the measurement probes. Over the regions measured, electrons scatter predominantly from the microfabricated channel sidewalls and retain phase coherence for distances well beyond the junction region itself. Clearly, in this regime the voltage measurement is non-local. Even if no net current flows into the voltage probes used in the measurement, the probes are strongly coupled to the junction region. We thus expect *a priori* that this coupling will be an intrinsic part of the underlying physics. The physical confinement of the electrons in these narrow conduction channels creates appreciable energy separation between electronic levels. For $100nm$ wide channels, energy level separations, ΔE, are of order $10K$. At low temperatures, $k_B T < \Delta E$, electron transport involves only a few of these one-dimensional "modes" at the Fermi energy, E_F. If the intermode scattering is infrequent over the length of the conduction channel these narrow structures constitute "waveguides" for the electrons.

In the situation described electron transport can be viewed as a scattering phenomenon, where a T-matrix, describing the junction, ultimately determines the macroscopically-measured "resistances".[13] Waves propagating in one lead towards the junction region are either reflected or

Science and Engineering of One- and Zero-Dimensional Semiconductors
Edited by S.P. Beaumont and C.M. Sotomajor Torres
Plenum Press, New York, 1990

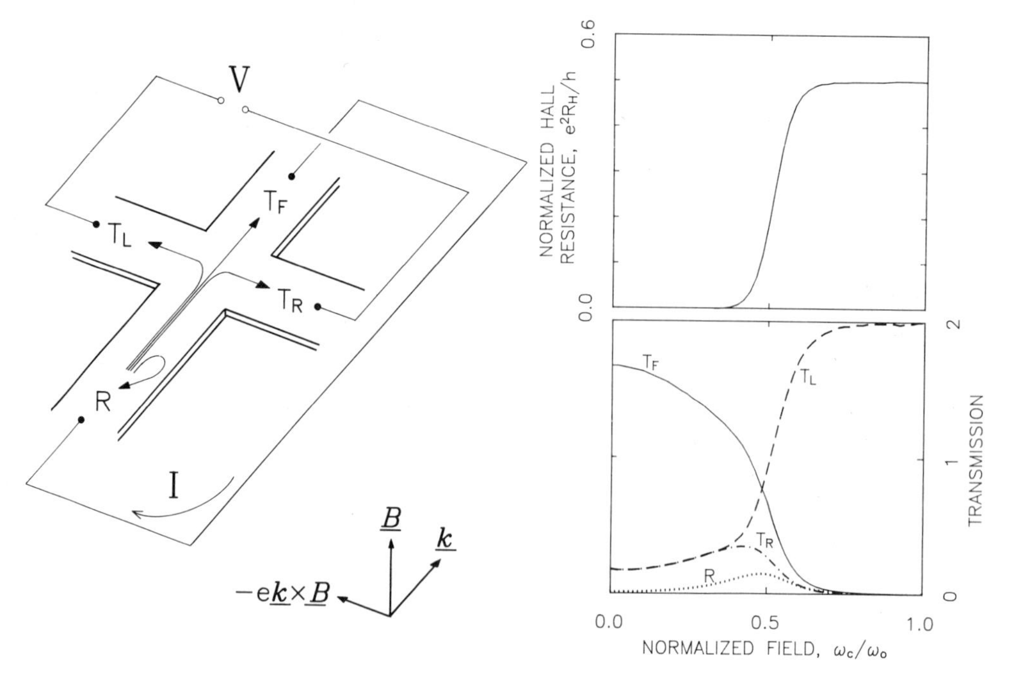

Figure 1. *left)* The Hall resistance, R_H, at a ballistic cross junction. *right)* R_H and corresponding transmission coefficients calculated[11] for the case of a "resonant quench" *(see text)*.

transmitted into outgoing waves in adjacent leads. The energy contours of the microfabrication-imposed confinement potential, $U(x,y)$, in the locale of the junction determine the matrix elements, T_{ij}, and, thus, how the (asymptotic) plane wave modes in each lead are coupled those in other leads. (Here, the z-direction is defined as normal to the heterojunction plane.) The presence of the leads *themselves* determine the form of this junction scattering potential. When many transverse modes are occupied, electron wavepackets can be constructed and transport can be described by semiclassical electron trajectories. For this multimode situation, the T_{ij}'s and, thus, the resistances have modelled experimental data in certain devices with a remarkable degree of accuracy by treating the electrons simply as "billiards".[14] In measuring transport through electron waveguide junctions we explore the interesting regime where, in general, resistances are sensed non-locally, yet *local* potential profiles of the junctions largely determine transmission into each of the samples' leads. In some sense, these investigations are the solid-state analogs of atomic scattering experiments.

Below, we present an abbreviated description of our recent measurements of magnetotransport at three types of "cross" junctions. Their conducting geometries are defined by the techniques of *self-aligned gates*, both with and without tapered leads, and by *pinched gates*. These techniques are described below. Our results with self-aligned gates represent the first reported measurements in electron waveguide structures in which electron density and the geometry of the conducting region can, to a large degree, be varied *independently*.[15,28] This permits a closer comparison with recent numerical simulations of quantum transport in small four-terminal devices. In these calculations confinement potentials have been modelled as fixed while E_F, and therefore the number of occupied modes, is varied.

MAGNETOTRANSPORT IN A BALLISTIC MULTITERMINAL CONDUCTOR

A intersecting pair of wires constitutes perhaps the simplest Hall bar. Consider conducting strips patterned from a 2DEG and crossed at right angles (Figure 1a) – if current is passed along one strip while voltage is measured across the other, only the transverse component of the voltage across the intersection, or junction region, is sensed. We've come to call the ratio of the voltage to the imposed current the *Hall resistance*, R_H. In what follows we'll continue to call it as such, although in very narrow ballistic conductors it will become readily apparent that little else, save for this operational definition of R_H just described, will remain. In *macroscopic* Hall geometry samples patterned from a 2DEG, except for deviations from ideal geometry, the Hall resistance vanishes for zero applied magnetic field. At finite B, R_H increases linearly with magnetic field, $R_H = B/n_S e$, and is antisymmetric with respect to field direction. At low temperatures and high magnetic fields, where Landau quantization is fully developed, quantum Hall plateaus are observed. This now familiar behavior is shown in Figure 2 *(left)* for a Hall bar comprised of 500 μm wide strips of 2DEG having mobility $\mu = 1.2 \times 10^6\ cm^2/V{\cdot}sec$ and electron sheet density $n_S = 3 \times 10^{11}\ cm^{-2}$ measured at 4.2K. In narrow Hall bars at high fields the behavior remains qualitatively the same but, at low-B, gross departures from the classically-expected linear behavior are evidenced. Figures 2b and 2c exemplify, for optimally defined narrow wires, the striking extent to which the Hall resistance at low-B can deviate from the linearity observed in wider samples. The 300*nm* wide sample shows the onset of the quenching phenomenon which, for 90 *nm* width wire, has become fully developed. At 90 *nm* width, R_H is seen to be quenched to a few parts in 10^3 of its classically-expected value. Similar results have been obtained subsequently in other laboratories.[5,16,17,18]

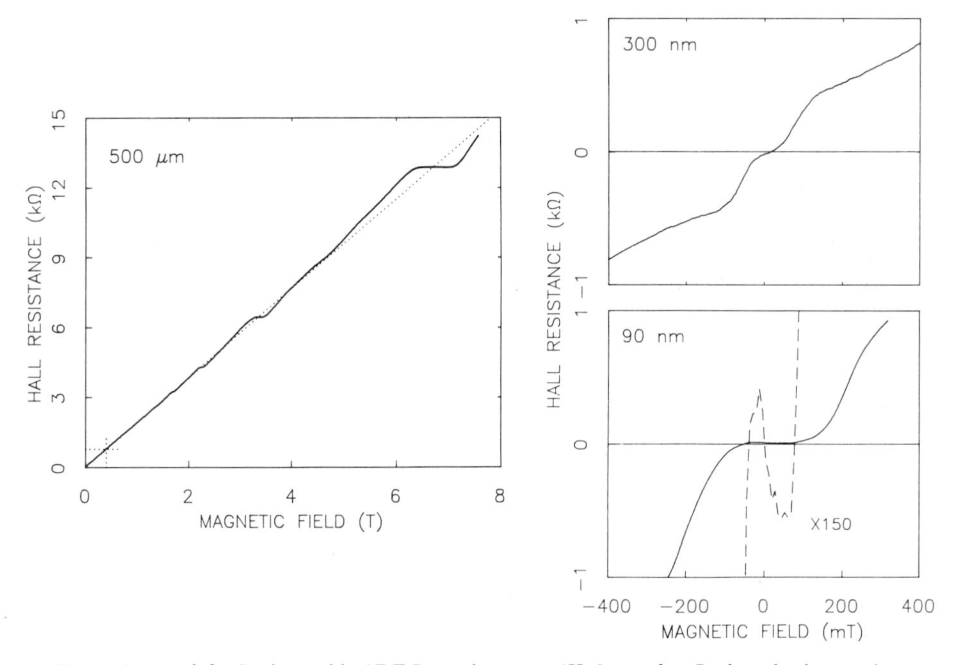

Figure 2. **a** *left)* R_H in a wide 2DEG conductor at 4K. **b,c** *right)* R_H in submicron wires.

In the last three years there have been great improvements in techniques for fabricating small structures from high mobility 2DEGs. These are briefly described below. With these advances it has been possible to suppress the role of random backscattering resulting from processing-induced defects. The reduction of this scattering allows systematic features arising from the microfabrication-imposed potentials to dominate electrical transport in these structures. Nonetheless, despite these improvements, some residual sample-specific randomness persists in all fabricated devices. Magnetotransport in these small structues is characteristically comprised of *both* systematic features arising from confinement and sample-specific fluctuations. In samples comprised of arrays of, e.g., narrow wires, "built-in ensemble averaging" helps to suppress the specific fluctuations intrinsic to each of the constituent wires. Except for the effect of temperature, no such inherent averaging is operative when isolated multiterminal ballistic devices are studied— the experimentalist here must face the onerous task of obtaining ensemble averages by measuring as many nominally-like devices as necessary to isolate the random from the systematic in his or her data. Obviously, advances in fabrication which reduce this random scattering are welcome, desired, and greatly simplify the task at the outset. As a concrete example of transport in the regime where the systematic is just emerging

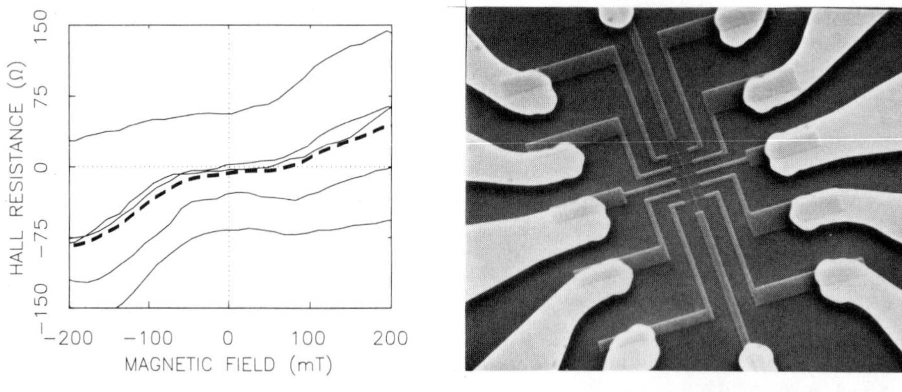

Figure 3. R_H traces from five junctions on a disordered sample and their ensemble average.

from the random, consider Figure 3. Five low-B Hall traces from each of five junctions on a single chip are displayed. These particular cross junctions were patterned using a somewhat over-vigorous ion exposure step (described below) which resulted in a sizable admixture of junction-specific fluctuations on each trace. As shown, the induced disorder causes offsets in R_H at zero field, deviations from the classically-expected antisymmetry, and adds a random "magneto-fingerprint" component to each junction's trace. The *systematic* feature inherent to all traces, however, a quench of $\sim 50mT$ width, survives ensemble averaging (dotted line) whereas the junction-specific fluctuations do not. Parenthetically, it is interesting to note that the traces shown previously in Figures 2b and 2c are actually *raw* data traces, each from a single junction without averaging. This dramatically underscores the importance of both optimizing fabrication to minimize random scattering, and, of measuring a *number* of samples to isolate fluctuation effects.

WHAT MAKES AN ELECTRON WAVEGUIDE?

There are two critical steps in the fabrication of small conducting structures from high mobility 2DEGS: *pattern definition* and *pattern transfer.*[1] In the first step, electron beam lithography is generally used to define patterns in insulating or metallic films deposited onto the surface of heterojunction material. These patterns will ultimately determine the conducting geometry of the narrow wires. A second critical step, however, must transpire before these *structural* features at the surface are transferred to the 2DEG below to define the actual *electrical* geometry. Successful transferral of submicron features has been achieved by etching[19,20], by very low energy ion beam exposure[1,21], by focussed ion beam-induced damage[22], or, using patterned metal surface films, by electrostatic depletion[23,16]. All result in a reduction of the conductivity of the 2DEG to zero over broad regions, to leave selected narrow paths of, hopefully, undisturbed material to form the conducting channels. Optimal pattern transfer generates conduction channels having electrical widths, w_{el}, which closely approximate the structural (lithographic) widths, w_{str}, of the patterned surface film. The panoply of techniques developed to achieve this transferral *(Figure 4)* reflect the difficulty in attaining this $w_{el} \sim w_{str}$ aspiration.

The difference between w_{el} and w_{str} arises from the depletion of carriers from beneath the patterned surface features at the edges of the electrical channel.[24,25] Let us first consider the case of channel definition using *pinched gates.* Here, the conductivity of narrow (ungated) regions of the 2DEG is preserved beneath narrow gaps or a patterned dielectric layer when sufficient gate potential, V_G, is applied to completely deplete broad (gated) regions *(Fig 4-iii).* Because the gate electrode is physically separated from the 2DEG, with donor and spacer layers of the 2DEG material intervening, screening effects and the divergent field from the gate electrode conspire to yield electrical channels that are *broadened* replicas of the gate geometry.[26] This lateral broadening is of order the total thickness of the intervening layers, which are typically 50-100 *nm,* and acts to reduce the abruptness of the imposed potential actually felt by the 2DEG. As a direct result of this, carriers beneath gaps smaller than ~ 200 *nm* tend to deplete out along with those under the broad gated regions as V_G is applied. From these considerations it is clear that *in pinched-gate devices the gate potential changes both the electrical geometry and the carrier density of the narrow conducting regions formed.* This confinement technique has been extensively employed in experiments involving point contacts.[e.g., 2]

Our early success in fabricating sub-100*nm* quantum wires was obtained using *very low energy ion beam exposure* to achieve pattern transfer.[1,21] This is the only technique which, to date, has demonstrated the capability of directly transferring surface features having lateral dimensions below $\sim 150nm$ into a high-mobility 2DEG. We have shown that by optimization of the ion exposure process, surface features (masks) smaller than 100 *nm* can still yield conducting channels below. This is possible because some fraction of charge control occurs closer to the 2DEG, through *physical* alteration of the 2DEG heterojunction material. In general, we find that structures patterned by ion beams; whether lightly exposed *(Fig.4-ii left)*, heavily damaged (as in the case of focussed ion beam exposure, *Fig. 4-ii right*), or physically etched *(Fig 4-i)*; have resulting "edge depletion" lengths that are *strongly* dependent on the particular choice of ion beam parameters, ion dose, and upon the exact configuration of the heterojunction. With precisely-controlled ion exposure this length has been reduced to as little as ~ 15 *nm*.[1]

We have developed the patterning of GaAs/AlGaAs heterojunctions into narrow wires having *self-aligned gates* in attempt to follow the evolution of scattering at junctions with *fixed* $U(x,y)$, in which subband occupancy *alone* is varied. In these structures a metal mask, used during the ion exposure step to achieve selective pattern transfer, subsequently serves as the top gate electrode *(Fig. 4-iv)*. Magnetic depopulation measurements[27] to determine the electrical width of such devices confirm the relative independence of w_{el} with changes V_G and, hence, changes in n_S.[28] This is to be contrasted with the behavior of pinched gate devices[16,18,32] in which *both* the junction geometry and carrier density change simultaneously with V_G. We have taken this approach in the hope of making closer comparison between experimental results and existing theoretical models which, in general, do not include self-consistent electrostatics.

In the early stages of experiments on electron waveguides, the successful fabrication of conducting wires with $w_{el} < 300$ *nm* was, of itself, enough to delight those working in the field. Taking the immediate next step, however, that of obtaining quantitative information about important properties determining the physics, has proven to be difficult. Perhaps most troublesome of the unknowns is the lack of information about the imposed confinement potential, $U(x,y)$, which determines the energy level spacing, the electrical width, and the number of modes occupied. From the discussion above, it is clear that the shape of the confinement potential, itself, is rather *sample-* (or, at best, *process-*) specific. In early work the number of occupied modes was estimated from rather loose determination of electrical widths and informed guesses for the transverse profile of $U(x,y)$. More recent work employing pinched gate junctions has allowed direct investigation of transport as a function of subband occupancy, at the expense, however, of *not* retaining fixed junction geometry[29]. Even in samples where $U(x,y)$ is essentially fixed, the exact geometry of the conducting paths, especially at junctions, still can only be approximately determined. As discussed below, this makes it difficult to make meaningful comparisons between experimental data and recent theoretical results which display a very high degree of sensitivity to the *precise* shape of $U(x,y)$.[8-12,14] These considerations also indicate that comparisons between experimental data obtained with junctions defined by different methods will, at best, serve to indicate only qualitative trends.

Finally, it is important to note that electron transport in a narrow conductor strongly depends on the quality of the boundaries imposed by the fabrication process. If the imposed lateral potentials are not smooth, the lifetime of electron eigenstates in the narrow conductor will be short. In the limit of very rough edges, level broadening can obscure the effects of confinement at low magnetic fields altogether. When many quasi one dimensional states are occupied, *intermode scattering* with small transfer of forward momentum can occur. This is analogous to small-angle scattering familiar in 2- and 3-dimensional systems containing a continuum of transverse modes. Transport phenomena that directly depend on the T_{ij} resulting from an imposed potential will tend to decay when the mean free path for intermode scattering becomes smaller than the length scale characterizing the imposed potential. In the multimode regime, intermode scattering need not result in a strong degradation of forward momentum; resistance and mobility, however, are strongly weighted for large-k scattering events – i.e. *back*scattering. Since scattering events with the requisite large momentum transfer occur only a fraction of the time, the transport mean free path in a 2DEG determined from the mobility, $\ell_0 = m^* v_F \mu / e$, can be quite large – exceeding 10 μm. The length scale for small-k intermode scattering in a narrow high mobility structure, however, primarily depends smoothness of the confining potential or upon residual background impurities and, in general, will be much shorter. This picture, of course, changes drastically when E_F is only slightly greater than the subband spacing, i.e. when only a few transverse subbands are occupied. In this regime *all* intermode scattering events are capable of significantly degrading the forward momentum.

Our recent investigations with narrow multimode wires have yielded a mean free path for small-k intermode scattering[30] and a length scale for diffuse boundary scattering[31] both exceeding 1 μm at low magnetic fields. Recent transverse electron focussing experiments, which involve specular collisions at low magnetic fields along a depletion boundary beneath a gate, have been demonstrated

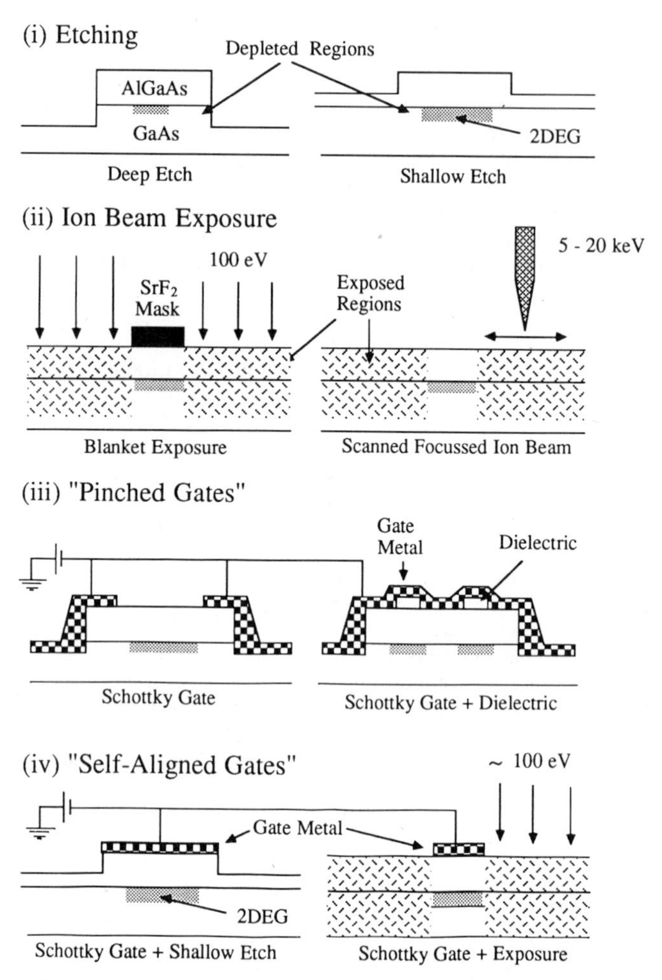

Figure 4. Techniques for imposing lateral confinement upon a 2DEG heterojunction.

over lengths of several μm.[32] These confirm that with optimal pattern transfer it is possible to create narrow conduction channels that act as electron waveguides over appreciable lengths, in contrast with a length ~ 0.3-$0.4\ \mu m$ previously deduced from the remote bend resistance in narrow wires patterned by physical etching *(Fig. 4-i right)*.[5] At high magnetic fields in the quantum Hall regime, the suppression of backscattering[33] in *all* devices tends to ameliorate the effect of potential fluctuations.

THE VARIOUS "FLAVORS" OF QUENCHING: *GENERIC, NEGATIVE, DISAPPEARING, RE-ENTRANT, ANTI- , . . .*

Ungated Junctions: "Generic" Quenching.

The first experimental traces displaying quenching of the Hall resistance were obtained in ungated devices which, therefore, had fixed width and constant electron density.[4] Qualitative dependence of this phenomenon upon w_{el} was obtained by studying a family of devices of different widths in these early experiments. *It was found that the field extent of the region of suppressed R_H (the "quench width") grew with decreasing channel width.* Quenching appeared to be a rather "generic" phenomenon in these ungated samples having n_S in the range $2-5\times 10^{11}\ cm^{-2}$ when w_{el} was smaller than a few hundred *nm.*

Pinched-Gate Junctions: "Generic", "Negative", and "Disappearing" Quenching

In experiments which followed, quenching was subsequently investigated in pinched-gate devices where w_{el} and, simultaneously, n_S were continuously varied with V_G.[16] *In these devices the field-width of the region of suppressed R_H grew monotonically with increasing negative gate bias as the channels were depleted from the edges,* confirming the results we had obtained earlier with ungated channels. Junctions in these devices were formed beneath a dielectric layer in the shape of cross strips $1\ \mu m \times 1.3\ \mu m$ wide *(Fig. 4-iii, right)* which lifted the overall gate electrode away from the surface of the heterojunction material. Quenching appeared when the edges of the remaining channel were pinched down to leave an estimated electrical width of $w_{el} \sim 0.47\ \mu m$ at which point $n_S \sim 2.0\times 10^{11}\ cm^{-2}$. In retrospect, assuming that the remaining channel depleted inward a constant amount from each edge, it is clear that the resultant electrical geometry of the junction had *significant rounding,* rather than square corners, at the value of V_G where quenching first appeared. For $V_G \sim -2.75V$, yielding an estimated $w_{el} \sim 300\ nm$ and $n_S \sim 1.6\times 10^{11} cm^{-2}$, and at lower gate voltages, we expect this rounding to have been quite pronounced. In this regime near pinch-off a *negative Hall slope,* $dR_H/dB < 0$ at low-B was observed *(Ref. 16, Fig. 4).* This negative slope had also been observed in narrow ungated devices in our laboratory patterned by low energy ion exposure. In what we assumed to be our "best" devices, however, if Hall slopes were negative, then they were also quite small in magnitude. (See, e.g., the magnified trace in Fig. 2c).

We have recently fabricated pinched-gate devices in our laboratory having somewhat narrower gaps, 600 *nm,* than those described above. *With these devices we observe a non − monotonic growth of the quenched region with decreasing gate voltage,* as depicted in Fig. 5.[15,34] The estimated widths for the set of data shown in Fig. 6 range from $w_{el} \sim 350\ nm$ at $n_S \sim 1.8\times 10^{11} cm^{-2}$ to $w_{el} \sim 150\ nm$ at $n_S \sim 1.3\times 10^{11} cm^{-2}$ for the range of V_G spanning channel definition to pinch-off. In between these two extremes a quench appears, and dR_H/dB is seen to become strongly negative. At channel definition and near pinch-off the Hall slope is positive and tends towards its classical value $(n_S e) dR_H/dB = 1$; i.e., the quench has disappeared. *The quench in this device thus appears, then disappears again, as V_G is swept monotonically.* Similar density dependence has recently been observed in the self-aligned junctions of Chang and Chang[17], patterned by shallow etching *(Fig. 4-iv left).*

Self-Aligned Gated Junctions with Tapering: Unexpected Density Dependence

In Figure 6 we display data obtained from devices with self-aligned gates having leads which are quite grossly tapered outwards as they approach the junction.[15] Such tapering has been argued to induce collimation of the electron "beam" injected into the junction.[35,8] Away from the junctions the gate widths define leads of a constant width, $w_{str} \sim 300\ nm$, which, after the particular pattern transfer conditions employed, depleted down to $w_{el} \sim 100\ nm$. This was confirmed by magnetic depopulation measurements on 6 μm long sections of wire between two junctions. At the junctions themselves the leads taper outward yielding electrical widths which we estimate to be approximately two times greater, $w_{el} \sim 200\ nm$. The low-B behavior of three such devices displayed in Fig. 6 *(right)* shows *quenching which monotonically grows with increasing electron density, i.e. increasing mode number,* over the range of V_G which could be applied to these devices. *This is precisely the opposite behavior first conjectured*[36]*, where quenching was assumed to be most pronounced if only a few of the lowest-lying modes were occupied.* The (normalized) Hall slope at low densities descends from near its classical value ($=1$) to negative values which are almost as large in magnitude. The width of the quenched region grows approximately following an $(n_S)^{0.5}$ power law, indicating its direct

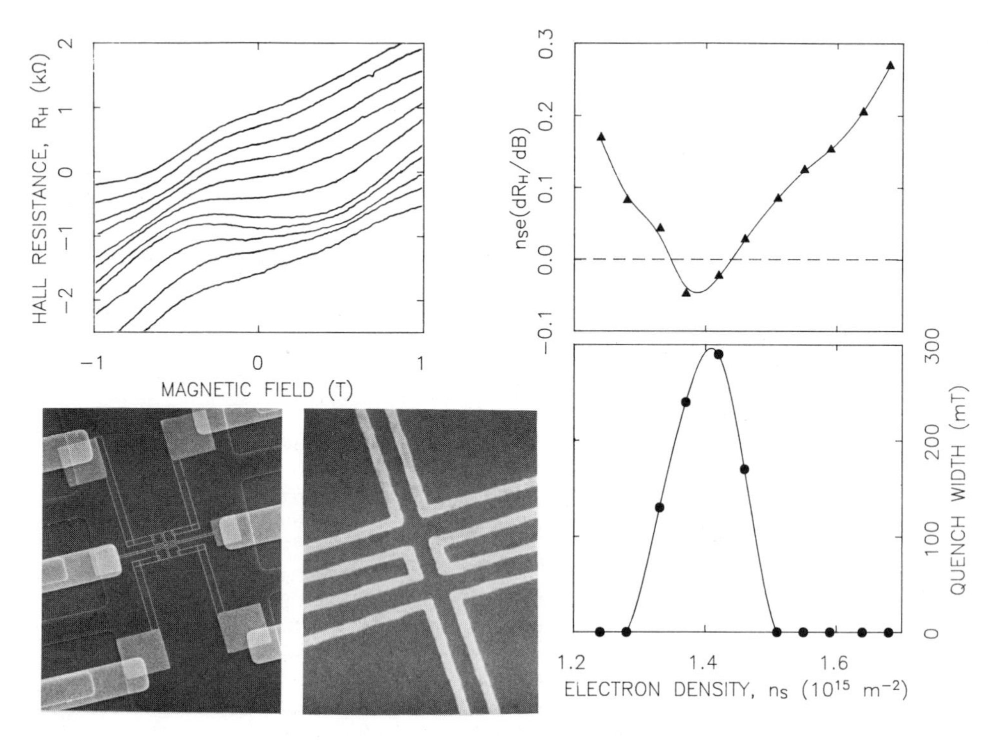

Figure 5. R_H in pinched-gate junctions.

proportionality with k_F in 2D, which, in the multimode limit, approaches k_F in a quasi 1D system. This unexpected dependence has recently been predicted from the "billiard" model of a ballistic junction with tapered leads.[14]

Straight Self-Aligned Gated Junctions: "Re-Entrant" Quenching

In Figure 7 we display data from a junction with self-aligned gates patterned by highly optimized low energy ion exposure *(Fig. 4-iv right.* Since edge depletion lengths from patterning were kept below ~ 30 *nm*, the $w_{el} \sim 200$ *nm* leads form electrical paths staying reasonably straight as they approach the junction. A number of similar devices in several widths were studied – the striking feature manifested in all of them was that *quenching occurred over a limited range at several specific values of electron density.* The "re-entrant" quenches generated by these junctions were considerably weaker, however, than that observed in junctions having a large degree of tapering. Here the width of the quenched region, bounded at either side by the points where dR_H/dB vanished, did not in any case exceed ~ 60 *mT*. Quench widths of several hundred *mT* are commonly seen in our laboratory, both with intentionally tapered devices and with pinched-gate junctions having the large "intrinsic tapering" discussed previously. We estimate, as a *lower* bound, that 4-5 modes are depopulated as n_S is decreased from the quench zone at $n_S \sim 4.3 \times 10^{11}$ cm^{-2} to the zone at $n_S \sim 2 \times 10^{11}$ cm^{-2}. At first sight, this appears to preclude explanations for the observed re-entrant behavior based on junction resonances, since the theory predicts quenching only for three narrow zones just preceding complete occupancy of the lowest three subbands.[10] In the experimental data the Hall slope varies from near its classical value to negative values equally large in magnitude, in contrast to behavior displayed in numerical simulations involving tapered junctions.[8] Similar trends, although much weaker in magnitude were displayed in junctions with self-aligned gates formed from ~ 400 *nm* wide straight wire segments.

Pinched-Gate Junctions: Dependence upon Geometry

We briefly mention the important recent work of Ford et al.[16] which elucidates the effect of junction geometry on low-field transport through pinched-gate junctions. Transport was measured in conducting paths beneath dielectric layers patterned in the form of cross junctions with various geometries. R_H was found to be suppressed to near zero for dielectric layers in the shape of a cross junction with *straight* leads. For a cross geometry with leads grossly *tapered* outward at the junction,

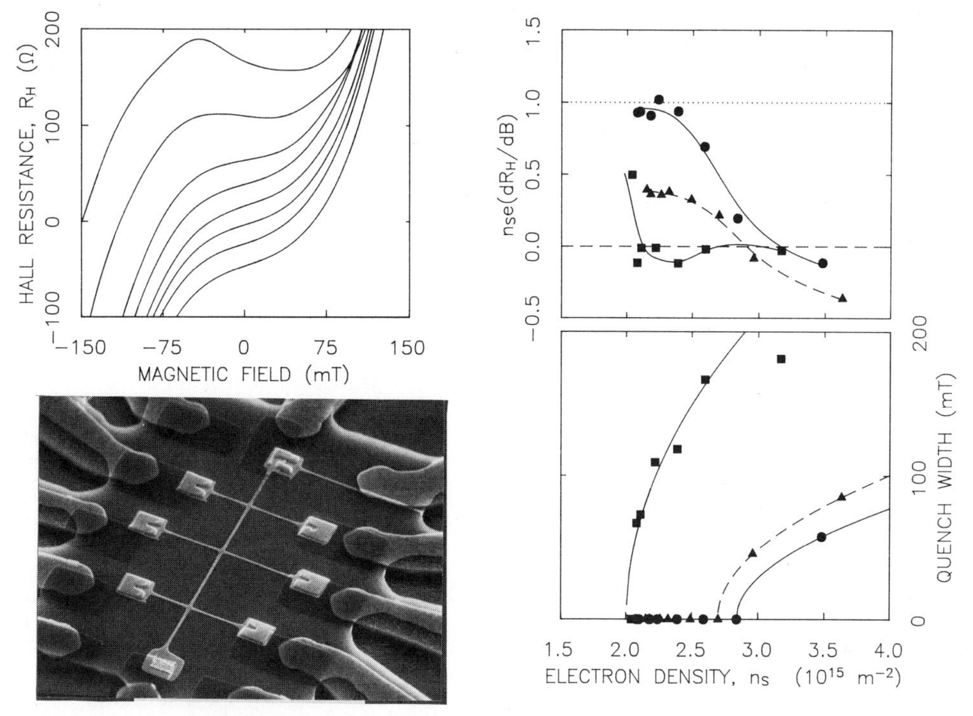

Figure 6. R_H in tapered junctions with self-aligned gates.

a quench with strongly *negative* dR_H/dB was observed. If an *obstruction* was imposed in the center of the junction region (a hole was opened in the dielectric causing carriers to deplete from the center of the junction), R_H did not quench at low field but actually "anti-quenched", *exceeding* its classically-expected value. It was argued that the imposed obstacle suppressed forward transmission, T_F, *(see Fig. 1a)* and favored scattering into the correct probe *(e.g., T_L in Fig. 1a)* at small magnetic fields. Inasmuch as these results were obtained in pinched-gate junctions, the electrical profiles actually obtained were broadened replicas of the patterned dielectric layers. In all likelihood, therefore, both collimation effects and the specific geometrical scattering in each type of device played important roles in these experiments.

THEORY vs. EXPERIMENT

Our early observations stimulated theoretical work following two distinct lines of investigation. Work reported first sought to explore the Hall effect *intrinsic* to a ballistic wire, in the limit where few 1D subbands are occupied. Here, by "intrinsic" we mean in the limit where probe scattering effects are negligible, either because no probes are attached or because they are somehow "weakly coupled". None of the experimentally observed anomalies were predicted.[7,37-39]

The second line of investigation sought to explicitly include and understand the effect of the probes by directly tackling the four-terminal junction scattering problem numerically. Before discussing these results it is helpful to consider how the Hall resistance develops in the symmetric four-probe geometry *(Fig 1., left)*. In the ballistic regime this is easy to conceptualize. The application of finite-B generates a Lorentz force on the incident electrons which enhances their transmission into a side probe, T_L, at the expense of forward transmission, T_F. Finite R_H arises when the zero-field symmetry existing between T_L and T_R is broken by the magnetic field. From the Buettiker formula we find that $R_H \propto T_L - T_R$, indicating that quenching implies equality of the "turning" probabilities at finite field; $T_L(B) = T_R(B)$. This can occur for two reasons – because the T's perform a suprising balancing act in the face of the Lorentz force, or because both turning coefficients vanish identically and the electrons avoid the side probes altogether. As an example, Fig. 1 *(right)* displays R_H and the corresponding behavior of the T_{ij}'s for one occupied subband while a "resonant" quench, described below, is occurring (*after Kirczenow* [11]). Here the balancing act occurs, $T_L = T_R \neq 0$, until, at sufficiently high field, $\omega_c/\omega_0 > 0.4$, the Lorentz force overcomes the junction potential and the $\nu = 2$ quantum Hall plateau is formed.

In the first theoretical consideration of quenching in the ballistic four-terminal configuration, it was argued semi-classically that geometric selection of electron trajectories at the cross caused suppression of the "transmission asymmetry" *(e.g.* $T_R \neq T_L$ *in Fig. 4)*, normally expected at finite B.[36] The qualitative predictions advanced did not agree with more detailed experimental results which followed. Computer work by Baranger and Stone[8] demonstrated that models of ballistic crosses with *strongly-coupled* probes show quenching of the Hall resistance. These are single-electron calculations, where the Schroedinger equation and the Buettiker formalism[13] are used to obtain transport coefficients. Both square[8,9] and parabolic[10,8] confinement potentials have since been considered by several groups. Detailed investigations have indeed confirmed that quenching *can* emerge from models involving untapered ("straight-wire") junctions, but only for select conditions (i.e. quenching is *not* "generic").

Based on our early conjecture[4,40] that localized states in the cross region may account for the experimental phenomena, theoretical work followed confirming their existence at zero magnetic field.[41,42]. This binding occurs, despite the open geometry of the region, because local "geometrical relaxation" of the confinement potential lowers the energy levels slightly in vicinity of the junction.[41] Subsequent numerical work has demonstrated analogous *transmission resonances* at finite-B which dramatically affect calculated magnetotransport in the few-subband limit.[8-10]. Kirczenow[10], in particular, has described the evolution of magnetotransport features resulting from such resonances bearing strong similarity to features observed experimentally – magnetoresistance peaks, quenches, and "last plateaus". He has followed these as a function of subband filling, to link their appearance to localized resonances, associated with each subband, which arise from the potential contours of the cross. Further numerical investigations of this *non-generic* quenching have shown, however, that thermal averaging can suppress this resonant behavior at elevated temperatures.[8] At present, firm connection to existing experimental data has not been established.

Recent computer modelling has demonstrated that *generic* quenching can result from model potentials with gross tapering of the leads outward at the junction.[8] These findings are in qualitative agreement with recent experiments in multimode wires[17] and with the results with *tapered* junctions presented here. It has been argued that collimation[35] of electrons injected into the junctions explains this behavior, since, as a result of the collimation, both turning probabilities (T_L, T_R) may be suppressed.[8] Very recently, however, Beenakker and van Houten have performed *classical* trajectory calculations to model ballistic electrons in an imposed lateral potential.[14] Magnetoresistance curves are obtained having overall features *remarkably* similar to experimental data from tapered junctions in the multimode regime – quenches, a last plateau, magnetoresistance peaks, and negative "bend resistances" are predicted. These calculations appear to demonstrate that these low-field features originate predominantly from *classical "size" effects*. This classical model does not reproduce certain fine features in the data, however, nor can it explain the "disappearing" or "re-entrant" quenching phenomena described previously. Of special significance here is the fact that quenching emerges from these calculations when the junction tapering is significant enough to affect appreciable collimation but, interestingly, in this limit the turning probablilities are *not* found to be small. It is argued that multiple scattering within the junction region suppresses the turning asymmetry normally expected at finite B – i.e., after enough scattering in the junction the electron loses memory of "where it was intially supposed to go." It would thus seem that the temporary trapping of electrons into these complicated trajectories and the resultant "scrambling" of their vector momentum plays a more important role here than mechanisms involving the forward collimation of the electron beam, recently advanced by Baranger and Stone.[8]

CONCLUSIONS AND PROJECTIONS

Recent theoretical developments seem to indicate that low-field magnetotransport in junctions with adiabatically-varying potentials can be explained, for the most part, in terms of classical "size" effects. Contrary to initial conjectures, the "generic" quenching phenomena now observed in many miniature Hall bar samples appears to fall into this class of phenomena. In general, however, experimental investigations reported to date can be described as being carried out either – i) in the multimode regime, or ii) with adiabatically-varying potentials or iii) in exceedingly broad junctions. *All of these factors presumably enhance classical behavior.* A set of experiments representing a possible exception to this rather broad categorization are the those reported herein – carried out in the narrowest straight-wire, self-aligned gated junctions fabricated to date, patterned by very low energy ion exposure. Indeed, with these junctions "generic" quenching is *not* observed – instead, a weaker

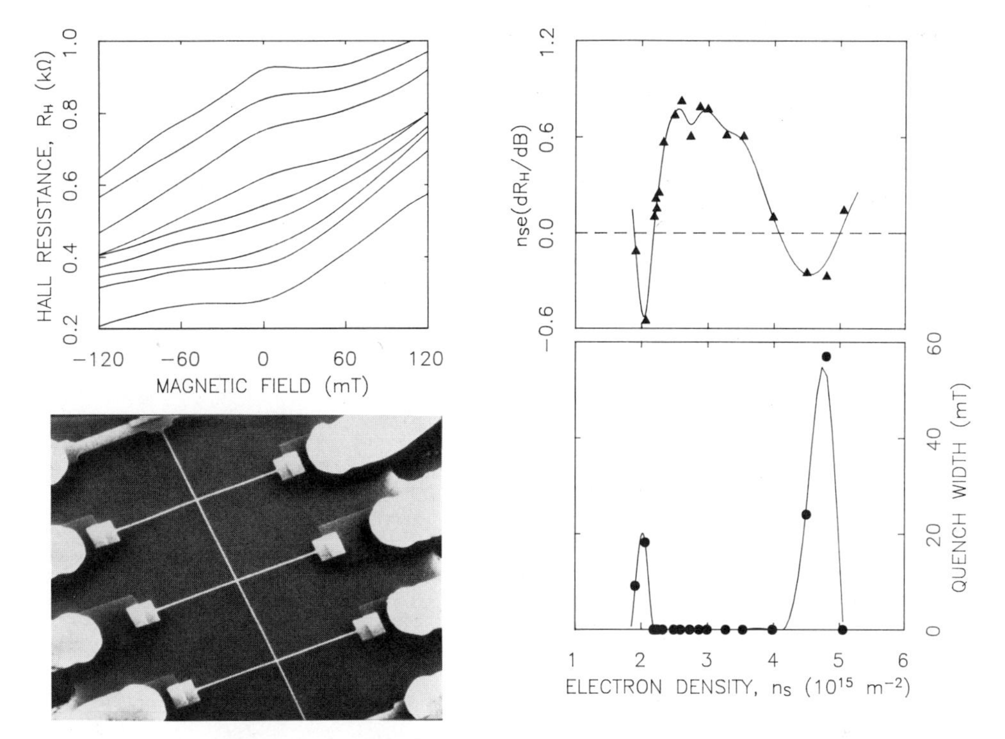

Figure 7. R_H in straight junctions with self-aligned gates.

phenomenon which we have called "re-entrant" quenching emerges. Numerical calculations of low-field transport in ballistic junctions, however, seem to display extreme sensitivity to the precise shape of the junction potential. Given this behavior, we cannot at this point summarily dismiss the possibility that, by allowing for very small electrostatic distortion of the junction potential with changing gate voltage, classical explanations may adequately account for this more delicate quenching phenomenon as well. Nonetheless, it is worth to note that the "classical" explanation proposed for quenching in the multimode regime – based on the brief trapping of electrons into multiply-scattered trajectories in the locale of the junction – bears interesting similarity to the concept of "resonant" quenching predicted in the few-subband regime where trajectory arguments are invalid. Further experimental and theoretical work will help sort the quantum from the classical.

It is a pleasure to acknowledge helpful discussions with S.J. Allen, Jr., H.U. Baranger, C.W.J. Beenakker, M. Buettiker, G. Kirczenow, F.M. Peeters, H. van Houten, and J.M. Worlock, and to thank R.J. Martin for his invaluable contributions to this work.

REFERENCES

†current address: Dept. of Elect. Eng., Princeton Univ., Princeton, NJ

1) A. Scherer and M.L. Roukes, Appl. Phys. Lett., **55**, 377 (1989).
2) B.J. van Wees, H. van Houten, C.W.J. Beenakker, J.G. Williamson, L.P. Kouwenhoven, D. van der Marel, and C.T. Foxon, Phys. Rev. Lett., **60**, 848 (1988). / D.A. Wharam, T.J. Thornton, R. Newbury, M. Pepper, H. Ahmed, J.E.F. Frost, D.G. Hasko, D.C. Peacock, D.A. Ritchie, and G.A.C. Jones, Solid State Comm., **68**, 715 (1988).
3) M. Buettiker, Y. Imry, R. Landauer, and S. Pinhas, Phys. Rev. **B 31**, 6207 (1985). / Y. Imry in "*Directions in Condensed Matter Physics*", G. Grinstein and G. Mazenko, eds. (World Scientific, Singapore, 1986), vol. 1, p. 102.

4) M.L. Roukes, A. Scherer, S.J. Allen, Jr., H.G. Craighead, R.M. Ruthen, E.D. Beebe, and J.P. Harbison, Phys. Rev. Lett. **59**, 3011 (1987).
5) G. Timp, H.U. Baranger, P. deVegvar, J.E. Cunningham, R.E. Howard R. Behringer, and P.M. Mankiewich, Phys. Rev. Lett. **60**, 2081 (1988).
6) Y. Takagaki, K. Gamo, S. Namba, S. Ishida, K. Ishibashi, and K. Murase, Solid State Comm. **68**, 1051 (1988).
7) F.M. Peeters, Phys. Rev. Lett. **61**, 589 (1988).
8) H.U. Baranger, *Proc. 4th Int. Conf. on Superlattices, Microstructures and Microdevices*, Trieste (1988). / H.U. Baranger and A.D. Stone, Phys. Rev. Lett., *to be published* (1989). / H.U. Baranger and A.D. Stone, *these proceedings* (1989).
9) D.G. Ravenhall, H.W. Wyld, and R.L. Schult, Phys. Rev. Lett. **62**, 1780 (1989).
10) G. Kirczenow, Phys. Rev. Lett. **62**, 1920 (1989). / G. Kirczenow, Phys. Rev. Lett. **62**, 2993 (1989).
11) G. Kirczenow, Solid State Comm., *to be published* (1989).
12) Y. Avishai and Y.B. Band, Phys. Rev. Lett. **62**, 2527 (1989).
13) M. Buettiker, Phys. Rev. Lett. B57, 1761 (1986).
14) C.W.J. Beenakker and H. van Houten, Phys. Rev. Lett. *to be published*, (1989).
15) A preliminary report of this work appears in lecture notes of "*Workshop on Quantum Electrical Engineering*", Theoretical Physics Institute, University of Minnesota (TPI-WORKSHOP-88/2), October 20-22, 1988.
16) C.J.B. Ford, T.J. Thornton, R. Newbury, M. Pepper, H. Ahmed, D.C. Peacock, D.A. Ritchie, J.E.F. Frost, and G.A.C. Jones, Phys. Rev. **B 38**, 8518 (1988).
17) A.M. Chang and T.Y. Chang, Phys. Rev. Lett., *to be published*, (1989).
18) C.J.B. Ford, S. Washburn, M. Buettiker, C.M. Knoedler, and J.M. Hong, Phys. Rev. Lett. **62**, 2724 (1989).
19) H. van Houten, B.J. van Wees, M.G.J. Heijman, and J.P. Andre, Appl. Phys. Lett., **49**, 1781 (1986).
20) T. Demel, D. Heitmann, P. Grambow, and K. Ploog, Appl. Phys. Lett. **53**, 2176 (1988).
21) A. Scherer, M.L. Roukes, H.G. Craighead, R.M. Ruthen, E.D. Beebe, and J.P. Harbison, Appl. Phys. Lett **51**, 2133 (1987) / T.L. Cheeks, M.L. Roukes, A. Scherer and H.G. Craighead, Appl. Phys. Lett. **53**, 1964 (1988).
22) T. Hiramoto, K. Hirakawa, Y. Iye, and T. Ikoma, Appl. Phys. Lett. **54**, 2103 (1989).
23) T.J. Thornton, M. Pepper, H. Ahmed, D. Andrews and G.J. Davies, Phys. Rev. Lett. **56**, 1198 (1986) / C.J.B. Ford, T.J. Thornton, R. Newbury, M. Pepper, H. Ahmed, D.C. Peacock, D.A. Ritchie, J.E.F. Frost, and G.A.C. Jones, Appl. Phys. Lett. **54**, 21 (1988).
24) K.K. Choi, D.C. Tsui, and K. Alavi, Appl. Phys. Lett. **50**, 110 (1987).
25) J.H. Davies, Semicond. Sci. Technol., **3**, 995 (1988).
26) A. Kumar, S.E. Laux, and F. Stern, Bull. Am. Phys. Soc. **34**, 589 (1989) / J.H. Davies, Bull. Am. Phys. Soc. **34**, 589 (1989).
27) K.F. Berggren, T.J. Thornton, D.J. Newson, and M. Pepper, Phys. Rev. Lett. **57**, 1769 (1986).
28) T.J. Thornton, M.L. Roukes, A. Scherer, and B.P. Van der Gaag, *these proceedings* (1989).
29) G. Timp, R. Behringer, S. Sampere, J.E. Cunningham, and R.E. Howard, in "*Nanostructure Physics and Fabrication*", M.A. Reed and W.P. Kirk, eds., Academic Press, New York (1989).
30) M.L. Roukes, A. Scherer, and B.P. Van der Gaag, *to be published* (1989).
31) T.J. Thornton, M.L. Roukes, A. Scherer, and B.P. Van der Gaag, *to be published* (1989).
32) H. van Houten, B.J. van Wees, J.E. Mooij, C.W.J. Beenakker, J.G. Williamson, and C.T. Foxon, Europhys. Lett. **5**, 721 (1988).
33) B.I. Halperin, Phys. Rev. **B 25**, 2185 (1982) / H. van Houten, C.W.J. Beenakker, P.H.M. Loosdrecht, T.J. Thornton, H. Ahmed, M. Pepper, C.T. Foxon and J.J. Harris, Phys. Rev. **B 37**, 8534 (1988) / M. Buettiker, Phys. Rev. **B 38**, 9375 (1988).
34) J.A. Simmons, Ph.D. Thesis, Princeton Univ. (1989). *unpublished.*
35) C.W.J. Beenakker and H. van Houten, Phys. Rev. **B 39**, 10445 (1989).
36) C.W.J. Beenakker and H. van Houten, Phys. Rev. Lett. **59**, 2406 (1988).
37) G. Kirczenow, Phys. Rev. **B 38**, 10958 (1988).
38) A. Devenyi (and Y. Imry), M.S. Thesis, Weizmann Institute (1988), *unpublished.*
39) H. Akera and T. Ando, Phys. Rev. **B 39**, 5508 (1989).
40) M.L. Roukes, Bull. Am. Phys. Soc. **33**, 600 (1988). / M.L. Roukes, Elect. Eng. Colloquium, Univ. of Illinois, Urbana, IL, *April 21, 1988.*
41) F.M. Peeters, *Proc. 4th Int. Conf. on Superlattices, Microstructures and Microdevices, Trieste (1988)*, Superlattices and Microstructures (1989). / F.M. Peeters, *these proceedings* (1989).
42) R.L. Schult, D.G. Ravenhall, and H.W. Wyld, Phys. Rev. **B 39**, 5476 (1989).

SOME AD-HOC METHODS FOR INTRODUCING DISSIPATION TO THE SCHRÖDINGER EQUATION

M. Cemal Yalabık

Department of Physics, Bilkent University
Bilkent 06533 Ankara, Turkey

ABSTRACT

Considerable interest has developed in the quantum mechanical simulation of electronic devices with the fabrication of structures whose geometric feature sizes are comparable to the quantum mechanical wavelengths of the carriers in these devices. The inclusion of dissipative effects to the study of quantum mechanical transport phenomena is a difficult fundamental problem. Although considerable progress has been made in the formal theory, these formalisms are computationally difficult to implement in numerical simulations of charge transport in realistic device structures.

Integration of the time-dependent Schrödinger equation has been utilized in various studies to describe the dynamics of charges in small devices. Although the simplicity of the method is promising, the method is applicable only to those systems in which the dissipative effects can be assumed to be negligible. In this study, a number of ad-hoc approximations that result in equations which contain dissipative effects will be discussed. These approximations lead to relatively simple integration procedures which may be useful in the simulation of devices where dissipation effects are weak, but not negligible.

I.INTRODUCTION

The study of numerical simulation of charge transport in electronic devices is important from a fundamental point of view as well as for practical reasons. The ongoing trend in the microelectronics field towards higher device densities has resulted in the fabrication of devices with geometric feature sizes comparable to the quantum mechanical wavelengths of the charge carriers in these devices. Modeling of the quantum transport effects in these geometries is an interesting problem. Although considerable progress has been made in the formal theory of quantum transport,[1-4] applicability of these methods to realistic problems with detailed geometrical structure is very limited.

If the size of the active region of the device is small compared to the mean free

Science and Engineering of One- and Zero-Dimensional Semiconductors
Edited by S.P. Beaumont and C.M. Sotomajor Torres
Plenum Press, New York, 1990

path of the carriers, the significant transport effects can in most cases be satisfactorily modeled by non-dissipative ballistic transport. However, there remains a large class of cases where dissipation cannot be completely ignored. In the absence of computationally feasible rigorous methods for including such effects, one naturally looks for approximations which hopefully will help in modeling important aspects of the phenomena. It should also be pointed out that there are cases (such as the modeling of the contact regions which act as charge reservoirs) in which the end results of the dissipative effects (i.e. thermal equilibrium) is more important than the details of how such effects take place.

There are a number of methods available for the simulation of ballistic quantum mechanical transport in devices. Integration of the time dependent single particle Schrödinger equation is one of these methods, allowing a detailed description of the state of the system as a function of time. The use of the single particle equation is justified in cases where the particle density is sufficiently low so that quantum degeneracy may be neglected. The complexity of the computation increases relatively slowly with increasing dimensionality, in contrast to, for example, Wigner function methods where one has to work with a $2d$ dimensional mesh for a d dimensional device. On the other hand, there are very basic problems in introducing realistic boundary conditions and dissipation effects to the Schrödinger equation.

An approximation for the electron phonon scattering effects within the context of the simulation of a GaAs MESFET using the time dependent Schrödinger equation has been reported by Yalabik et. al.[5,6] The method involves updating of the momentum-space amplitudes of the wave function consistent with the corresponding scattering cross sections. This procedure then guarantees a correct equilibrium state and a two-point time correlation function consistent with the results of field theoretic methods for a homogeneous system. However, the method is computationally very expensive, 95% of the total simulation time being used up in the handling of the dissipation effects. It may also result in undesirable non-locality in some applications because of the inherently non-local treatment in the momentum space. In this paper we will report the preliminary results of a study of two methods through which one can introduce energy and momentum dissipation to the time dependent Schrödinger equation. The dissipative effects are introduced through a real, local, and time dependent potential term. The procedure is expected to be useful in the simulation of electronic devices in which the dissipative effects are not very significant in some small active region of the device, but the overall effect over a larger length scale is not negligible.

II. THE METHOD

Energy relaxation

Consider a scaled form of the Schrödinger equation

$$i\frac{\partial \Psi}{\partial t} = -H\Psi \tag{2.1}$$

with

$$H = -\nabla^2 + V$$

where V is the potential. The formal solution to the equation may be written as

$$\Psi(t+\Delta t) = \mathrm{e}^{-iH\Delta t}\Psi(t). \tag{2.2}$$

One way of integrating this equation numerically is to use the approximation

$$\Psi(t+\Delta t) \approx \mathrm{e}^{-iV\Delta t}\, \mathrm{e}^{-i\Delta t\nabla^2}\, \mathrm{e}^{-iV\Delta t}\Psi(t). \tag{2.3}$$

The break-up of the Hamiltonian H preserves the time reversal symmetry and is exact in the limit Δt approaches zero. The integration then involves multiplication of the "old" wave function by the appropriate exponentials (note that the $\mathrm{e}^{-i\Delta t\nabla^2}$ term becomes a scalar in momentum space) to obtain the "new" wave function. We are interested in the addition of a dissipation term to the potential so that $V \rightarrow V + V_{diss}(t)$. The time dependent potential $V_{diss}(t)$ is to be chosen so that it tends to decrease the energy. Equation 2.3 shows clearly that V_{diss} will introduce some additional phase to the wave function at each time- step. The variation of the expectation value of the energy E with respect to the phase $\theta(x)$ of the wave function $\Psi(x)$ at a certain point x is given by

$$\frac{\delta E}{\delta\theta(x)} = 2\mathrm{Im}\ \Psi^*(x)\nabla^2\Psi(x). \tag{2.4}$$

Hence, the presence of a term of the form

$$V_{diss} = \alpha \mathrm{Im}\ \Psi^*\nabla^2\Psi \tag{2.5}$$

then results in a modification of the phase of the wave function in a direction that corresponds to a reduction in the local contribution to the energy. The parameter α is a constant that controls the rate of the relaxation. One point that needs attention is the possibility of generating very short wavelength instabilities in the simulation due to the large values that will be produced by the Laplacian operation at these short wavelengths. This possibility is eliminated in practice by the existence of a finite size mesh (hence a high frequency cutoff in momentum space).

The potential in Equation 2.5 is equal to zero for any Ψ that has a constant phase, and in particular for all eigenfunctions of the original Hamiltonian operator. This implies that the eigenfunctions of the original Hamiltonian correspond to states whose energies will not change with the introduction of the term in Equation 2.5. However, our numerical experiments indicate that all eigenfunctions except the ground state eigenfunction correspond to unstable equilibrium points from the point of view of energy relaxation. Any small perturbation around a pure excited eigenstate will carry the wave function to the ground state in the presence of V_{diss}. In fact, the procedure can be used very effectively for determining ground-state wave functions in general.

Another point to note is that if periodic boundary conditions are used in the simulation and if the coherence of the wave function extends over the whole system, the energy of the system cannot be reduced to that of the ground state by the addition of V_{diss}. This is because in this case, the phase of the wave function changes by an integral number of 2π radians from one side of the system to the other, and this integral number cannot be reduced by making small local changes of the phase at each time-step. Therefore, for the procedure to work with periodic boundary conditions, the coherence of the wave function from one end of the system to the other must be destroyed either by the geometry of the original potential, or through a thermal noise term as will be discussed below.

The relaxation mechanism described above will reduce the energy of the system until the ground state is reached. To model a realistic dissipation process, one also needs an additional mechanism so that at equilibrium the particle energies are distributed in proportion to $e^{-E/kT}$, where k is the Boltzmann constant and T is the temperature. (The use of the Boltzmann factor in place of the Fermi distribution is based on the basic assumption that the particle density is low, and that the particle energies are sufficiently larger than the Fermi energy.) This distribution can be obtained by the inclusion of a noise term in V_{diss}, so that the system is simultaneously excited in energy as well as relaxed. There is no unique way to introduce this thermal noise term, whose effect would physically correspond to the loss of coherence of the wave function due to dissipative scattering from a heat bath. One simple way is to construct a V_{diss} of the form

$$V_{diss} = \alpha \mathrm{Im}\ \Psi^* \nabla^2 \Psi + (3\alpha kT)^{\frac{1}{2}} R(x,t) \tag{2.6}$$

where $R(x,t)$ is a random (uncorrelated in space and time) function which changes uniformly between $\pm \frac{1}{(\Delta t)^{\frac{1}{2}}}$. Other possibilities include a random function that has sinusoidal correlation in space, which will correspond to more local transitions in the momentum space. An important point to note here is that this procedure generates an ensemble of wave functions whose energies are expected to be distributed as $e^{-E/kT}$ for a sufficiently large number of "sample" wave functions. This does not in general mean that the distribution of the eigenstate occupation numbers corresponding to a particular wave function at a certain time will be a Boltzmann distribution. However, this will happen if the phase coherence of the wave function does not extend over all parts of the simulated system so that different parts of the wave function may be treated as different "samples".

Momentum relaxation

A relaxation in the momentum of a particle can be achieved through the introduction of a potential term into the Schrödinger equation corresponding to a force that is in opposite direction to the local momentum density:

$$-\nabla v_{diss} = \beta(\Psi^* i \nabla \Psi - \Psi i \nabla \Psi^*) \tag{2.7}$$

Here, β is a parameter that controls the speed of the relaxation. This equation then yields

$$\nabla^2 v_{diss} = 2\beta \mathrm{Im}\ \Psi^* \nabla^2 \Psi. \tag{2.8}$$

Hence, the potential term that corresponds to momentum relaxation can be obtained from the solution to the Laplace equation with a source term that has the same form as the energy relaxation potential. This will bring only a small amount of computational load to "self consistent" simulations in which the Laplace equation is solved frequently in

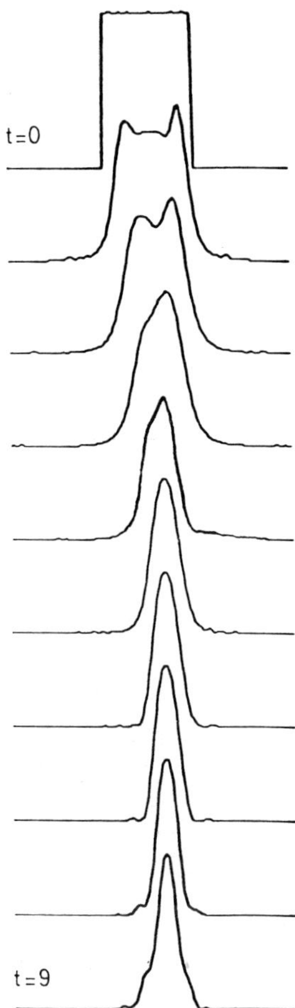

Figure 1. Plots of the magnitude of the wavefunction as a function of time in the presence of the energy dissipating potential (no noise term present). The limits of the displacement axis are $x = \pm 1$, the maximum initial magnitude is 1, the original Hamiltonian has a potential energy term x^2, and the initial state wavefunction has a phase factor $e^{3.13ix}$ corresponding to an initial momentum towards the right. The wavefunctions have been scaled to give equal maximum magnitudes. The parameter α equals 0.1.

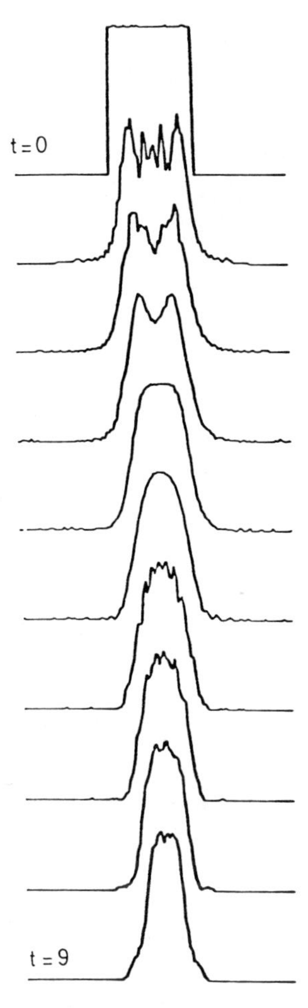

Figure 2. Plots of the magnitude of the wavefunction as a function of time in the presence of the momentum dissipating potential. The parameters are the same as those in Figure 1 except that $\alpha = 0$ and $\beta = 0.01$.

order to obtain the electrostatic potential which is affected by the particle distribution at that particular time instant. Note that due to the form of the potential v_{diss}, periodic boundary conditions will rarely be appropriate with this procedure.

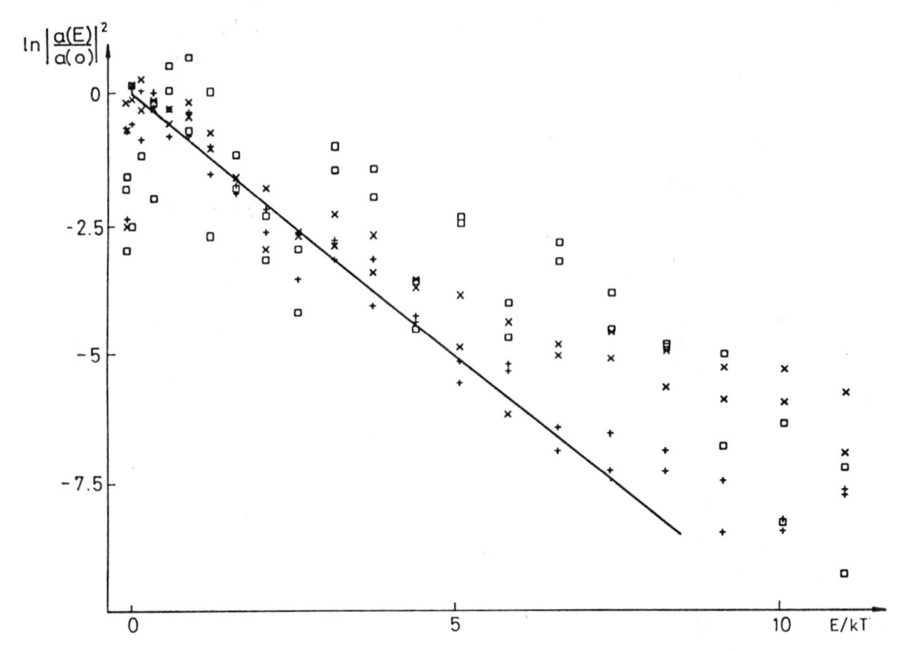

Figure 3. The occupation numbers in the momentum space in the presence of an energy dissipating potential for a homogeneous system. For this figure, the Schrödinger equation with parameters corresponding to an electron in the Γ conduction valley of GaAs was used. The vertical axis is the logarithm of the occupation number. The horizontal axis is scaled so that it corresponds to the energy divided by kT. Crosses correspond to T =100K, plusses to T =200K and squares to T =300K. The straight line corresponds to the theoretical Boltzmann distribution.

III.NUMERICAL SIMULATION RESULTS

Both the energy and momentum relaxation procedures were implemented (on one, two, and three dimensional meshes) numerically and applied to a variety of problems. Only the results pertaining to the relaxation procedures will be reported here. Figure 1 shows the relaxation of an initial wave function under the effect of a potential of the type given in Equation 2.5. Figure 2 shows the relaxation of the same initial wave function when the relaxation potential is of the type given in Equation 2.8. Both of these relaxations were carried out for an initial (non-dissipative) Hamiltonian with a quadratic potential energy term. Finally, Figure 3 shows the momentum space distribution of occupation numbers for a free particle under the influence of energy relaxation and thermal noise. The formal analysis of the effect of the thermal noise on the Schrodinger equation is quite complicated, work is in progress to understand which choice of randomness corresponds to a more realistic modeling.

Acknowledgement: I would like to thank Prof. Mehmet Baray for his help in obtaining the computer plots.

REFERENCES

1. G. D. Mahan, Quantum transport equation for electric and magnetic fields, Physics Reports 145:251 (1987).

2. K. K. Thornber and R. P. Feynman, Velocity acquired by an electron in a finite electric field in a polar crystal, Phys. Rev. B1:4099 (1970).

3. A. O. Caldeira and A. J. Leggett, Quantum tunneling in a dissipative system, Ann. Phys. 149:374 (1983).

4. A. P. Jauho and J. W. Wilkins, Theory of high-electric-field quantum transport for electron-resonant impurity systems, Phys. Rev. B29:1919 (1984).

5. M. C. Yalabik, J. D. Gunton, G. Neofostistos, and K. Diff, Simulation of charge transport in a GaAs MESFET using the time dependent Schrödinger equation, Superlattices and Microstructures 3:463 (1987).

6. M. C. Yalabik, G. Neofostistos, K. Diff, H. Guo, and J. D. Gunton, Quantum mechanical simulation of charge transport in very small semiconductor structures, IEEE Trans. on Elect. Dev. (in print, 1989).

TRANSPORT IN ELECTRON WAVEGUIDES: NON-LINEARITIES AND GATED RINGS

P.G.N. de Vegvar,[1] G. Timp,[2] P. M. Mankiewich,[3] and R. Behringer[2]

[1]AT&T Bell Laboratories
Murray Hill, New Jersey 07974

[2]AT&T Bell Laboratories
Holmdel, New Jersey 07733

[3]Boston University
Boston, Massachusetts 02215

Over the last few years it has become possible to construct electron waveguides using nano-fabrication techniques.[1-3] This contribution will discuss some recent advances in quantum transport occurring in these small heterostructure semiconductor systems.

To appreciate best the nature of these devices, it is instructive to introduce a hierarchy of length scales. How these lengths are determined is outlined below. The first one is the so-called electron dephasing length L_ϕ. This is the distance an electron travels before the phase of its wavefunction is upset by some kind of randomizing interaction with the "environment." An example of the latter would be an electron-phonon collision. In our samples at $T \simeq 300$ mK, $L_\phi \gtrsim 5$ μm. Another significant length is the electron mean free path or momentum relaxation distance L_e. This corresponds to how far an electron propagates before undergoing a large angle elastic scattering event. From zero-field mobility measurements L_e ranged from 1 to 5 μm. Of course, there are also the geometric length L and conducting width W of the devices. L varied from 0.9 to 6 μm depending on the lead configuration, and $W \simeq 900-1000$ Å. For comparison purposes, the electron Fermi wavelength for typical sheet densities was $\lambda_f \simeq 500$ Å. Thus there are only 2-3 transverse channels occupied below the Fermi energy, and the electrons propagate ballistically through the devices.

The nonlocal nature of quantum transport on a length scale below L_ϕ has been extensively discussed in the literature.[4-7] One useful picture in this so-called "mesoscopic" regime was developed by Buettiker.[8] According to this view, the 4-point resistance measured on such a length scale is determined by scattering probabilities between the leads used to make the measurement. Briefly, if one injects a current I into lead k, withdraws it from lead l, and measures the voltage difference V between leads m and n, then $R_{kl,mn} \equiv V/I = (h/e^2)(T_{mk}T_{nl} - T_{nk}T_{ml})/D$. Here T_{mk} is the transmission probability from lead k to m, and D is a combination of T's.

If mesoscopic resistance is basically a coherent scattering experiment, how can we controllably alter it? Here we discuss two methods of perturbing scattering of electrons in waveguides. The first involves using a small gate that intercepts one branch of an Aharonov-Bohm (AB) annulus in order to construct a tunable electron interferometer.[9] It also turns out that the current used to make the resistance measurement perturbs the mesoscopic system. Some aspects of this non-linear transport form the second part of this presentation.[10]

The electron waveguides were constructed by electron beam lithography and reactive ion etching techniques. The processing details are covered elsewhere.[11] To summarize, electron beam lithography is employed to define a 0.5 μm wide etch mask directly above an MBE grown modulation doped GaAs-AlGaAs heterostructure. These were then dry etched only down to the Si doping layer. This removed the underlying two dimensional electron gas (2DEG) everywhere except beneath the mask. Lateral depletion due to surface traps introduced by the etching further reduced the width $W \simeq 0.1$ μm of the conducting channel to less than the mask width. The small AuPd gates for the interferometer were fabricated using alignment marks and lift-off. Schematic views of a gated waveguide annulus are shown as insets to Figure 1. The resistance between the electrons in the waveguide and the metal gate was in excess of 1 GΩ.

Materials parameters were determined by several different techniques. All resistance measurements were performed at $T \simeq 300$ mK using a 4-point ac method at 11 Hz. Shubnikov-de Haas and Hall data on a wide (300 μm) bar were used to extract the starting 2DEG sheet density and zero-field mobility. In the narrow samples, the high-field quantum Hall effect was used to estimate the sheet density from the plateau with $R_{xy} = h/ie^2$ where $i = 2$. L_ϕ and the conducting width W were obtained from analysis of Fourier spectra of AB oscillations in ring samples.[12] For example, the gated annuli discussed below were constructed from material with an initial 2DEG sheet density $N_{2D} = 3.9\times10^{11}$ cm^{-2} and mobility $\mu_{2D} = 3\times10^5$ cm^2/V$-$sec, implying $L_e = 1.6-2$ μm. Non-linear transport was investigated in two materials. The first had $N_{2D} = 4.5\times10^{11}$/cm^2, and $\mu_{2D} = 3.5\times10^5$ cm^2/V$-$sec, with $L_e = 2.7$ μm. At high fields in narrow wires this gave $N(i=2) = 2.95\times10^{11}$/cm^2 and $\lambda_f = 460$ Å. The parameters describing the second starting material were $N_{2D} = 4.2\times10^{11}$/cm^2 and $\mu_{2D} = 8.7\times10^5$cm^2/V$-$sec, corresponding to $L_e = 6.4$ μm. Similarly, $N(i=2) = 1.9\times10^{11}$/cm^2, and $\lambda_f = 575$ Å. Fourier decomposition of the AB effect in ring samples without gates yielded the values for channel width and dephasing length quoted above.

Since only a few (2-3) transverse sub-bands are occupied in these electron waveguides, an annulus displays AB magnetoresistance oscillations whose peak-peak amplitude may be up to 10% of the total ring resistance. By fabricating a gate over one arm of such a ring and biasing it, we modify the interference causing the AB effect. According to one picture, a negative gate bias locally depletes electrons from beneath the gate. This alters the density and Fermi velocity in one arm relative to the other. The AB oscillations are thereby shifted in much the same manner as placing a dielectric slab into one branch of a photon interferometer causes optical fringe displacement.

Figure 1 shows the tunable AB effect in such a device. The magnetoresistance data have been digitally filtered to include only the Fourier components in the AB pass-band. For gate voltages below -1800 mV, one branch of the interferometer is pinched-off, and the AB oscillations are quenched.

Recently, Washburn and coworkers[13] have demonstrated a tunable AB effect by placing an Sb ring inside a capacitor. The gated ring discussed here differs from that experiment in that the applied field or scalar potential in the gated annulus perturbs one branch of the ring while a ring in a capacitor will have both arms affected. The two techniques also differ in some important

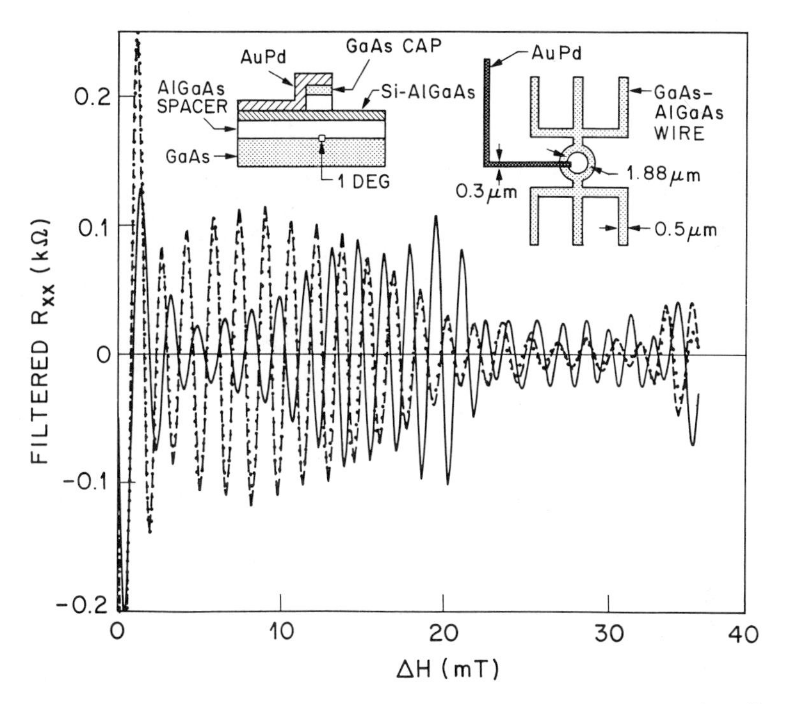

Fig. 1. Digitally filtered Aharonov-Bohm oscillations in a locally gated waveguide annulus. The dashed trace was obtained first with $V_g = 0$ mV. The solid curve refers to $V_g = -300$ mV, and the dotted one to cycling back to $V_g = 0$ mV. The insets are schematics of the devices, and are not to scale.

materials properties. Screening lengths are 200 Å in GaAs, but are 100 times shorter in Sb. Also the Sb wires are 3-dimensional conductors with about 5600 transverse channels, in contrast to 2-3 such channels in the GaAs waveguides. The ring in a capacitor experiment could not distinguish between a gross shift of Feynman paths (wavefunction) contributing to the measured resistance and the so-called electrostatic AB effect. The latter arises from contact of the wavefunction with a scalar potential rather than with the more conventional vector potential (magnetic flux). This phenomenon may also be viewed as an adiabatic distortion of the wavefunction by the external potential. It is predicted to give rise to changes periodic in the gate voltage V_g since the electron propagator acquires a phase $\exp(\frac{ie}{\hbar}\int V\,dt)$ in the presence of a potential V. A gross change in wavefunction gives rise to changes generally aperiodic in applied potential. One may use these gated annuli to check for such periodicities.

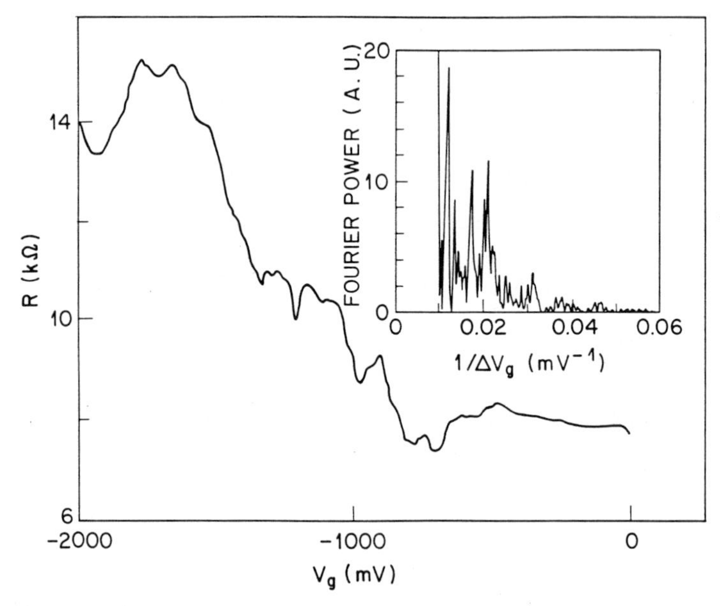

Fig. 2. Ring resistance vs. V_g in zero magnetic field. Inset: Fourier decomposition. There is no clear signature of an electrostatic Aharonov-Bohm effect.

One way to search for such effects is through the magnetoresistance. Is there a phase shift of the AB oscillations periodic in V_g? It turns out experimentally that there is no well-defined rigid phase shift of the oscillations. The phase change is only locked over 10 or so cycles for a given change in V_g. This contrasts with the behavior of an optical interferometer in which the fringes exhibit a global shift when a phase delay is introduced into one arm. In the absence of such a well-defined "fringe" displacement in the electronic interferometer, another method must be used to search for the electrostatic AB effect. Figure 2 shows the ring resistance in zero magnetic field as a function of V_g. To look for periodic effects, a Fourier decomposition of these data was performed and is shown in the inset. No clear evidence of periodic phenomena are apparent.

The lack of observation of an electrostatic AB effect may not be so surprising in view of two facts. In the magnetic AB effect for the case where the applied flux threads only the hole of the annulus, the wavefunction is exactly periodic under changes of a single electron flux quantum h/e. Thus an infinite number of AB cycles will be observed as a function of magnetic field until the field penetrating the wire has a large effect. Such a gauge transformation is absent in the electrostatic analog. The number of cycles possible in the electrostatic AB effect is limited in the gated AB annuli to the number of

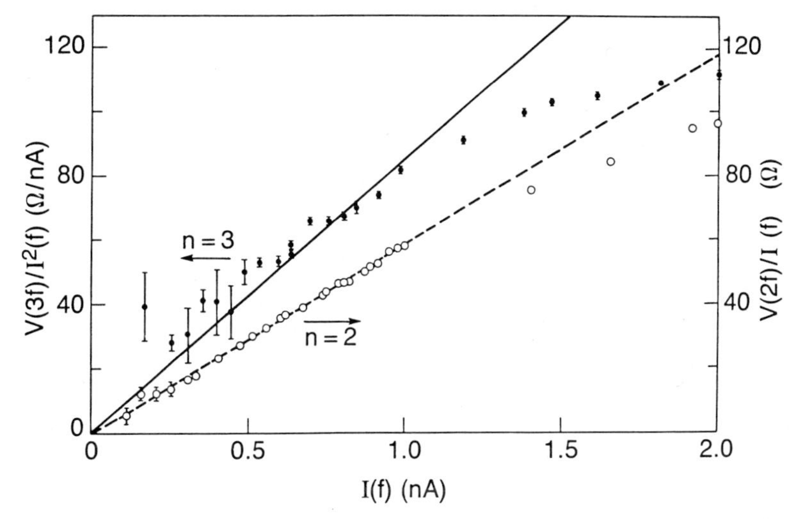

Fig. 3. Current dependence of V(2f) (open circles) and V(3f) (solid circles). The dashed and solid lines are best fits to the power law: $V(nf) \propto I^n(f)$. Deviations for $I > 1$ nA are due to higher order mixing effects.

electron wavelengths under the gate. Since this is only 5-6 cycles for these devices, a clear Fourier peak will be at best elusive. One can also estimate the change in potential experienced by an electron necessary to generate one electrostatic AB period to be about 2 mV. This is comparable to energies required to raise transverse sub-bands through the Fermi level, and thus cannot be considered as an adiabatic deformation of the wavefunction.

So far we have only discussed the way a gate alters the scattering probed by a mesoscopic 4-point resistance measurement. The finite current used to do this will also perturb the system being measured. To this end, consider expanding the voltage response V to a current excitation I in a Taylor series:

$$V = R^{(1)}I + R^{(2)}I^2 + R^{(3)}I^3 + \cdots$$

In a homogeneous macroscopic sample, T-invariance requires $V(I) = -V(-I)$. This means the even order R's must vanish. On the other hand, when L approaches L_ϕ the electron transport is nonlocal and sensitive to the scattering potential it experiences while traversing the sample. The system is then far from being homogeneous. As discussed below, this can lead to non-vanishing even order "resistances." In particular, if one injects a sinusoidal current at a frequency f, the even order R's will produce a voltage response at 2f. Because heating is proportional to the square of the current, such a 2f signal cannot arise from heating effects. We have observed such second harmonic generation (SHG) in our mesoscopic electron waveguides. As a control, we note that the 2f signal was immeasurably small at 4 K and also in large samples (300 μm wide Hall bars) at all temperatures. The SHG is weak: $V(2f)/V(f) = 3\times10^{-4}$ for I = 1 nA. Harmonics up through 10f have been detected.

Figure 3 shows the detailed current dependence of the nonlinearities. V(nf) for n = 2, 3 is plotted against I(f) is such a way that $V(nf) \propto I^n(f)$ appears as a straight line. This behavior is indeed found to hold for currents below 1 nA. In the range 1-2 nA higher order terms in the Taylor series mix down to give deviations from this simple power law behavior. Similar harmonic generation has also been observed in metallic mesoscopic systems. There, however, V(nf) was found to be proportional to I for all n at the lowest currents employed.[14]

A demonstration of the mesoscopic origin of the phenomena is presented in Figure 4. Here both the linear (Ohmic) resistance as well as the second order response coefficient $R^{(2)}$ are shown as a function of magnetic field for a ring geometry. Both display the AB effect periodic in the single electron flux quantum. The cross correlation coefficient between the oscillations is only $.33\pm.15$, indicating that $R^{(1)}$ and $R^{(2)}$ probe different physics. $R^{(2)}$ was also observed not to obey the same lead-field symmetries as the linear resistance.[8,15]

There are several ways to picture the physical mechanisms responsible for these nonlinearities. The first is a simple one dimensional model. Consider two electron reservoirs with chemical potentials μ_1 on the left and μ_2 on the right separated by a wire of length $L < L_\phi$ in which there is some scattering potential V(x) in the absence of injected current. Here x is measured from the midpoint of the wire. Suppose a current I is injected from the left reservoir. Taking the electron's charge as positive, this raises μ_1 relative to μ_2. Now, what is the scattering potential experienced by one additional electron transversing the wire from left to right? To first order, it will see V(x) plus the linear voltage drop from μ_1 to μ_2. If one now reverses the sign of the injected current, the reservoirs exchange roles. An extra electron coming from the right will see V(x) now tilted down to the left. It is clear that the scattering potential experienced in the first case is NOT the same as in the second provided $V(x) \neq V(-x)$. Thus the transmission probabilities $T(I>0) \neq T(I<0)$, and $R^{(2)} \neq 0$. What happens if $L>>L_\phi$? Then we add many L_ϕ sized sections in "series". But as the electron propagates across these boxes, it losses phase coherence and "forgets" where x=0 is. The resistance is no longer sensitive to the fluctuations or inhomogenities in the scattering potential on a length scale less than L_ϕ. The mesoscopic inhomogenities are averaged out, and one recovers $T(I) = T(-I)$. This picture tells us that the non-linear response measures the voltage or energy dependence of the transmission probabilities, in contrast to the linear response which probes the transmission directly.

An alternative point of view is the following. As shown by Larkin and Khmelnitskii,[16] the voltage response V to a current I is a random function of I for $V \geq V_c \sim \dfrac{\hbar}{e(L_\phi/v_f)} = 21\ \mu V$. The idea is that the injected current generates electric fields that act to alter the electrons' energy. If the change in

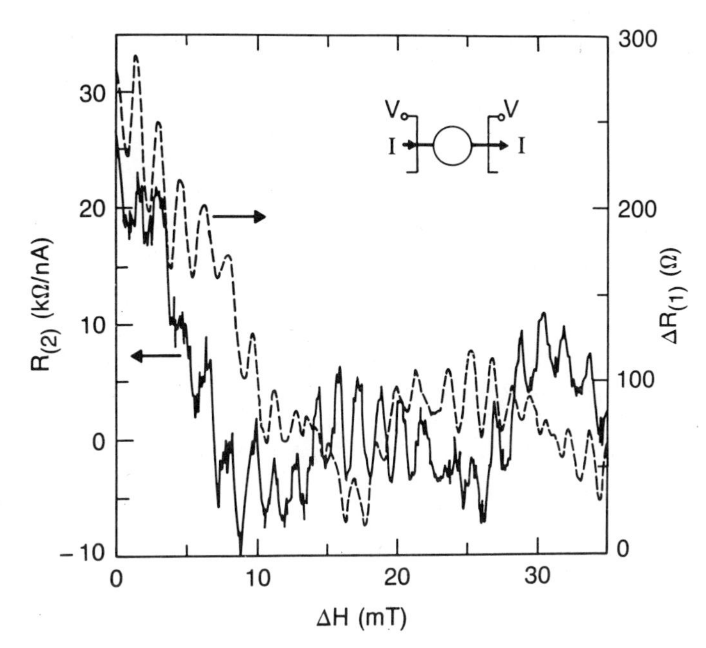

Fig. 4. Change in linear resistance (dashed) and second order response (solid) vs. magnetic field in an annulus. Inset: Lead geometry used to perform the measurement.

energy is eV_c or larger, then the Feynman paths contributing to the resistance are modified. This is a fully deterministic process, and the electrons do not heat unless L_ϕ degrades. A shift of paths is of course the same physically as a distortion of the wavefunction by the injected current or by the chemical potential difference between the reservoirs. The nonlinearities are then a mesoscopic version of the atomic Stark effect.

The current scale of the non-linearities implied by the above picture may be inferred as $I_c \sim V_c/R \sim 4$ nA. An experimental estimate for I_c can be taken as the current level at which higher order mixing occurs. This is at 1-2 nA in our samples. So the theory and experiment are in qualitative agreement.

In conclusion, we haved turned some experimentally accessible knobs available in mesoscopic systems in order to change their scattering behavior. Tunable AB oscillations were demonstrated in gated waveguide annuli, but no signature of an electrostatic AB effect was observed. Scattering was also found to be sensitive to current. The higher order resistances $R^{(n)}$ probe different physics than the conventional resistance. At low currents $V(nf) \propto I^n(f)$, in contrast to metallic systems. Theory and experiment give roughly the same current scale for the non-linear effects. These non-linearities hold promise as new tools with which to learn more about mesoscopic systems.

REFERENCES

1. G. Timp, A.A. Chang, P. Mankiewich, R. Behringer, J.E. Cunningham, T.Y. Chang, and R.E. Howard, Phys. Rev. Lett., 59: 732 (1987).
2. G. Timp, H.U. Baranger, P. de Vegvar, R. Behringer, J. Cunningham, P. Mankiewich, and R.E. Howard, Phys. Rev. Lett., 60: 2081 (1988).
3. M.L. Roukes, A. Scherer, S.J. Allen Jr., H.G. Craighead, R.M. Ruthen, E.D. Beebe, J.P. Harbison, Phys. Rev. Lett., 59: 3011 (1987).
4. R.A. Webb, S. Washburn, C.P. Umbach, and R.B. Laibowitz, Phys. Rev. Lett., 54: 2696 (1985).
5. S. Washburn and R.A. Webb, Adv. Phys. 35: 375 (1986).
6. G. Timp, A.M. Chang, J.E. Cunningham, T.Y. Chang, P. Mankiewich, R. Behringer, and R.E. Howard, Phys. Rev. Lett., 58: 2814 (1987).
7. C.J.B. Ford, T.J. Thornton, R. Newbury, M. Pepper, H. Ahmed, C.T. Foxon, J.J. Harris, and C. Roberts, J. Phys. C, 21: L325 (1988).
8. M. Buettiker, Phys. Rev. Lett., 57: 1761 (1986).
9. P.G.N. de Vegvar, G. Timp, P.M. Mankiewich, R. Behringer, and J. Cunningham, submitted to Phys. Rev. B.
10. P.G.N. de Vegvar, G. Timp, P.M. Mankiewich, J.E. Cunningham, R. Behringer, and R.E. Howard, Phys. Rev. B Rapid Comm., 38: 4326 (1988).
11. R.E. Behringer, P.M. Mankiewich, and R.E. Howard, J. Vac. Sci. Technol. B, 5: 326 (1987).
12. G. Timp, A.M. Chang, P. de Vegvar, R.E. Howard, R. Behringer, J.E. Cunningham, and P.M. Mankiewich, Surf. Sci., 196: 68 (1988).
13. S. Washburn, H. Schmid, D. Kern, and R.A. Webb, Phys. Rev. Lett., 59: 1791 (1987).
14. R.A. Webb, S. Washburn, and C.P. Umbach, Phys. Rev. B, 37: 8455 (1988).
15. A.D. Benoit, S. Washburn, C.P. Umbach, R.B. Laibowitz, and R.A. Webb, Phys. Rev. Lett., 57: 1765 (1986).
16. A.I. Larkin and D.E. Khmelnitskii, Zh. Eksp. Teor. Fiz., 91: 1857 (1986) [Sov. Phys. JETP, 64: 1075 (1986)].

THEORY OF BALLISTIC QUANTUM TRANSPORT THROUGH A 1D CONSTRICTION

E. Tekman and S. Ciraci

Department of Physics, Bilkent University
Bilkent 06533 Ankara, Turkey

ABSTRACT

In this paper the ballistic quantum transport through a 1D constriction in a 2D electron gas is investigated using a refined formalism. The quantization of the conductance at integer multiples of $2e^2/h$ is found to be the main property of these quantum point contacts. The agreement and discrepancies with the experiments and the existing theories are summarized. Special emphasis is given to the effects of the constriction geometry and finite temperature. Some unresolved aspects of the experiments are clarified on this basis.

I.INTRODUCTION

In the last two decades much interest has been raised in the electronic structures and the transport properties of the lower dimensional systems owing to the improvements in fabrication of such devices. Among these, the quasi-one-dimensional (Q1D) systems have important transport characteristics, and they have been the subject of several studies[1-3]. Earlier, the effcts of the quantization of the transversal momentum in the infinite 1D electron waveguides were investigated theoretically[3] and the conductance of the constriction, G_c, was found to be quantized as $G_c = 2e^2N_c/h$, N_c being the number of the occupied subbands.

Most recently two experimental groups[4-5] independently achieved the measurement of the conductance G, through a narrow constriction between two reservoirs of two dimensional electron gas (2DEG) in a high mobility GaAs-GaAlAs heterostructure. The constriction they made by a split-gate was sufficiently narrow so that its width w was comparable with the Fermi wavelength ($w \sim \lambda_F$), and also significantly short ($d < l_e$ electron mean free path) so that electrons can move ballistically. The resulting conductance of the transport through this constriction was found to change with w (or gate voltage) in steps of $\sim 2e^2/h$. These pioneering experiments[4-5] were interpreted in terms of the quantization of the constriction conductance G_c. However, in the devices used for measuring the conductance the electron mean free path is large enough to result in a ballistic transport

Science and Engineering of One- and Zero-Dimensional Semiconductors
Edited by S.P. Beaumont and C.M. Sotomajor Torres
Plenum Press, New York, 1990

even in the 2DEG. Apparently, the measured G has to differ from the conductance arising from the constriction alone, G_c. Although the existing studies on the subject[6-10] clarified some of the main features of the phenomenon, most of the crucial aspects are not resolved yet.

In this paper we present a thorough analysis of the quantum conductance through a constriction by using a refined formalism. The expression we derived provides exact calculation of G and allows the investigation of the effects of temperature and contact geometry. In Section II we derive a conductance expression for a uniform constriction in 2DEG. The results emerging from the application of our theory to various cases are presented in Section III. Concluding remarks are given in Section IV.

II.FORMULATION OF THE CONDUCTANCE

The most general constriction system can be visualized as two reservoirs adjacent to the 2DEG which represent the external circuit connections, and a constriction placed (along z-direction) between two 2DEG. The form of the confining potential depends on the split-gate geometry, gate voltage and the material parameters. We assume that the transversal momentum(p_y) in the constriction region is quantized.

To formulate the conductance in a finite and abrupt constriction we first consider the system consisting of a semiinfinite constriction of uniform width w for $z \leq 0$, and a 2DEG for $z \geq 0$. For an electron incident from the constriction side the wavefunction of energy E is written as:

$$\begin{aligned} \psi_n(y, z \leq 0) &= e^{i\gamma_n z}\phi_n(y) + \textstyle\sum_{m=1}^{\infty} e^{-i\gamma_m z}\phi_m(y) r_{mn} \\ \psi_n(y, z \geq 0) &= \textstyle\int_{-\infty}^{\infty} d\kappa \; A_n(\kappa) e^{ik_z(\kappa)z} e^{i\kappa y}, \end{aligned} \tag{2.1}$$

where $\gamma_n^2 = 2m(E - \epsilon_n)/\hbar^2$, $k_z^2(\kappa) + \kappa^2 = 2mE/\hbar^2$ and ϕ_n (ϵ_n) is the nth eigenstate (eigenenergy) of the infinite-well constriction potential. The reflection coefficient r_{mn} and the transmission function $A_n(\kappa)$ are determined from the boundary conditions at $z = 0$. By introducing the transversal Fourier transform $\Phi_n(q) = (2\pi)^{-1/2} \int_{-\infty}^{\infty} dy \; e^{-iqy}\phi_n(y)$, the continuity equations can be expressed in q-space, therefrom $A_n(q)$ is eliminated to yield the relation $\sum_m[\gamma_n + k_z(q)]\Phi_m(q) r_{mn} = [\gamma_n - k_z(q)]\Phi_n(q)$. This equation is multiplied by $\Phi_j^*(q)$ and integrated over q. The resulting equations can be cast in a matrix form:

$$\mathbf{r} = (\mathbf{\Gamma} + \mathbf{K})^{-1}(\mathbf{\Gamma} - \mathbf{K}), \tag{2.2}$$

where $\mathbf{\Gamma}$ has the elements $\Gamma_{ij} = \gamma_i \delta_{ij}$. γ_i is either real or imaginary, and therefore $\mathbf{\Gamma} = \mathbf{\Gamma_R} + i\mathbf{\Gamma_I}$, $\mathbf{\Gamma_R}$ and $\mathbf{\Gamma_I}$ being diagonal matrices with positive real elements. $\mathbf{K}$ is given by $K_{ij} =< \Phi_i \mid k_z(q) \mid \Phi_j >$. The expression for $\mathbf{r}$ is reminiscent of the reflection coefficient obtained for 1D step potential, $r = (k' - k)/(k' + k)$, k and k' denoting incident and transmitted wave vectors, respectively. To formulate the incidence from the 2DEG we consider the mirror image of the above geometry. Following similar steps we obtain the transmission coefficients corresponding to the incident wave vector $\vec{k} = (k_o, \kappa_o)$:

$$\mathbf{t}_{\vec{k}} = \sqrt{2\pi}(\mathbf{\Gamma} + \mathbf{K})^{-1} 2k_o \mathbf{\Phi}^\dagger. \tag{2.3}$$

Here the vectors, $\mathbf{t}_{\vec{k}}$ and $\mathbf{\Phi}$ have elements $t_{n\vec{k}}$ and $\Phi_n(\kappa_o)$, respectively. The equation for $\mathbf{t}_{\vec{k}}$ is analogous to the 1D semiinfinite barrier transmission coefficient $t = 2k/(k + k')$.

Finally, to obtain the conductance for a constriction of finite length d and width w between two reservoirs of 2DEG as described in the inset of Fig. 1, we consider the incident wave $\vec{k}$. This gives rise to right-going constriction states with amplitudes, $\mathbf{t}_{\vec{k}}$, which are reflected back at the right boundary with amplitudes, $\mathbf{r}$ and the phase factors, $e^{i\mathbf{\Gamma}d}$. Then the wavefunction including multiple reflections in the constriction region can be written as:

$$\mathbf{\Phi}(y)[e^{i\mathbf{\Gamma}z}\mathbf{\Theta}(\vec{k}) + e^{-i\mathbf{\Gamma}z}\mathbf{\Delta}(\vec{k})], \tag{2.4}$$

where $\mathbf{\Theta}(\vec{k}) = [\mathbf{I} - (\mathbf{r}e^{i\mathbf{\Gamma}d})^2]^{-1}\mathbf{t}_{\vec{k}}$ and $\mathbf{\Delta}(\vec{k}) = e^{i\mathbf{\Gamma}d}\mathbf{r}e^{i\mathbf{\Gamma}d}\mathbf{\Theta}(\vec{k})$. We next evaluate the current operator for the state $\vec{k}$ and integrate over all the states at the Fermi level to obtain the conductance:

$$G = \frac{e^2}{\pi h}\int_{-k_F}^{k_F}\frac{d\kappa}{k_z(\kappa)}\{[\mathbf{\Theta}^{\dagger}(\vec{k})\mathbf{\Gamma_R}\mathbf{\Theta}(\vec{k}) - \mathbf{\Delta}^{\dagger}(\vec{k})\mathbf{\Gamma_R}\mathbf{\Delta}(\vec{k})] + 2\mathrm{Im}[\mathbf{\Theta}^{\dagger}(\vec{k})\mathbf{\Gamma_I}\mathbf{\Delta}(\vec{k})]\}. \tag{2.5}$$

In this expression the first and second terms in the parentheses are related to the right-going and left-going states, respectively. The contribution of the evanescent states are expressed by the third term. At small d the sharp rises in conductance corresponding to the opening of a new channel are smoothed out by the evanescent states. In contrast to methods[4-8] proposed earlier, the contributions of various types of states are explicitly given in the present formalism. This provides a better description of the quantum phenomena taking place in the constriction, and thus leads to a thorough understanding of the detailed structure of G. Since the formalism of quantum conductance presented in this work uses a mixed basis set consisting of the plane waves and constriction states, the numerical results converge rapidly. For example, the conductance showing 7-8 steps can be handled with a reasonable accuracy by using (10×10) matrices for any d. Starting from the same type of basis set, Kirczenow[9] and the authors[10] have independently arrived at the similar expressions of G. The analogy with a 1D potential problem is established in the present study, however. Moreover, our method is extended to study more realistic situation, namely a non-uniform or smooth constriction geometry as described in Fig. 2 and 3; and the roughness of constriction as well. In this case the constriction is described by closely spaced uniform constrictions with different widths, and a transfer matrix method is used for multiple boundary matching. The theory can also be extended to investigate various aspects such as temperature, scattering in the constriction and magnetic field. Based on the calculations performed by using the above formalism we will discuss the effects of the constriction geometry, roughness and temperature in the next section.

III.RESULTS

Fig.1 illustrates the conductance $G(w/\lambda_F)$ calculated for the uniform constriction corresponding to $d = 0$, λ_F and $5\lambda_F$ at $T = 0^o$K. The first curve ($d = 0$) corresponds to the Sharvin[11] conductance, which was known to be $G_s = (2e^2/h)\ 2w/\lambda_F$. One sees that the whole curve is displaced, and weak oscillations are superimposed owing to the quantum interference effects even at $d = 0$. However, the longer is d, the sharper are the quantum jumps and the flatter the plateaus. Moreover, for $d \gtrsim 5\lambda_F$ steps occur exactly at the integer multiples of $2e^2/h$. The resonance structure which originates from the interference of multiple reflected waves, becomes more pronounced at large d. Within the approximation of real and diagonal $\mathbf{r}$, the position of the mth resonance on the nth plateau is estimated to be $w_{mn} \simeq n\lambda_F[4 - (m\lambda_F/d)^2]^{-1/2}$, and the number of resonances on the nth plateau $M_n \simeq 2d/\lambda_f[(2n+1)/(n+1)^2]^{1/2}$. Based on these approximate expressions we arrive at the conclusion that the number of resonances on a plateau is proportional to d, but decreases with increasing w. The same approximation also yields that δG_{mn} (i.e. the difference in conductance between the mth resonance and the next anti-resonance on the nth plateau) decreases as either m or n increases. As for fixed m and n, the larger is d, the greater becomes δG_{mn}. Furthermore, we examined the effect of the intersubband mixing by neglecting the off-diagonal elements in $\mathbf{K}$ and found that δG_{mn} is not affected in any essential manner. This is at variance with earlier results[6].

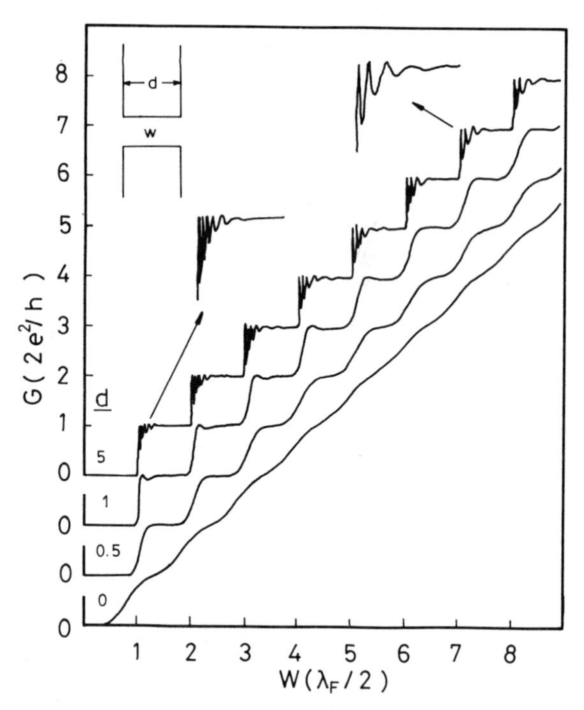

Fig.1 Quantum conductance G of a uniform constriction calculated for various lengths, d. The unit of length is λ_F, and $T = 0^o$K. The constriction geometry is shown in the inset. The resonance structure of the first and seventh channels are magnified by the inset.

In Fig.2 we present the results obtained for the wedge-like constriction shown in the inset. Up to a certain value of wedge-angle ($\alpha \sim 50^o$), the conductance curve does not deviate from that of Sharvin contact geometry corresponding to $\alpha = 0^o$, but beyond this value, the quantum steps become more pronounced. $\alpha = 90^o$ corresponds to the uniform constriction shown in Fig.1. At a given angle ($\alpha \gtrsim 60^o$), the quantum effects are emphasized as d increases. In spite of the apparent step structure at large α, the resonance structure does not occur owing to the phase incoherence caused by the mixing among different subbands in the aperture. As an extension of this contact geometry we consider also the smooth entrance to a uniform constriction described in Fig.3. The constriction geometry used by Khmelnitskii[12] can be compared to the form with $\alpha = 75^o$. Owing to the negligible band mixing and reflection in this geometry the quantization of conductance is not affected in any essential manner. However, because of insufficient phase coherence

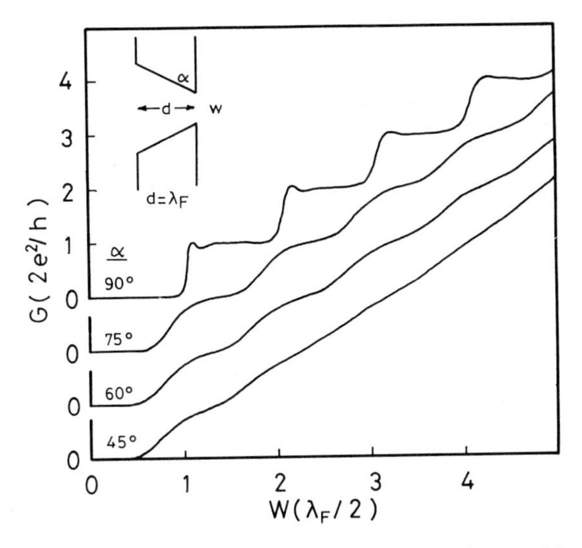

Fig.2 Quantum conductance G of a wedge-like constriction (shown in the inset) calculated for various wedge angles, α. The constriction length $d = \lambda_F$, and $T = 0^o$K.

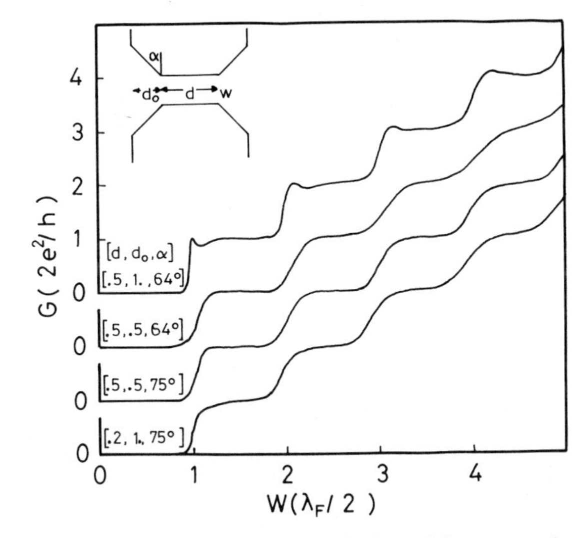

Fig.3 Quantum conductance G of a constriction with a smooth entrance calculated for various d_o, d and wedge angle α as described by the inset. The unit of length is λ_F, and $T = 0^o$K.

the resonance structure is weakened. In the present study the representation of a smooth tapering by a sequence of discrete steps may give rise to the reflections. These artificial reflections are, however, eliminated by using large number of steps. It is found that the effect of the smooth entrance is insignificant for $d_o/d \ll 1$ (d_o and d are denoted by the inset), but becomes crucial if $d_o \sim d \sim \lambda_F$. In the latter case, additional structure superimposed on the plateaus, and thus the step structure is deformed.

We have studied also the effect of the roughness along the constriction by a random modulation of the width. Preliminary results are illustrated in Fig.4, in which the shaded areas correspond to the fluctuations in G depending on the form of the roughness. It is seen that the heights of the steps deviate from the ideal value of $2e^2/h$. Moreover, the resonance structure is also affected, and becomes either weaker or stronger. That the observed conductance curve[4,5] is lacking resonances can be attributed to the roughness of the constriction.

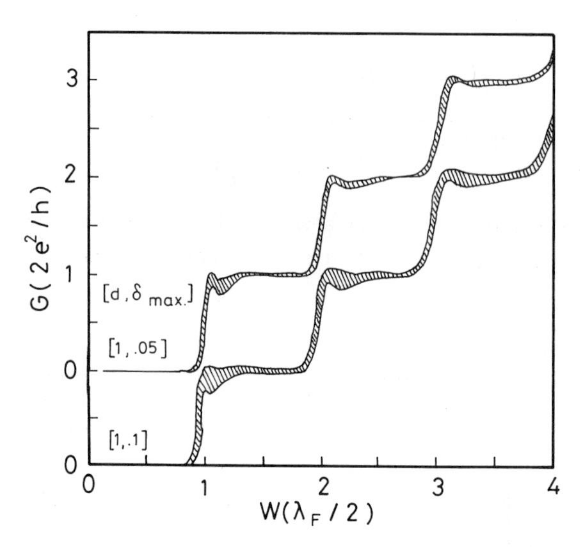

Fig.4 Quantum conductance G for a rough constriction for maximum modulation amplitude $\delta_{max} = 0.05\lambda_F$ and $0.1\lambda_F$. The length $d = \lambda_F$ and $T = 0^o$K.

Since carrier mobilities are decreased and the sharp Fermi distribution is smeared at finite but small temperatures, the quantum conductance is expected to deviate considerably from that at $T = 0^o$K. Ignoring the effects of inelastic scattering and assuming that l_e is still larger than d we studied the effect of temperature in the range from $T = 0^o$ to 5^oK within the ballistic transport regime. The quantum conductance $G(w/\lambda_F)$ was calculated for T =0, 0.6 and 5^oK and is shown in Fig.5. At $T = 5^o$K the resonance structure completely disappeared. While the higher lying steps are smeared out, the steplike structure of the first ten channels are still maintained. In the range of temperature ($\lesssim 0.6^o$K) where the experiments were performed[4,5], the resonance structure is maintained for small constriction length ($d \lesssim 2.5\lambda_F$). At small d the resonance peaks are so widely separated that they persist in spite of the energy spreading due to temperature. In contrast to this, for long constrictions ($d \gtrsim 10\lambda_F$) closely spaced resonances can easily be destroyed. An important predict of this result is that the experimental data[4,5] which do not display a clear resonance structure, ought to be obtained from an effectively long constriction ($d \simeq 0.5\mu$m).

IV.CONCLUSIONS

In this paper, we investigated the ballistic quantum transport through a 1D constriction in a 2DEG using a refined formalism. We have found that: **i)** The weak oscillations around the classical Sharvin conductance evolve into the quantized steps for finite d. A resonance structure is superimposed on the plateaus, which becomes pronounced as d increases. **ii)**

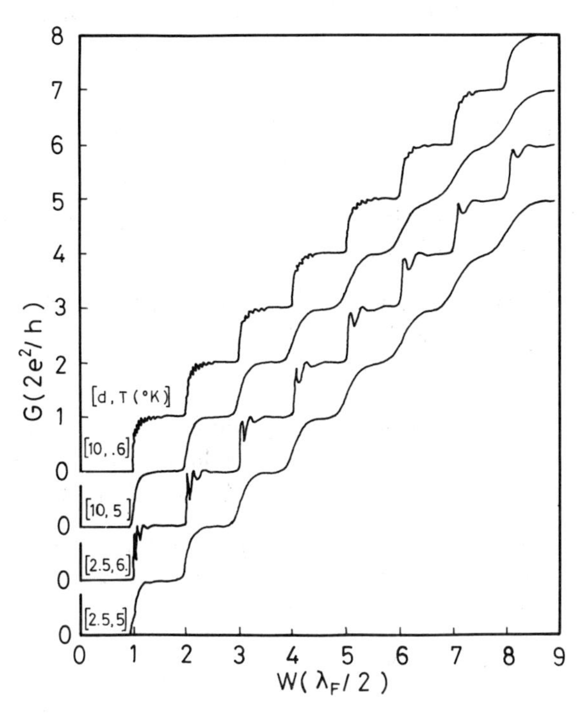

Fig.5 Quantum conductance G of uniform constrictions of $d = 2.5\lambda_F$ and $10\lambda_F$ calculated for $T = 0.6^o$ and 5^oK. The Fermi wavelength λ_F is taken to be 42 nm as in Ref. 4.

For a wedge-like constriction the step structure becomes significant only for large wedge-angle, α. Tapering of the constriction can result in the reflectionless transport. **iii)** In a rough constriction quantization deviates from the ideal values and resonance structure gets weaker or stronger. **iii)** At a typical experimental temperature ($T \lesssim 0.6^o$K) the resonance structure disappears only for large d ($d > 5\lambda_F$). For even higher temperatures ($T \sim 5^o$K) the steplike structure is smoothed, but is still recognizable.

ACKNOWLEDGEMENTS

Part of this work has been carried out while one of the authors (E.T.) was visiting ICTP, Trieste. We acknowledge valuable discussions with Professors A. Baratoff, N. García and Dr. M. Büttiker. We wish to thank E. Haanapel and D. van der Marel for providing us their work prior to its publication, and Prof. C. Yalabık for his careful examination of the manuscript.

REFERENCES

1. R. Landauer, IBM J. Res. Develop. **1**, 233 (1957); Z. Phys. **B68**, 217 (1987).
2. P. W. Anderson, D. J. Thouless, E. Abrahams and D. S. Fisher, Phys. Rev. **B22**, 3519 (1980).
3. M. Büttiker, Phys. Rev. **B33**, 3020 (1986); Y. Imry, in *Directions in Condensed Matter Physics*, ed. G. Grinstein, G. Mazenko (World Scientific, Singapore, 1986), vol.1, p. 102.
4. B. J. van Wees, H. van Houten, C. W. J. Beenakker, J. G. Williamson, L. P. Kouwenhoven, D. van der Marel, and C. T. Foxon, Phys. Rev. Lett. 60, 848 (1988).
5. D. A. Wharam, T. J. Thornton, R. Newbury, M. Pepper, H. Rithcie and G. A. C. Jones, J. Phys C21, L209 (1988).
6. A. D. Stone, "Theory of Quantum Conductance of a Narrow Constriction", Working Party on Electron Transport in Small Systems, held in Trieste, 1988, unpublished; A. Szafer, A. D. Stone, Phys. Rev. Lett. 62, 300 (1989).
7. D. van der Marel, "Oscillations in the Sharvin Point Contact Resistance", Working Party on Electron Transport in Small Systems, held in Trieste, 1988, unpublished; E. G. Haanapel, D. van der Marel, Phys. Rev. B39, No. 11 (1989).
8. N. García, "Oscillatory Quantum Elastic Resistances of Small Contacts: Holes, Constrictions, Tube Precursors ", Working Party on Electron Transport in Small Systems, held in Trieste, 1988, unpublished.
9. G. Kirczenow, Solid State Commun. 68, 715 (1988).
10. E. Tekman, S. Ciraci, Phys. Rev. B39, No. 11 (1989).
11. Yu. V. Sharvin, Zh. Eksp. Teor. Fiz. 48, 984 (1965) (Sov. Phys.–JETP 21, 655 (1965)).
12. D. E. Khmelnitskii, "Reflectionless Quantum Transport and Fundamental Steps of Ballistic Resistance in Microconstrictions", Working Party on Electron Transport in Small Systems, held in Trieste, 1988, unpublished.

BOUND AND RESONANT STATES IN QUANTUM WIRE STRUCTURES

François M. Peeters

University of Antwerp (U.I.A.)
Department of Physics
Universiteitsplein 1
B-2610 Antwerpen

I. INTRODUCTION

In recent years[1,2] there has been a growing interest in systems of reduced dimensionality. A variety of phenomena exhibiting quantum interference between alternative carrier paths has been studied in such systems including the Aharonov-Bohm effect[2,3], 'universal' fluctuations[1,4], and resonant phenomena. New quantum-size effects have been found such as: non-local bend resistance[5,6], the quenching of the quantum Hall effect[7], quantized point contact resistance[8], the oscillatory behavior of the capacitance[9] and of the dc-conductivity[10,11],... It has also become clear that the behavior of nanostructures may resemble, in many ways, properties of waveguides. This notion has recently found clear experimental verification[5]. Nanostructures offer the possibility of achieving device functions[12] by the use of quantum interference effects. Such functions would be analogous to those achieved in two-terminal resonant tunneling devices[13,14].

In the present paper we will study wires with perfect sharp boundaries in the absence of impurities and free of defects. Such systems are realized, to a very good approximation, by lateral confinement of a two-dimensional electron gas (2DEG) at a $GaAs/Al_xGa_{1-x}As$ heterojunction. The elastic mean free path l_ϵ may be larger than a few micron and the phase-breaking length l_ϕ is typically an order of magnitude larger. The lateral confinement can define wires of width W down to the fermi wavelength λ_F. When the measuring probes are placed within a phase-breaking length the electron motion between the measuring probes is coherent and the electron motion is *ballistic*. This is the regime which will be of interest to us.

Two different quantum wire structures will be investigated. The *bend* structure which is a quantum wire with a bend at right angle. Such a system exhibits[15] one bound state which is localized at the bend. This state has a pure quantum mechanical origin and is a consequence of the relaxing of the boundary conditions at the bend, which reduces the zero-point energy of the electron. I will show how this bound state offers the possibility of creating a new type of quantum wire which has the prospect of easy integration features. The second system which will be investigated is another waveguide-like structure, schematically shown in Fig. 1. The structure consists of a local thickness increase of the quantum wire. This cross-shaped system is assumed to be infinitely long in the x-direction.

Science and Engineering of One- and Zero-Dimensional Semiconductors
Edited by S.P. Beaumont and C.M. Sotomajor Torres
Plenum Press, New York, 1990

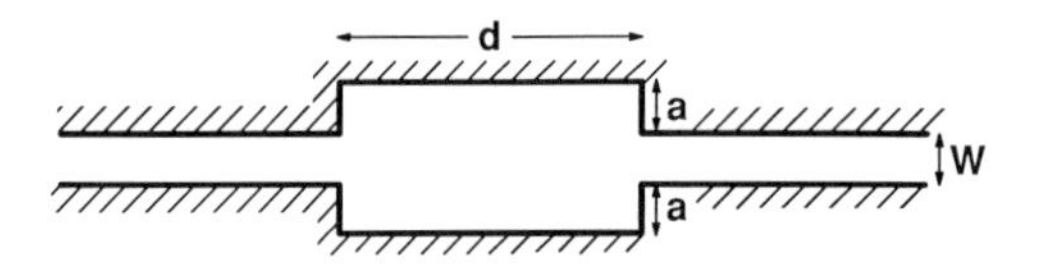

Fig. 1. The cavity studied in this paper.

while the transverse (taken as the y-direction) arms are of finite length which are bounded by a hard wall. Both arms are taken of equal size. The idea of device function is based on the fact that the potential which confine the electrons inside the nanostructure, can be changed directly by using external voltages and this will vary the electron interference patterns. In conventionally metal waveguides this is achieved by mechanically moving a piston inside a resonant cavity to change the interference patterns of the electromagnetic field. Note that if the length of the side arms of the system of Fig. 1 is much larger than the inelastic mean free path, this structure represents just two intersecting wires. Hall measurements are done on such cross structures and recently[16–19,6] it has been show that the geometrical details of the cross are of fundamental importance in such an experiment and may lead to quenching of the Hall resistance[7,20] and even a negative Hall resistance[21] can be induced. Here I will study such a cross in the absence of a magnetic field and I will investigate bound and scattered states. This work is a generalization of Ref. 22 to a cross system in which the voltage probes and the current carying probes have not the same width. For completeness I want to mention that recently Sols *et al*[12] solved the T-shaped structure by using tight-binding Green's function techniques.

The quantum mechanical problem corresponding to the structure depicted in Fig. 1 is a difficult boundary value problem even if one assumes that the electron potential is flat and delimited by hard walls. Nevertheless it is possible to solve it exactly (numerically). Different theoretical approaches have been persuit to solve such quantum mechanical problems:

1) *Wavefunction matching.* This technique will be applied in the present paper. For the particular geometry of Fig. 1 we benefit from the exact knowledge of the eigenstates in the different sections of the leads. The wavefunction is expanded in a complete set of such eigenstates which are solutions of the Schrödinger equation satisfying the boundary conditions. The coefficients in this expansion are determined by requiring continuity of the wavefunction and its first derivative at the boundaries between the different sections. This leads to a set of coupled linear equations, for these expansion coefficients, which has to be solved numerically. This involves a single matrix inversion of infinite order. The numerical results in the present paper were obtained for a finite number of modes, but care was taken that upon increasing the number of modes the result does not change with more than 0.1%.
2) *Tight binding model*[6,12,23] on a 2D lattice with nearest neighbor hopping parameter t. The advantage of this method is that it is relatively easy to include: i) a magnetic field, ii) impurity scattering, and iii) more general geometries (boundary conditions). The disadvantages are: i) it is a pure numerical technique, and ii) no bound states are obtained, only scattered states are found.

We found that the transmission and reflection coefficients for the structure of Fig. 1 show a very sensitive dependence on the incident transverse mode index N, the wavelength for longitudinal motion λ, the wire width L and the dimensions a, d of the cavity. Even for the simple geometry of Fig. 1 the variety and structure of scattering amplitudes as a function of incident quantum numbers and device dimensions are remarkable. A relative small change in the dimensions of the cavity can induce dramatic changes in the transmission.

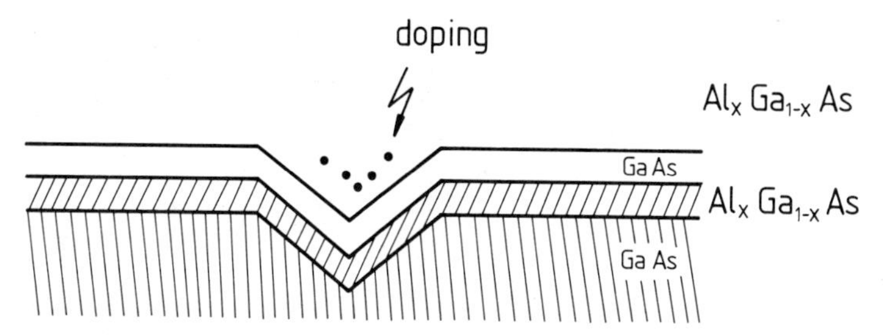

Fig. 2. At the bottom of the crevice the electron motion is 1D.

In the present work, as in most discussions up to know, confinement by hard walls is assumed. Limitations of the hard wall model can be partially overcome by a parabolic confining potential. In principle one has to treat the finite potential barrier case as occurs at a $GaAs/Al_xGa_{1-x}As$ interface. The penetration of the wavefunction in the barrier is not important at low energy (the electron energy is close to the bottom of the well). Thus this effect is only relevant at high energies (i.e. higher modes) and thin wires (large zero point energy).

II. THE BEND

The bound state in the bend can be used to create a new quantum wire. A proposal for such a system is schematically shown in Fig. 2. First one grows a thick $GaAs$ layer. In this layer a crevice is made, which may have a valley of sidewalls at an angle of 45^0. Subsequently one covers the surface with a few atomic layers of $Al_xGa_{1-x}As$. On top of this one grows the quantum well of $GaAs$. Next a thick layer of $Al_xGa_{1-x}As$ is added with some remote doping which provides the electrons in the quantum well. The 2D quantum well looks as shown in Fig. 3a: it is a 2D plane which has been bend and both surfaces are at right angle. The energy spectrum of this structure is shown in Fig. 3b. The z-axis is choosen along the bend of the 2D quantum well and the x-axis is the other coordinate along the quantum well. Quantum confinement is present along the y-direction. For convenience we have taken the x-axis and the minus x-axis perpendicular to each other. The spectrum consists of 2D parabolas with spacing $E_n = n^2 E_0$ which correspond to the well-known quantum well states in which the electrons are free to move in the x-z plane.

Because of the presence of the bend a localized state will exist around $x = 0$. This state is localized in the x-direction but is free to move along the z-direction. This state is a result of the small increase in distance between the two quantum well barriers at $x = 0$. As a result the zero point energy of this state is smaller than E_0 and we obtain the one-dimensional parabola with energy $E(k_z) = E_0' + \hbar^2 k_z^2/2m$, where $E_0' < E_0$ with $(E_0 - E_0')/E_0 = 7.8\%$. In order to get quasi-one-dimensional (Q1D) behavior the Fermi energy has to be below E_0. This can be achieved by a right choice of doping.

The main advantage of this new type of quantum wire is that the sidewalls of the quantum well are well-defined: it are $GaAs - Al_xGa_{1-x}As$ interfaces which are know to be smooth on an atomic scale. Furthermore no electric charges are present on the walls or near the walls. This new technique provides the possibility for integration of quantum wire structures.

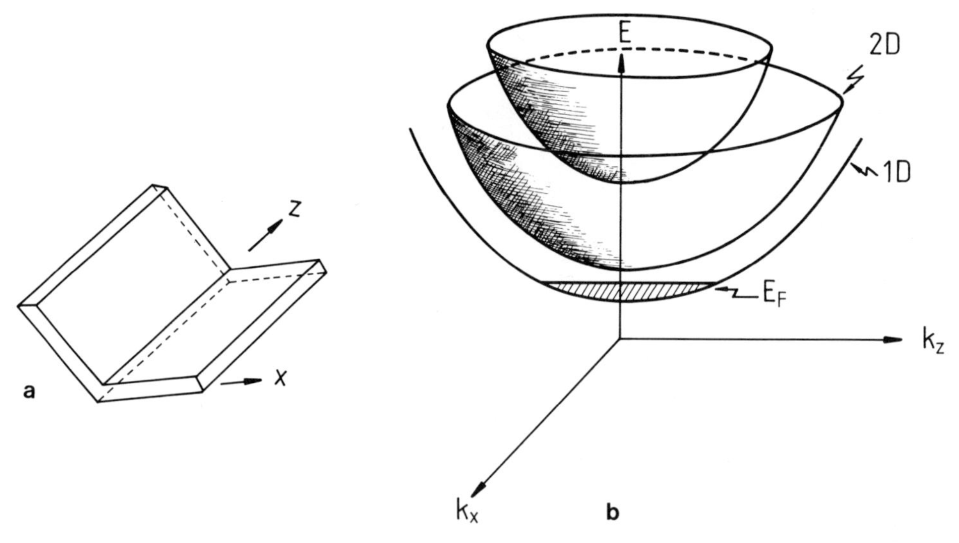

Fig. 3. Configuration of the 2D quantum well bend (Fig. 3a) and energy spectrum (Fig. 3b) of the quantum well structure of Fig. 2.

In practice there may be problems to create a perfect crevice at right angle. At present such bends have been realized mainly with the aim of creating laser action[24]. No free electrons were present because of the absence of doping near the quantum well. At present bends of 2D systems have been created with rounded bends. This is not necessarily a disadvantage, in fact it will make it easier to have a bound state.

III. CAVITIES

Resonant cavities have been very important in microwave physics. Here I will discuss a cavity for a solid state electronic waveguide structure. Such a cavity may lead to mode selection, resonances, and thus can be considered for transistor action.

The structure under study is depicted in Fig. 1. In order to limit the number of parameters I will only consider the symmetrical cavity, i.e. the thickness change is the same on both sides of the quantum wire. Furthermore the cross is a special case of this structure in which the sidewalls of the cavity are at infinity, i.e. $a \to \infty$. Cross structures are present in Hall bars and they are of crucial importance for the quenching of the quantum Hall effect. Note that we will also be able to discuss the asymmetric cross, i.e. $d \neq W$ (such a system has still $x \to -x$, $y \to -y$ symmetry but no rotational symmetry about 90°).

Note that the present structure is the inverse of a point contact[8,25] structure. In a point contact the electrons move through a narrow constriction from a 2DEG to another 2DEG. This was studied by numerous people and there is no need to dwell on it. Such a system has $W + 2a \ll W$. Very recently, a cavity connected on both ends through point contacts to 2DEG's was studied theoretically[26] and experimentally[27]. In the present study we consider a cavity connected by quantum wires. After completion of this work it came to my attention that the T-shaped analog of this structure was studied recently by Sols *et al*[12] using thight-binding Green's function techniques.

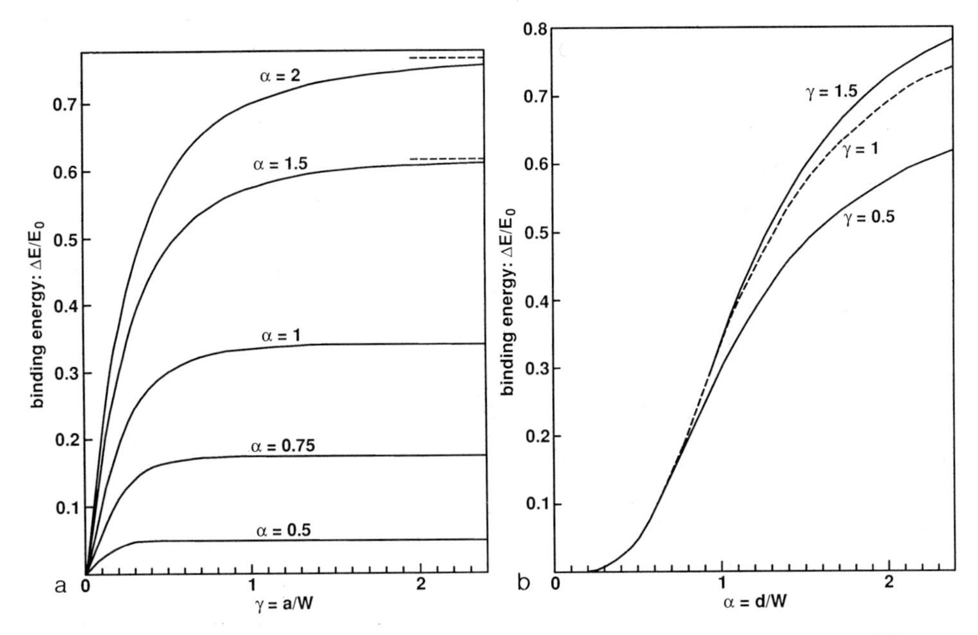

Fig. 4. Bound state energy of the cavity as function of the thickness $\gamma = a/W$ (Fig. 4a) for different values of the width $\alpha = d/W$. Fig. 4b is the same but now as function of the width and for different values of the thickness.

III.1. Bound states in the cavity

Due to the local thickness increase of the quantum wire the electron will have a lower zero-point energy when it is in the cavity. Thus in the cavity the electron can occupy a state which has an energy lower than the zero point energy in the leads and will thus not be able to propagate into the leads. In the following we take W as our unit of length and we define the dimensionless quantities $\gamma = a/W$, $\alpha = d/W$ and use $E_0 = \hbar^2\pi^2/2mW^2$ as our unit of energy.

For a long cavity ($d \gg W + 2a$) a simple estimate of the binding energy can be made. The zero-point energy in the cavity is approximately given by

$$E = \frac{\hbar^2\pi^2}{2m(W+2a)^2} = \frac{E_0}{(1+2\gamma)^2} \quad , \tag{2}$$

and consequently the binding energy with respect to motion in the quantum wire becomes

$$\Delta E = E_0 - \frac{E_0}{(1+2\gamma)^2} \quad , \tag{3}$$

and thus $\Delta E/E_0 \to 4\gamma$ when $\gamma = a/W \to 0$. In the limit of $d \gg a$ the binding energy is independent of $\alpha = d/W$. For general values of α and γ a numerical calculation has been performed whose results are shown in Figs. 4. For $\alpha = 1.5$ and $\alpha = 2$ the $\gamma \to \infty$ value is indicated by the horizontal line in Fig. 4a. For large values of γ the binding energy becomes independent of γ. The reason is that in the $\gamma \to \infty$ limit one obtains the cross structure. The bound state in the cavity no longer feels the sidewalls. Remember that the electron wavefunction is an exponential decaying function from the center of the cavity. From Fig. 4b it is apparent that the binding is independent of γ for $\alpha \ll 1$. For small α the binding energy quickly decreases to zero.

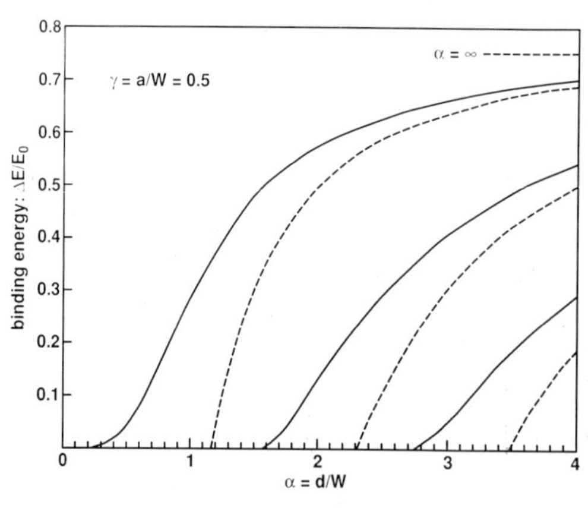

Fig. 5. The binding energy of the bound states (solid curves) in a cavity as a function of the length of the cavity for fixed value of the thickness. The results from the quantum box approximation are shown as dashed curves.

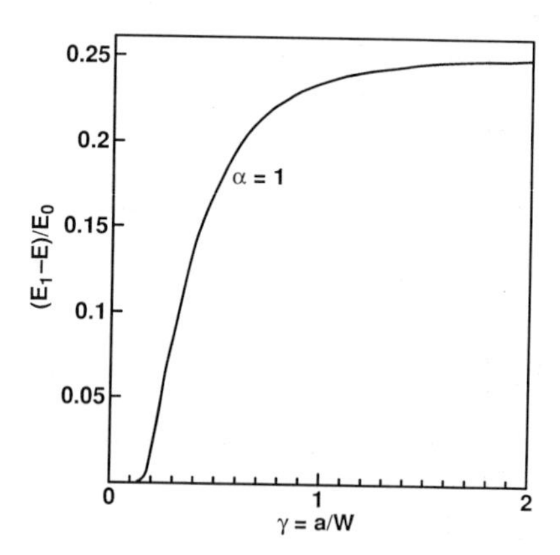

Fig. 6. Binding energy of the lowest odd parity state as function of the thickness of the cavity for $\alpha = 1$.

When α is sufficiently large it is possible to have several bound states. This is depicted in Fig. 5 where the binding energy of all the bound states are shown as function of the length of the cavity (d) for a fixed value of γ ($\gamma = 0.5$ corresponds to a cavity which is twice as wide as the quantum wire). With increasing α the number of bound states increases. The behavior observed in Fig. 5 is analogous to the one expected for a 1D quantum well in which the depth or the width is varied. A 1D quantum well, irrespective of its width or depth, has always *one* bound state. This is also true for the cavity structure although, for small values of α the binding energy of this bound state becomes extremely small. The fundamental differences between such a 1D quantum well and the cavity under study are: 1) the present problem is 2D, and 2) there is no attractive potential. Classically the electron will not be bound. The binding is a pure quantum mechanical effect.

The $\alpha \to \infty$ behavior in Fig. 5 is well approximated by considering the cavity as a 2D quantum box. The energies in such a box are

$$E = n_1^2 \frac{\hbar^2\pi^2}{2m(W+2a)^2} + n_2^2 \frac{\hbar^2\pi^2}{2md^2} \quad . \tag{5}$$

Refering this energy to the groundstate energy in the quantum wire we obtain the binding energy

$$\frac{\Delta E}{E_0} = 1 - n_1^2 \frac{1}{(1+2\gamma)^2} - n_2^2 \frac{1}{\alpha^2} \quad . \tag{6}$$

which is a lower bound to the exact binding energy of the cavity. To obtain Eq.(6) we have added hard walls at the two places where the cavity is connected to the quantum wire which will decrease the binding energy. These energy levels are shown as dashed lines (only $n_1 = 1$ leads to binding for $\gamma = .5$) in Fig. 5.

Shult *et al*[22] have found that a cross with arms of equal width exhibits two bound states; the first is equivalent to the lowest one just studied and the other is in the continuum and has energy E_1' such that $E_0 < E_1' < E_1 = 4E_0$ with $(E_1 - E_1')/E_1 = 6.8\%$. This state has the following symmetry: its wavefunction is odd about the x,y center lines and is even

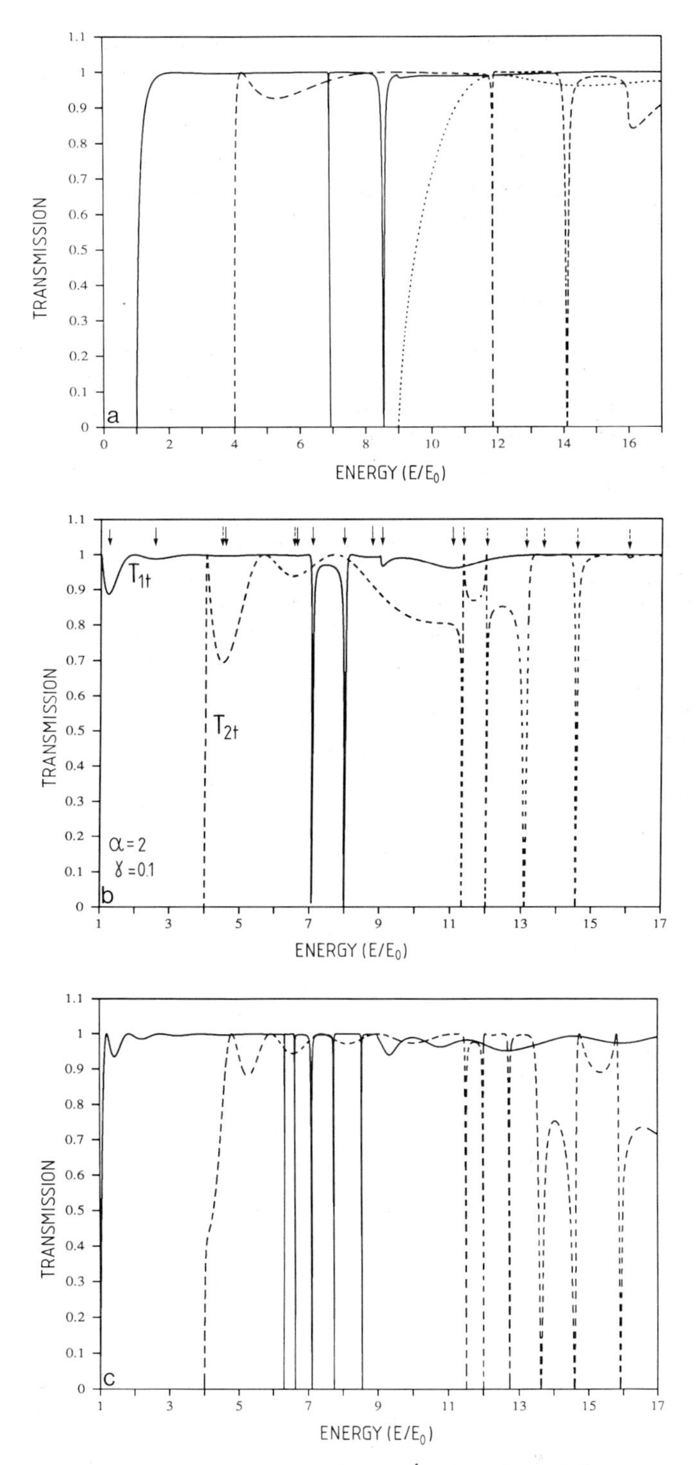

Fig. 7. The total transmission coefficient as function of the energy for an electron impinging on a cavity with thickness $W + 2a = 1.2W(\gamma = 0.1)$ and different values of the length: $\alpha = 1$ (Fig. 7a), $\alpha = 2$ (Fig. 7b) and $\alpha = 3$ (Fig. 7c). The solid curve is for an electron in the lowest mode $(N = 1)$, the dashed curve for $N = 2$ and the point curve (only in Fig. 7a) for the $N = 3$ mode.

about the x=y line. I have looked for a similar state in the cavity and found that such a state can exist *only* for $\alpha = 1$. The γ-dependence of the binding energy of this state is shown in Fig. 6. The unstability of this bound state against a variation in α away from 1 will make it almost impossible to observe it. Also the fact that the state is localized in the continuum will imply that any geometrical imperfection, or the presence of some potential fluctuations (e.g. impurities) will lead to a decay of this bound state into a free propagating scattered state.

III.2. The scattered states

For energies $E > E_0$ the electron wavefunction is not bound in the x-direction and the cavity will act like a scattering center. Resonances in the transmission coefficients are expected. Note that the fundamental difference with a cross is that in this case there is no sideways transmission. In this respect there is a fundamental difference between $\gamma = \infty$ and $\gamma \neq \infty$. This is different for the boundstates where a continuous behavior is found with increasing γ.

Fig. 7 shows the total transmission coefficient for $\gamma = 0.1$ for three different values of $\alpha = 1, 2, 3$. $\gamma = 0.1$ corresponds to a 20% increase of the thickness of the wire. The solid line is the total transmission coefficient when the electron is injected in mode $N = 1$, the dashed line for mode $N = 2$ and in Fig. 7a also the transmission coefficient for mode $N = 3$ is shown. Two different types of resonances are found: 1) weak resonances such that $T_{Nt} \neq 0$, and 2) strong resonances such that $T_{Nt} = 0$. With increasing α the number of such well-defined resonances increases and their width in energy decreases. In Fig. 7b the position of all the minima in T_{Nt} are indicated by arrows.

Our numerical results are summerized in Figs. 8. The position of the resonances are shown as function of α for fixed $\gamma = 0.1$ and for the modes $N = 1$ (Fig. 8a) and $N = 2$ (Fig. 8b). The solid curves give the position of the resonances which lead to $T_{Nt} = 0$. The weak resonances are shown as dashed curves. Note that the $T_{Nt} = 0$

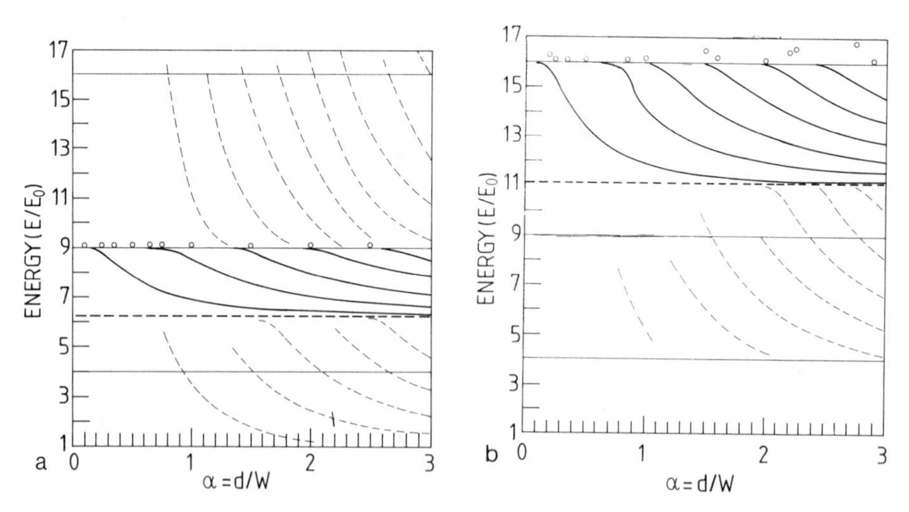

Fig. 8. The position of the resonances in the total transmission coefficient as function of α for fixed $\gamma = 0.1$. The resonances for the lowest mode are shown in Fig. 8a and the $N = 2$ mode in Fig. 8b. The solid curves are resonances such that $T_{Nt} = 0$ while the thin dashed curves (and circles) are the position of the weak resonances.

resonances are concentrated into energy windows. A simple estimate of the position of the well-defined resonances can be made in case of $\alpha \to \infty$. If we approximate the system by a 2D quantum box the energy levels are given by Eq.(5). In the present case for $\gamma = 0.1$ we find $E/E_0 = 0.694n_1^2 + n_2^2/\alpha^2$. For $N = 1$ the $T_{Nt} = 0$ resonances are located between $6.25 < E/E_0 < 9$ which corresponds to the states $(n_1 = 3, n_2)$. For $N = 2$ we found the band $11.11 < E/E_0 < 16$ corresponding to $(n_1 = 4, n_2)$. This can be understood as follows: the $N = 1(2)$ state which is even (odd) can most effectively resonante with an even (odd) state in the cavity and thus n_1 has to be odd (even). As a consequence for $N = 1$ we have the bands $0.694 < E/E_0 < 1$; $17.35 < E/E_0 < 25, \ldots$ But $n_1 = 1$ would lead to an energy window which is below the cutoff energy E_0 where no scattered states can exist. The same for $N = 2$, we have the bands $2.776 < E/E_0 < 4; 11.104 < E/E_0 < 16; 14.184 < E/E_0 < 36$ where the first one is again below the energy cutoff $E_1 = 4E_0$ and is thus not relevant.

From Figs. 8 we notice that changing the dimensions of the cavity will change the position of the resonances. At fixed energy a small change in e.g. α can drive the system from total transmission $T_t \sim 1$ to no transmission $T_t = 0$. This feature may be used to build transistors based on these quantum principles. In practical systems several channels may be occupied and the transmission coefficient will be very complex. Nevertheless it is evident that for $a = 0$ the device will be transparent while, with increasing a the device becomes opaque in certain energy windows. Experimentally the present ideas can be tested by measuring the conductivity of this device. Landauer[28] and Büttiker[29] have considered the conductivity problem as a quantum scattering problem. In such a formalism the above calculated transmission coefficients, at the Fermi energy, are the essential ingredients. Thus resonances in the transmission coefficients will be reflected in resonances in the conductivity.

IV. CROSS

IV.1. Symmetric cross

In the limit $a \to \infty$ (i.e. $\gamma \to \infty$) the cavity of Fig. 1 becomes a cross. The symmetric cross, i.e. $d = W$ $(\alpha = 1)$ was studied by Schult *et al*[2]. For completeness I will review the results for the transmission coefficients of such a structure. An electron wave impinging on the cross in mode N will be reflected into mode n with a probability R_{Nn}, it may be transmitted into mode n of the side arms of the cross with probability S_{Nn}, or it may go straight through and be scattered into mode n with a probability T_{Nn}. It is also convenient to consider the total probability for reflection $R_{Nt} = \sum_n R_{Nn}$, sideways transmission $S_{Nt} = \sum_n S_{Nn}$ and straight-through transmission $T_{Nt} = \sum_n T_{Nn}$. For leads with unequal widths it is important to realize that the above probabilities in fact refer to current probabilities. Therefore we will have that $T_{Nt} + R_{Nt} + S_{Nt} = 1$.

The above mentioned total probabilies are shown in Figs. 9 as function of the energy for an electron in mode $N = 1$ (Fig. 9a) and mode $N = 2$ (Fig. 9b). The different resonances can be understood through Figs. 10 where the separate probabilities into the different modes are shown for $N = 1$ and $N = 2$. For $N = 1$ we found that $T_{1,2n} = R_{1,2n} = 0$. The transmission or reflection of an even wavefunction into an odd wavefunction is forbidden. The cross, as a scattering centre, preserves the parity of the wavefunction. The reason is that the cross is invariant under a parity transformation. For sideways scattering, no such a selection rule is present because the wavefunction with even parity along the x-axis may be scattered into a wavefunction with even or odd parity along the y-axis. Note also that just above threshold the electron is completely reflected. This is a typical quantum mechanical effect. Such a reflection is relatively easy because reversal of the

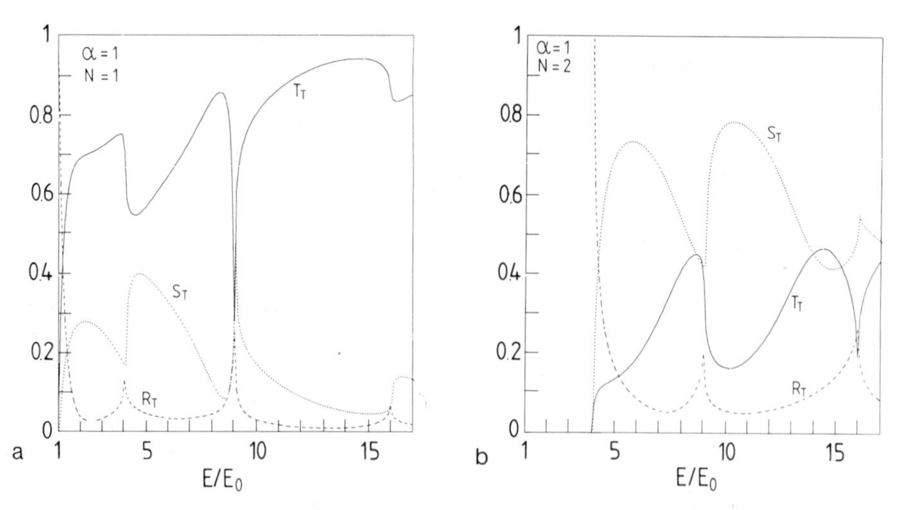

Fig. 9. The total probability for transmission (T_t), reflection (R_t) and sideways transmission ($\overline{S_t}$) for a cross with leads of equal width. The result for mode $N = 1$ (Fig. 9a) and mode $N = 2$ (Fig. 9b) are shown.

wavevector corresponds only with a very small wavevector exchange Δk. With increasing energy the reflection coefficient R drops very fast. Each time when the electron energy approaches a new threshold, i.e. $E = E_n = n^2 E_0$, a dip in the transmission is observed and a peak in the reflection. This is due to: 1) the opening up of another sideways channel, and 2) for odd (even) N, reflection and transmission into odd (even) n channels of the same wire are possible.

From Figs. 9 we observe that for $N = 1$ most of the electrons are not scattered and go straight-through the cross. On average T_{1t} increases with energy. For $N = 2$ (see Fig. 9b) on average most of the electrons turn into the side leads. The cross act as a strong scattering centre. This is even more pronounced for higher mode numbers. The results are summarized in Table 1 were the approximate energy averages of the total transmission, reflection and side ways transmission coefficients are liten. Table I shows that for $N \geq 3$ most of the electrons are scattered into the side leads. Electrons in higher modes have a larger momentum in the y-direction and, classically, have many more collisions with the sidewalls per unit distance in the x-direction. At the center of the cross there are no sidewalls and the electron will be able to move freely in the y-direction and disappears into one of the side arms of the cross.

Table I. The symmetric cross ($\alpha = 1$)

	$N = 1$	$N = 2$	$N = 3$
$T(\%)$	80	30 – 40	5 – 10
$R(\%)$	5	10	< 5
$S(\%)$	20	50	80 – 90

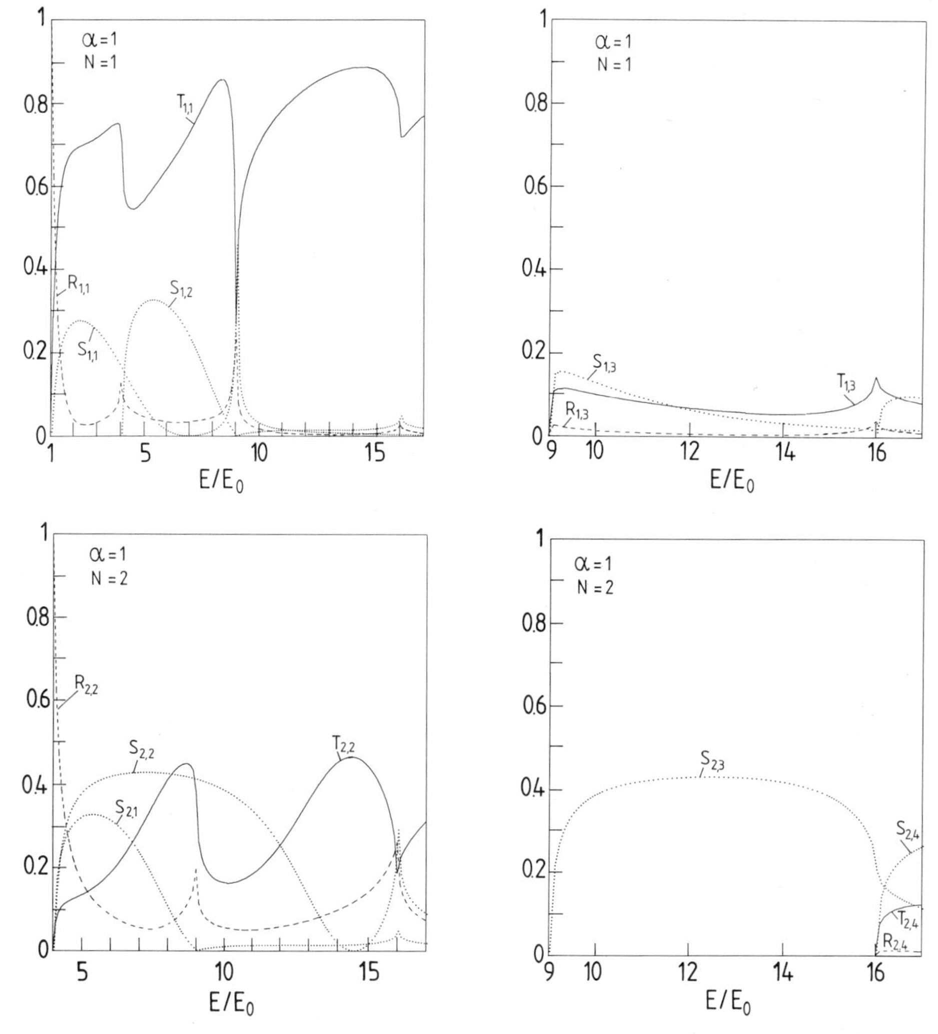

Fig. 10. The probability for scattering into the individual modes n is shown for an electron impinging on the cross when the electron is in mode $N = 1$ and $N = 2$.

IV.2. Asymmetric cross

In case the side leads are different in width (d) from the injection leads (W) the energy levels in both leads will no longer match. This influences the T, R, S coefficients considerably. Those coefficients are shown in Fig. 11 for $N = 1$ and $N = 2$ with $\alpha = 0.5$ and $\alpha = 1.5$ respectively. For $N = 1(2)$ weak structure is observed at $E = E_{2n+1} = (2n+1)^2 E_0 (E_{2n} = (2n)^2 E_0)$ due to the onset of additional scattering channels $T_{1,2n+1}(T_{1,2n})$ and $R_{1,2n+1}(R_{1,2n})$. The side channels induce resonant structure at the energies $n^2 E_0/\alpha^2$. As intuitively expected narrow side leads (e.g. $\alpha = .5$) result in larger transmission coefficients: it is harder for the electrons to turn the corner. Wide side leads (e.g. $\alpha = 1.5$) result in large S values and thus small T values. The average values for T, R and S corresponding to Figs. 11 are summarized in Table II for the modes $N = 1, 2, 3$. These values are only indicative because T, R and S are function of the energy.

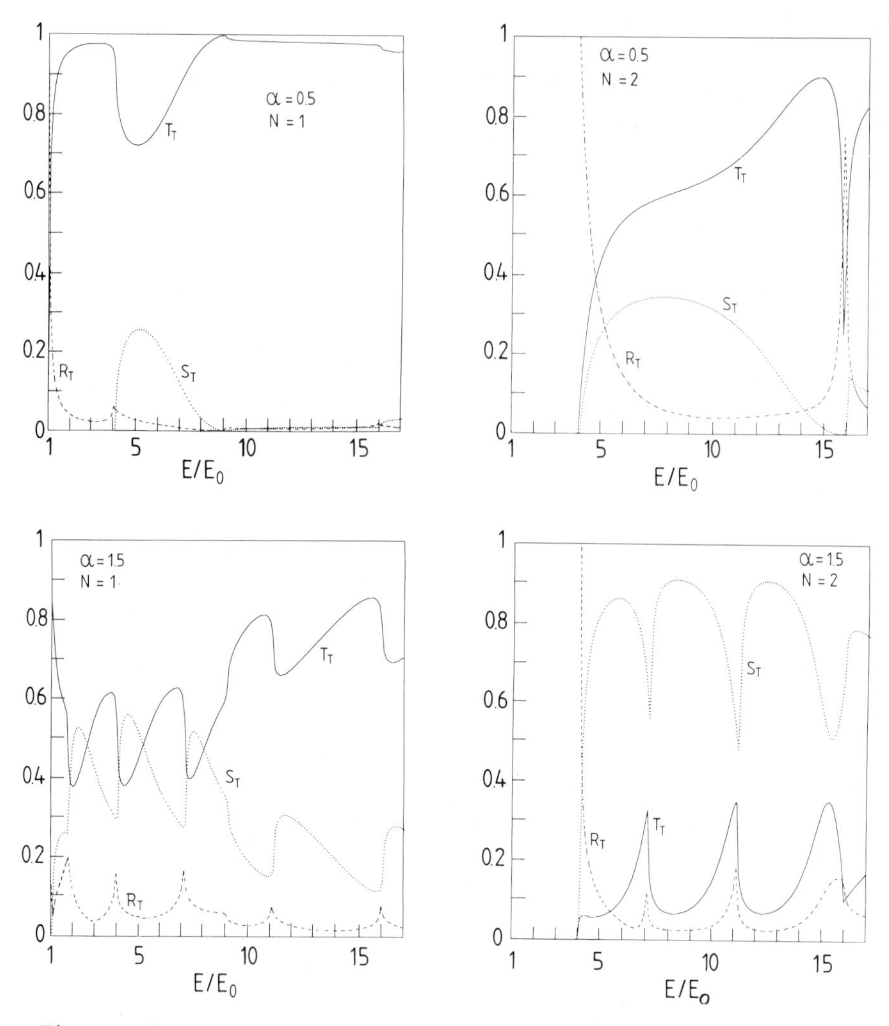

Fig. 11. The total probability for transmission (T_t), reflection (R_t) and sideways transmission (S_t) for a cross with leads of unequal width. The result for mode $N = 1$ and mode $N = 2$ are shown for narrow ($\alpha = 0.5$) and wide ($\alpha = 1.5$) leads.

Table II. Transmission coefficients for the asymmetric cross

		$\alpha = 0.5$			$\alpha = 1.5$	
	$N = 1$	$N = 2$	$N = 3$	$N = 1$	$N = 2$	$N = 3$
$T(\%)$	> 90	70	50	$50 - 70$	$10 - 20$	10
$R(\%)$	< 5	10	10	$5 - 10$	5	< 5
$S(\%)$	< 10	20	40	$30 - 40$	80	90

V. CONCLUSION

Ballistic transport of electrons through quantum wire structures was investigated. Two examples of nanostructures were considered. It is shown how in the case of the bend the bound state can be exploited to create a new type of quantum wire. In fact the bound states of the cavity may also be used to create a similar quantum wire.

The resonances of the scattered states of the cavity may be used for device applications. It offers the possibility of the action of a tunnel structure without having to tunnel through potential barriers. Here, the geometry is essential and leads to variations in the geometrical confinement which, for example may result in bound states. The studied phenomena are quantum mechanical and no simple classical analogon exist for the bound states. Extension of this work to include a magnetic field would be interesting. It is expected that such a magnetic field perpendicular to the cavity will be able to shift[30,31] the position of the resonances.

ACKNOWLEDGEMENTS

During the course of this work I have benefited from stimulating discussions with M.L. Roukes, S.J. Allen, Jr., E. Kapon and H.U. Baranger. A large part of this work was done while I was visiting Bellcore. This work was supported by the Belgian National Science Foundation (NFWO).

REFERENCES

1. R. Landauer, IBM J. Res. Develop. **32**, 306 (1988).
2. See e.g. *Physics and Technology of Submicron structures*, Eds. H. Heinrich, G. Bauer and F. Kuchar (Springer-Verlag, N.Y., 1988).
3. C.P. Umbach, C.V. Haesendonck, R.B. Laibowitz, S. Washburn and R.A. Webb, Phys. Rev. Lett. **56**, 386 (1988).
4. C.P. Umbach, S. Washburn, R.B. Laibowitz and R.A. Webb, Phys. Rev. **B30**, 4048 (1984).
5. G. Timp, H.U. Baranger, P. de Vegvar, J.E. Cunningham, R.E. Howard, R. Behringer and P.M. Mankiewich, Phys. Rev. Lett. **60**, 2081 (1988).
6. See also H.U. Baranger and A.D. Stone (in this proceedings).
7. M.L. Roukes, A. Scherer, S.J. Allen, Jr., H.G. Craighead, R.M. Ruthen, E.D. Beebe and J.P. Harbison, Phys. Rev. Lett. **59**, 3011 (1987).
8. B.J. van Wees, H. van Houten, C.W.J. Beenakker, J.G. Williamson, L.P. Kouwenhoven, D. van der Marel and C.T. Foxon, Phys. Rev. Lett. **60**, 848 (1988).
9. T.D. Smith III, H. Arnot, J.M. Hong, C.M. Knoedler, S.E. Laux and H. Smid, Phys. Rev. Lett. **59**, 2801 (1987).
10. T. Ohshima, M. Okada, M. Matsuda, N. Yokoyama and A. Shibatomi, Superlattices and Microstructures (1989).
11. P. Vasilopoulos and F.M. Peeters (to be published).
12. F. Sols, M. Macucci, U. Ravaioli and K. Hess, Phys. Rev. Lett. **54**, 350 (1989).
13. T.C.L.G. Sollner, W.D. Goodhue, P.E. Tannenwald, C.D. Parker and D.D. Pech, Appl. Phys. Lett. **43**, 588 (1983).
14. F. Capasso, in *Semiconductors and Semimetals* (Academic Press, New York, 1987), Vol. 24, Ch. 6.
15. F. Lenz, J.T. Londergan, E.J. Moniz, R. Rosenfelder, M. Stingl and K. Yazaki, Ann. Phys. **170**, 65 (1986).

16. F.M. Peeters, Phys. Rev. Lett. **61**, 589 (1988).
17. G. Kirczenow, Phys. Rev. Lett. **62**, 1920 (1989).
18. R.G. Ravenhall, H.W. Wyld and R.L. Schult, Phys. Rev. Lett. **62**, 1780 (1989).
19. H.U. Baranger and A.D. Stone, Phys. Rev. Lett. (to be published).
20. A.M. Chang and T.Y. Chang (to be published).
21. C.J. Ford, S. Washburn, M. Büttiker, C.M. Knoedler and J.M. Hong, Phys. Rev. Lett. , (1989).
22. R.L. Schult, D.G. Ravenhall and H.W. Wyld, Phys. Rev. **B39**, 5476 (1989).
23. D. van der Marel and E.G. Haanapel, Phys. Rev. **B39**, 7811 (1989).
24. E. Kapon, D.M. Hwang, R. Bhat and M.C. Tamargo, Superlattices and Microstructures **4**, 297 (1988); E. Kapon, M.C. Tamargo and D.M. Hwang, Appl. Phys. Lett. **50**, 347 (1987); E. Kapon, J.P. Harbison, C.P. Yun and N.G. Stoffel, Appl. Phys. Lett. **52**, 607 (1988).
25. A.D. Stone and A. Szafer, Phys. Rev. Lett. **62**, 300 (1989).
26. D. van der Marel *et al*, in proceedings of the *Symposium on Nanostructure Physics and Fabrication*, Eds. W.P. Kirk and M. Reed (Academic Press, N.Y., 1989).
27. R.J. van Wees, L.P. Kouwenhoven, C.J.P.M. Harmans, J.G. Williamson, C.E. Timmering, M.E.I. Broekaart, C.T. Foxon and J.J. Harris (to be published).
28. R. Landauer, IBM J. Res. Develop. **1**, 233 (1987) and Z. Phys. **B68**, 217 (1987).
29. M. Büttiker, Phys. Rev. Lett. **59**, 1761 (1986).
30. F.M. Peeters, Superlattices and Microstructures (1989).
31. G. Kirczenow, Solid Stat. Commun. **68**, 715 (1988).

SELECTIVE POPULATION OF MODES IN ELECTRON WAVEGUIDES:

BEND RESISTANCES AND QUENCHING OF THE HALL RESISTANCE

Harold U. Baranger

AT&T Bell Laboratories 4G-314
Holmdel, NJ 07733

A. Douglas Stone

Applied Physics, Yale University Box 2157
New Haven, CT 06520

1. INTRODUCTION

In determining the salient characteristics of transport in small structures, the size of the sample compared to the important length scales in the material— the fermi wavelength, the elastic mean-free-path, the phase coherence length, and the localization length— defines several different regimes. In the last five years a great deal of attention has been devoted to studying the conductance fluctuations present in the coherent diffusive regime for which the system size is of the order of the phase-breaking length and much larger than the elastic scattering length.[1] More recent work has established that in certain materials electrons can travel without significant scattering or dephasing over a surprisingly large distance. The material of choice in this regard is the two-dimensional electron gas which is created at the GaAs/GaAlAs interface in modulation-doped heterostructures for which the elastic mean-free-path can be greater than 1 μm and the phase-breaking length can be greater than 10 μm. Furthermore, one can define wires in this material with a width of order the fermi wavelength.[2] Studies of transport in these "quasi-one-dimensional ballistic microstructures" have revealed many novel features.

As a first step in describing experiments in this regime, it is useful to neglect scattering within the wires completely. Thus, wires with widths of order the fermi wavelength will act as solid-state electron waveguides.[2] The wave-function in the transverse direction will be quantized and we will refer to the resulting subbands as modes or channels. Each mode has a threshold energy, and for typical fermi energies several modes are populated. It is therefore natural to think about the population of the different modes. If one treats the wires as waveguides, it is clear from the analogy with microwave waveguides that the junctions between the wires are going to be strong scattering sites and therefore play an important role in determining the physics of these structures. We will focus our attention on understanding several important physical properties of these junctions and will emphasize their role in modifying the population of electrons among the modes.

Since we wish to address resistance measurements, we first must ask how to calculate the resistance for the phase-coherent multi-probe structures that are of interest here. This question was first addressed by Büttiker[3] who viewed the problem as a quantum scattering problem between ideal reservoirs in the tradition of Landauer.[4] He related the currents in the leads, I_m, to the voltages V_n, using the transmitted intensities between the reservoirs (T_{mn} for transmission from lead n to m),

$$I_m = \frac{e^2}{h}\sum_n T_{mn}(V_n - V_m) \ . \qquad (1)$$

Here the current in lead m is very naturally the sum of the pairwise currents between leads m and n where the pairwise current is the transmission coefficient (at the fermi energy) times the difference in voltage or chemical potential. Note that the dependence on the fermi velocity and density of states has cancelled, as usual for Landauer-type arguments. Eq. (1) was subsequently derived[5] [6] from the more traditional Kubo-Greenwood linear response approach, a derivation which shows that Eq. (1) is valid for arbitrary magnetic field and which gives useful Green function expressions for the conductance coefficients.

The calculations presented here were performed by discretizing the continuum problem onto a square lattice to yield a tight-binding Hamiltonian for which Green functions can be calculated recursively.[7] Transmission amplitudes between modes can be extracted by projecting the Green function onto the transverse states in the asymptotic part of the leads. Then, applying the appropriate constraints on the currents in the leads, one can solve Eq. (1) for the voltages and hence the resistances.[5] Throughout this work we include only one spin channel.

While there are important similarities between electron waveguides and microwave or optical waveguides, the physics of resistance measurements in electron waveguides differs from these more familiar cases in several important respects. First, the boundary condition that one can typically impose in the electron case is injection from a reservoir. This implies that all open modes in the waveguide (from mode 1 to N_M) carry equal amounts of current; in fact, electron waveguides are usually in the multi-moded regime. Second, it is natural in the electron case to measure the resistance rather than the transmission. From solving Eq. (1), we see that the resistance depends on a rather complicated combination of transmission coefficients (except in the two-probe case). Finally, the interaction of the electrons with a magnetic field which gives rise to the Hall effect has no analogue in the microwave or optical cases.

In the remaining sections of this paper we present calculations of the resistance of multiprobe waveguides. First, we find that scattering from a junction leads to a local bend resistance.[8] Furthermore, when several junctions are present, there is a non-local bend resistance[8] which can be explained from the selective population of modes by a junction. Then, we turn to consideration of the low-field Hall effect which is observed to be supressed or "quenched" in such systems.[9] The simplest model for a junction, a straight-wire cross, does not agree with the experiments in that the quenching is not seen.[10] Finally, we show that if the wires gradually widen as the junction is approached, quenching does result because of collimation of the electrons in the forward direction.[10]

2. LOCAL BEND RESISTANCE

Before calculating any resistances, we should first understand the scattering properties of a junction.[8] Thus we consider a junction of straight waveguides with infinite hardwall boundaries as shown in the inset to Fig. 1a.[11] The first question to ask is, suppose one injects an electron in a given mode, does the electron go around the bend or straight through? Let $T_B(\mu,\nu)$ be the transmission probability around a single bend from mode ν to mode μ and $T_F(\mu,\nu)$ be the probability to go forward through the junction. Fig. 1b shows the probability that an electron injected in a given mode ν either goes forward or turns left or right; that is, we plot $\langle\sum_\mu T_F(\mu,\nu)\rangle_E$ and $\langle\sum_\mu 2T_B(\mu,\nu)\rangle_E$ as a function of ν. The results are averaged over energies for which there are four modes open. Clearly, the low-lying modes have a high probability to go straight through, while the high-lying modes are most likely to turn the corner. This behavior is straight forward to understand semiclassically: low-lying modes correspond to classical particles with forward directed k-vectors which therefore have a low probability of "seeing" the bend into the side probe, while high-lying modes correspond to particles with a large transverse k-vector which therefore almost always go off into one or the other of the side probes.

We can also turn around the question asked in the last paragraph and ask what happens if one injects electrons from a reservoir with a fixed chemical potential as in a transport measurement. Fig. 1c shows the fraction of those electrons which come out of the junction in a certain mode given that they went either straight through or around the bend. More precisely, we show $<\sum_{\nu} T_F(\mu,\nu)/T_F>_E$ and $<\sum_{\nu} T_B(\mu,\nu)/T_B>_E$ as a function of μ where, as before, we average over energies for which there are four modes open. Thus, this is the distribution (normalized to one) of current among the outgoing modes. After going straight through, the distribution of current is weighted towards low-lying modes, while after going around the bend, the occupation of the lowest mode is severely supressed. The junction, then, is a crude filter for the modes of the electron waveguide.

These scattering properties cause the junction to have a resistance[8][12] (even in the absence of impurity scattering) as was first observed in GaAs/AlGaAs structures by Timp, et al..[8] A four-probe measurement which shows this clearly is indicated schematically in the inset to Fig. 1a. Using Eq. (1), the resistance for this situation, which we call the local bend resistance, is

$$R_B = \frac{h}{e^2} \frac{T_F - T_B}{4 T_B (T_F + T_B)} . \tag{2}$$

The calculated bend resistance shown in Fig. 1a is particularly large (>10kΩ) when only one channel is occupied since in this case it is difficult to make the electron wave turn the corner. Note that the fermi energy in Fig. 1a is normalized to the threshold energy of the lowest mode, $E_1=\hbar^2\pi^2/2mW^2$ where W is the width of the infinite-square-well waveguide. R_B is still substantial (≈few kΩ) in the few channel number regime, but decays as the number of channels becomes large. The energy dependence of the bend resistance within a subband (subband thresholds are indicated by the dashed lines) follows directly from the scattering properties above. As the energy increases from just above threshold to just below the next threshold, all of the modes become more low-lying. Semiclassically, the k-vector of the electrons becomes more forward directed as the energy increases for fixed mode number. Hence T_F increases compared to T_B and the bend resistance increases.

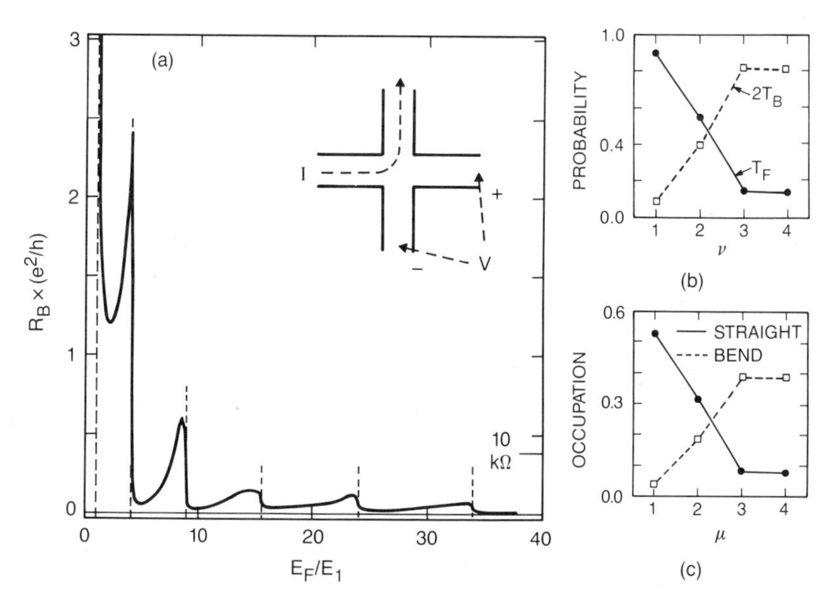

Fig. 1. (a) The bend resistance of the four-probe cross shown in the inset as a function of the fermi energy normalized to the lowest subband energy. The dotted lines indicate the thresholds of the subbands. (b) The probability that a particle injected in mode ν either goes straight through (T_F) or turns either corner ($2T_B$), summed over outgoing modes. (c) The fractional occupation of mode μ after either going straight through the junction or around a bend given that all modes are injected.

We have concentrated on the scattering states in this discussion of the properties of a junction because these are the states important for transport. However, it is interesting to note that true bound states exist in the junction region for these classically unconfined systems: states with energies between the two-dimensional zero of energy and E_1 will be bound, and for symmetry reasons higher energy bound states may exist. Such states were studied in a waveguide with a right-angle bend by Lenz, et al.[13] and more recently in a four-probe junction.[14] [15]

3. NON-LOCAL BEND RESISTANCE

More surprising consequences of the scattering properties of junctions can be obtained by considering more complicated structures. We find, for instance, that the resistance depends on the path which the current follows through the sample,[8] a result which is not expected from macroscopic experience. In the structure shown in Fig. 2b,[11] suppose we feed the current along either path 1 or path 2. The difference in resistance as a function of fermi energy shown in Fig. 2a is substantial when there is more than one mode present at the fermi energy and decays as the number of modes present becomes large. The rich fine structure in the resistance difference is presumably caused by the interference from scattering from the three junctions.

The non-local bend resistance can be explained in terms of selective population of modes by considering the modal distribution of electrons in the wire segment between the lowest and middle junctions. For current path 1, after going straight through the lower junction the low-lying modes are preferentially populated (see Fig. 1c). On the other hand, for current path 2, the electrons are preferentially in the mid/high modes after turning the corner. The upper part of the structure presents a different resistance for the different modes, and thus the two current paths result in different resistances. In particular, for path 1 the middle junction siphons off a smaller fraction of the electrons than for path 2 (low-lying modes go straight through) so that the flux of electrons going up the two voltage probes is more equal for path 1 than for path 2, causing R_2 to be larger than R_1. Since it is the differing population of the modes which causes the non-local bend resistance, when there is only one mode open the effect should be zero as in Fig. 2.

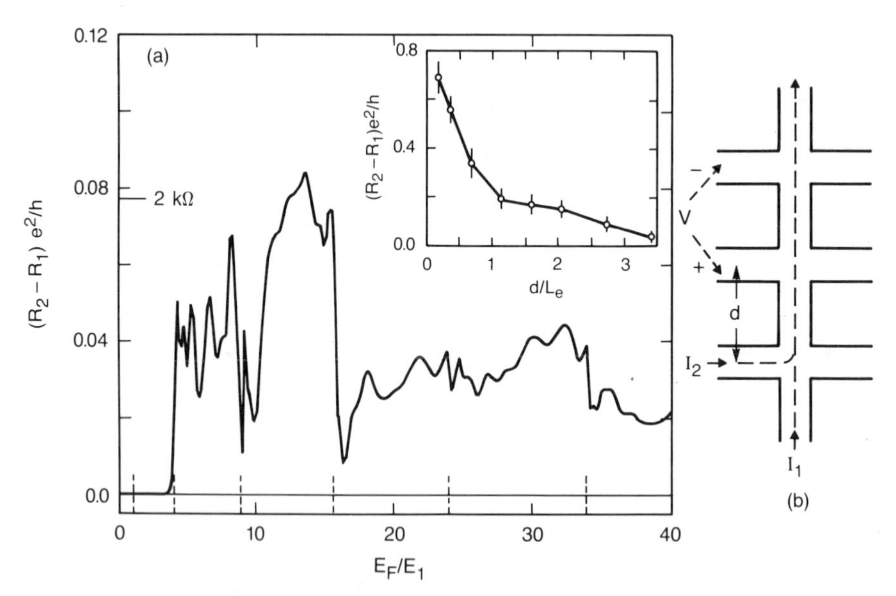

Fig. 2. The difference in resistance (panel (a)) between current path 2 and path 1 for the structure in panel (b) without disorder, showing a substantial non-local bend resistance. The inset shows that the decay of this non-local effect in the presence of disorder occurs over approximately an elastic mean-free-path. The distance between the current bend and the closest voltage probe, d, is normalized to the mean-free-path, L_e.

We now consider briefly the effects of elastic scattering on the bend resistances. One expects the non-local effect to be particularly sensitive to elastic scattering since it involves the distribution of electrons among the modes of the waveguide which will be disturbed if any elastic scattering occurs. On the other hand, the local bend resistance depends only on properties of a single junction and will be only weakly affected by elastic scattering until the mean-free-path is of the order of the width of the junction. We show the ensemble averaged non-local bend resistance as a function of the distance between the junction where the current path is bent and the nearest voltage lead in the inset to Fig. 2a. Disorder is included as fluctuations in the onsite energies of the tight-binding Hamiltonian (the Anderson model),[7] and the transport mean-free-path is approximately five times the width of the wires. The scale of the non-locality of the bend-resistance is of the order of the elastic mean-free-path, a result consistent with the expectation that average resistance effects are non-local over an elastic scattering length while it is only the resistance fluctuations which are non-local over a phase-coherence length.

The theoretically predicted non-locality of the bend resistance has been observed experimentally.[8] However, the length scale of the non-locality observed experimentally is considerably smaller than the transport mean-free-path. This may be caused by the large amount of small-angle impurity scattering which is known to exist in the electron gas formed by modulation doped AlGaAs/GaAs heterostructures.[16] Such small-angle scattering implies that the mode-mixing length of relevance to the non-local bend effect would be much smaller than the back-scattering length of relevance to the transport mean-free-path. In the calculations reported in Fig. 2, the disorder causes s-wave scattering which implies that these two lengths are the same.

4. HALL RESISTANCE USING STRAIGHT WIRES

An important difference between electron waveguides and microwave or acoustic waveguides is the sensitivity of electron transport to magnetic fields. It was observed experimentally by M. L. Roukes, et al.[9] that the low-field Hall resistance measured at intersections of narrow wires (≈100nm) is suppressed below the expected two-dimensional value, an effect called the quenching of the Hall resistance. Subsequent measurements by a wide variety of groups[8][17] [18] [19] [20] showed that quenching is *generic* in that it occurs in many differently-fabricated structures— those made using edge depletion[17][19][20] or using self-aligned gates[8][18][20]— and for a broad range of densities— certainly corresponding to a span of from one or two modes up to about ten modes at the fermi energy.

A previously proposed explanation of the quenching of R_H by Beenakker and van Houten[21] noted that there is a field scale at which the character of the electron states at the fermi surface changes from being edge-like to traversing the entire width. Then, they argued that R_H would be quenched for fields less than this threshold field because traversing trajectories hit both sides of the wire with equal frequency and hence are nearly as likely to go into one Hall probe as the other; however, they provided no calculation that this was the case. Subsequent microscopic calculations by Peeters[22] in the limit of weakly coupled Hall probes failed to show any quenching; in fact, his calculations showed that R_H tends to be larger than the two-dimensional value in this limit. It was suggested that the lack of quenching in these systems might be due to the weakly coupled nature of the probes.

Thus we are naturally led to consider the low-field Hall effect in the kinds of waveguides that we have been discussing.[10] In particular, we consider a four-probe cross of the type shown in Fig. 1a (the simplest structure in which one can measure a Hall effect) in which each branch has a constant cross-section;[11] we refer to these structures as *straight-wire crosses*. Very recently, two other groups— Ravenhall, et al.[23] and Kirczenow[24] — have also considered the Hall effect in this type of structure. While both of these other groups analyzed in detail the Hall effect at arbitrary magnetic field, we have concentrated on the low-field regime in an attempt to explain the quenching measurements. As we shall see, the behavior of straight-wire crosses does not explain the low-field experiments[10] and we consider them mainly as a foil for the more complicated structures discussed in the next section which do show quenching.

The Hall resistance of a symmetric cross structure can be expressed in terms of the transmission intensities to turn right (T_R), turn left (T_L), and go forward (T_F) as

$$R_H = \left(\frac{h}{e^2}\right)\frac{T_R - T_L}{2T_F(T_F+T_R+T_L)+T_R^2+T_L^2} \; . \tag{3}$$

Hall resistance traces for straight-wire crosses show several kinds of behavior depending on the fermi energy; examples are shown in Ref. 10. We find that there are energies at which R_H is definitely quenched before rising to the n=1 or 2 quantum Hall plateau. However, the more typical behavior, and the only behavior for higher energies, is for R_H to oscillate around the expected two-dimensional value and never approach zero.

To present these results more systematically, Figure 3 shows the slope of $R_H(B)$ at B=0 normalized to the 2D value as a function of fermi energy for several different cases. (For a 100 nm wire, an energy of 10 E_1 corresponds to about 5 meV.) First, for a symmetric cross with infinite hard walls, as used in Fig. 1,[11] the solid line shows that there is rich structure in R_H (in particular, cusps at the threholds for the modes) and quenching for a restricted range of energies for which the number of modes open, N_M, is less than two, $N_M \leq 2$. However, there is not generic quenching since the normalized slope mostly oscillates about one. The dashed lines shows our calculation of the slope of the Hall resistance in the limit of weakly coupled Hall probes. In this case, we insert a tunneling barrier between each Hall probe and the current carrying channel as in the geometry considered by Peeters.[22] The slope tends to be above the 2D value as found by Peeters; however, just above the threshold for each mode, the slope is suppressed by a substantial amount, leading to a striking sawtooth behavior of S/S_{2D}.

The qualitative features of the results presented in the last paragraph are not sensitive to the shape of the confining potential used. In Figure 3b we present our results for a junction between waveguides with harmonic confinement, as originally introduced by Kirczenow:[24] $U(x,y)=\theta(|x|-|y|)ay^2+\theta(|y|-|x|)ax^2$. Again, while there is quenching for energies with $N_M \leq 2$, there is no quenching for higher energies, only oscillations near the 2D value. Temperature averaging (simply by convolving with the derivative of the fermi function) reduces the deviations from the 2D value and, specifically, reduces the amount of quenching.

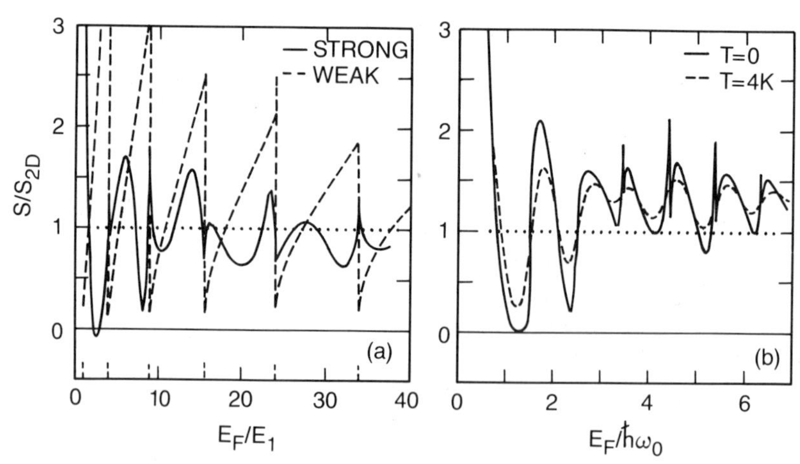

Fig. 3. The slope of the Hall resistance normalized to its two-dimensional value as a function of fermi energy for various four-probe cross structures. (a) The confining potential in the wire is an infinite hard-wall barrier (see Fig. 1b). The dashed line is for weakly coupled voltage leads; the solid line is for four identical leads. (b) The confining potential is harmonic and all leads are identical. Temperature averaging (T=0 solid, T=4K dashed) smooths the oscillations and shows clearly that the average is near 1. The lack of quenching for $N_M \geq 3$ shows that these straight-wire models cannot explain the generic quenching seen experimentally.

The straight-wire crosses considered in this section, then, cannot explain the range of experimental results. Furthermore, the fact that these structures do not show generic quenching demonstrates that quenching is not intrinsic to one-dimensional ballistic structures and rules out mechanisms for quenching[21] (based on a non-interacting model) which do not explicitly take into account the junction geometry.

5. HALL RESISTANCE USING GRADED WIRES

The important physical property absent from the straight-wire crosses is the rounding of the corners in the experimental junctions. In GaAs/GaAlAs structures the electrostatic confinement causing the conducting channels is due to a delicate balance between positive ions in the modulation doped layer, negative charges in the damaged surface layer in the case of the self-aligned structures, negative charges on the depleting gates in the case of the edge-depletion structures, and, of course, the electrons in the conducting channels itself. Because of the high mobility needed, the charges causing the conducting channel are of necessity removed from the channel itself: typical spacer layers are 20-30 nm and the surface charges are even further away (50-80 nm). It is difficult to see how the corners could be sharper than approximately the spacer-layer distance. In fact, in many structures, there may be more rounding: in the edge-depletion structures simply because as one narrows the channels through gate depletion the relative amount of rounding will increase, and in the self-aligned structures because of the difficulty of doing sharp lithography as well as the possibly different behavior of the etching agents near a junction. Very recently, self-consistent calculations of the potential of wires[25] and constrictions[26] have been performed which indicate that the gate-induced structures do indeed have substantially rounded corners. Thus we conclude that rounded corners, while not an essential feature of the quasi-one-dimensional ballistic limit, are very likely to be present in the structures made to date.

To see how the rounded corners at the junction introduce a crucial new physical effect, consider the rounded-corner junction in Fig. 4a, which we will refer to equivalently as a *graded-wire cross*. Suppose one injects a wave-packet from the left in the horizontal arm; what is the behavior of this wave-packet at later times? We show in Fig. 4b the thresholds for the transverse subbands at the two points x_1 and x_2 assuming infinite strips of the same width as the graded wire. Since at x_1 the wire is narrow, the transverse subbands are widely spaced, while at x_2, the subbands are much closer together. If the widening of the wire is gradual, the electrons will travel adiabatically from point 1 to 2, and thus will conserve their transverse mode number. That is, the injected wave packet sees a time-varying confinement potential, $U(y,t) \approx U(y,x_1+v_g t)$ for short times. In general this will generate transitions between the subbands; however, if the time variation is slow enough, the wavepacket will largely remain in the subband in which it is injected (adiabatic behavior). Thus, particles injected from a reservoir such that the current is evenly distributed across all the subbands below E_F (solid lines at x_1 in Fig. 4b) end up in the low-lying transverse subbands as they approach the junction (solid lines at x_2). A non-equlibrium momentum distribution has been created in the junction region.

Near the junction, the adiabatic approximation breaks down since the equi-potential contours must turn through 90 degrees. Here the wave-packet is scattered into different modes and into the different probes. The fact that the scattering properties of a junction are very different for the different modes (as shown above) implies that the physical result of injecting a non-equilibrium momentum distribution could be quite different from injecting the current in all the modes. Indeed, by analyzing the transmission properties of a straight-wire cross, we find that the low-lying modes make a much smaller contribution to R_H than the high-lying modes.[10] Because of the non-equlibrium modal distribution caused by the graded wires, then, quenching results.

The above quantum-mechanical scenario can be viewed classically in a manner analogous to that used by Beenakker and van Houten[27] in the case of a constriction. Classically, a particle injected with a large transverse momentum, $k_\perp$, has its k-vector rotated by the gradually widening of the wire so that it is injected into the junction region with a large component of its k-vector parallel to the injecting lead, $k_{||}$ (see the trajectory in Fig. 4a). These are the classical states which correspond to the low-lying modes of the quantum description. Thus particles injected in a hemispheric distribution far from the junction reach the junction collimated into a forward cone: graded wires produce a collimated electron beam. Quantum-mechanically, collimation comes about because $k_\perp$ tends to be conserved in the non-adiabatic scattering at abruptly terminated interfaces, as has been shown in the case of abruptly terminated constrictions.[28] van Houten and Beenakker have pointed out that additional collimation may be caused by a potential gradient perpendicular the wires.[29] Such a gradient is likely to be present in the experimental structures because the wider conducting channel near the junction will probably be accompanied by a deeper poential well.

The possible effects of collimation on R_H can be made clear by writing $R_H=\alpha T_{RL}/D$ where $\alpha=(T_R-T_L)/(T_R+T_L)$ is the relative asymmetry between left-turning and right-turning electrons, $T_{RL}=T_R+T_L$ is the total probability to turn a corner, and $D=(e^2/h)[2T_F(T_F+T_{RL})+T_{RL}^2(1+\alpha^2)/2]$ is relatively insensitive to whether the electrons go forward or turn either corner. R_H can be suppressed either by (1) decreasing the asymmetry while the total turning probability remains large ($\alpha \to 0$ while $T_{RL} \approx$constant) or by (2) decreasing the total turning probability quite apart from the asymmetry ($T_{RL} \to 0$ while $\alpha \approx$constant). It is clear that collimation, and hence grading, will reduce T_{RL} simply because more classical trajectories will impinge on the opposite lead when collimation is present. The effect of grading or collimation on the asymmetry is less clear *a priori*[30] though we find numerically that grading reduces α as shown below.

Is there numerical evidence for the quenching mechanism described above? We calculated the Hall resistance of the structure in Fig. 5b which consists of a gradual widening of the wires fed into a hard-wall cross (all four leads are identical).[11] The Hall resistance for an energy at which four modes are open (Fig. 5a) shows a fine quenched region bounded by a region in which R_H overshoots the two-dimensional value.[31] The slope of the Hall traces obtained from

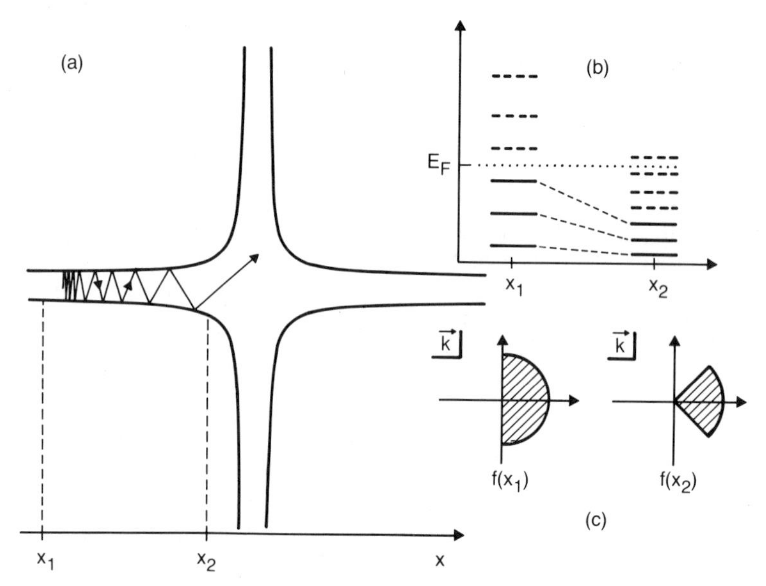

Fig. 4. Schematic showing the physics of collimation in a cross structure with graded wires and rounded corners as shown in panel (a). (b) The threshhold energy of the transverse subbands at two places in the structure. For gradual grading, the electrons conserve their mode number (dotted lines). Current initially injected in all modes below the fermi energy (solid lines at x_1) is carried only by low lying modes near the junction. Modes near the junction which are below the fermi energy but which are adiabatically connected to modes above the fermi energy (dashed lines) in the narrow region do not carry current. (c) Distributions of classical particles at two places in the structure. The grading rotates the k-vector of the classical particles into the forward direction, as indicated by the trajectory in part (a). Thus particles injected with a hemishperic distribution of k-vectors emerge collimated into a cone.

fitting R_H to a straight line in an interval about B=0 is shown in Fig. 6. In contrast to the result for the straight-wire cross (Fig. 3), the slope in this graded-wire case is near zero for a large range of fermi energies for which $N_M \geq 3$; that is, the quenching in this structure is indeed *generic*. The influence of grading on the asymmetry between the left and right transmission intensities can be evaluated through the slope of $\alpha(B,E_F)$ near zero field. Averaging over energies for which 3-5 modes are open, we find that the ratio of the slope of α in the graded-wire cross normalized to that in a straight-wire cross is 0.39. A careful analysis of the transmission properties of the graded junction[10] indicates that both this reduction in α and the increase in forward transmission discussed above contribute to the supression of R_H.

To emphasize the importance of the gradualness of the widening, we also present results for two abruptly widened cases: a sudden increase in the width by a factor of two (Fig. 6b) and a widening through 45 degree cuts at the junction (Fig. 6c). Neither of these structures shows

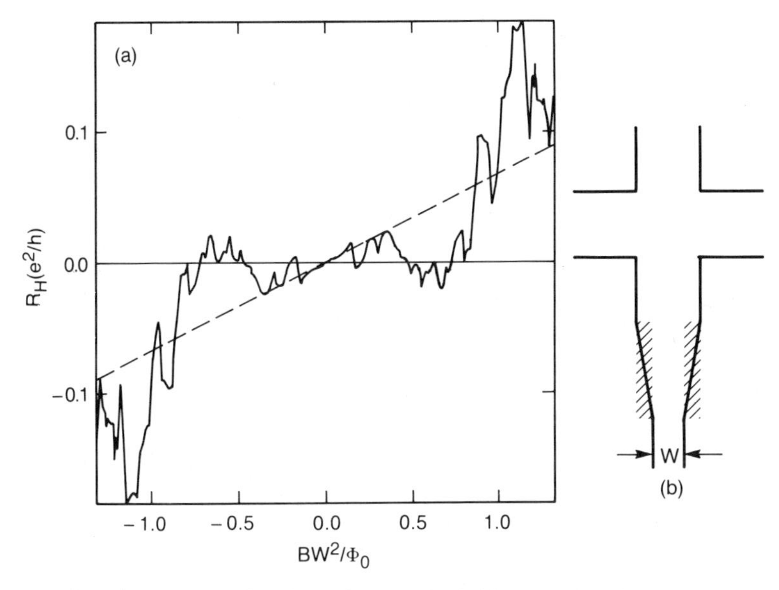

Fig. 5. The Hall resistance as a function of magnetic field (panel (a)) for the structure in panel (b) (all leads are identical). R_H is clearly below the two-dimensional value (dashed line) for a substantial range of B fields.

quenching for $N_M \geq 3$; in fact, both tend to give values somewhat above the two-dimensional value. In addition to the results for the structure shown in Fig. 5, we have considered several other ways of grading using both different cross-sectional potentials and different ways of changing the cross-section as the junction is approached; generic quenching is seen in all the adiabatically graded structures and not in the others. The degree of adiabaticity can be determined from a two-probe calculation on a single lead of the junction. More precisely, for electrons injected from the narrow end, we look at the ratio, ρ, of the intensity in the high-lying levels (which adiabatically would be empty) to the intensity expected if the current were evenly distributed across all the modes. If the grading is sufficiently gradual so that $\rho \leq 0.1$ (which is the case for the structure in Fig. 5), we find generic quenching in all geometries at which we have looked. These numerical results provide strong support for our explanation of the quenching through collimation.

To summarize our discussion of the quenching of R_H, the grading of the wires produces a selective population of the low-lying modes near the junction region which corresponds to a collimated electron beam. The subsequent scattering of this collimated beam off of the non-adiabatic part of the junction produces quenching both because collimation reduces the total probability to turn the corner and because grading reduces the asymmetry between left and right turning electrons.

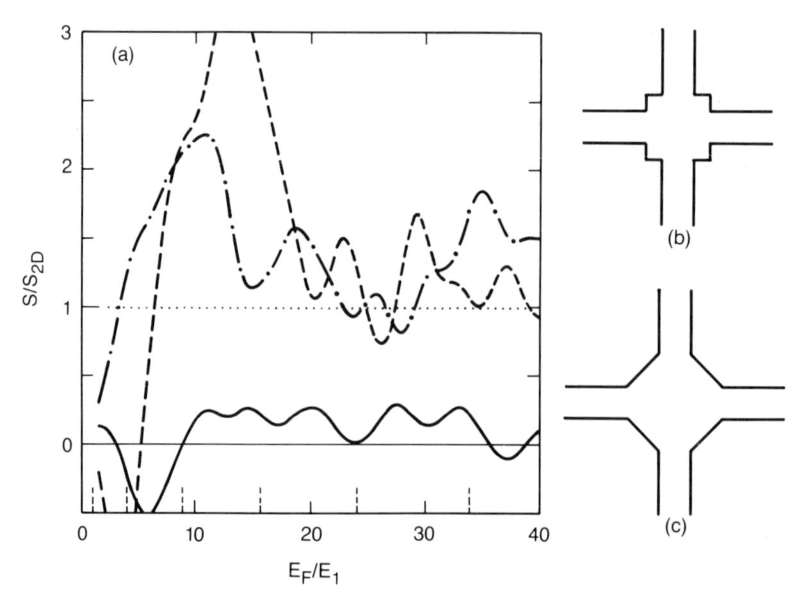

Fig. 6. The slope of the Hall resistance as a function of fermi energy at $T=E_1$ ($\approx$5K for a 100 nm wire). Generic quenching behavior is found for the graded wire structure of Fig. 5b (solid) but not for an abruptly widened structure (panel (b)) or for the "flattened" square structure of panel (c). The slopes are obtained by fitting $R_H(B)$ to a straight line in the interval $[-0.45,0.45]\Phi_0/W^2$.

6. CONCLUSION

We have taken the point of view that the quasi-one-dimensional ballistic microstructures which have been a subject of great interest in the last few years can be viewed as electron waveguides.[2] The junctions between waveguides inherent in multi-probe structures act as strong scattering sites which affect the different modes of the waveguide quite differently. Resulting effects on transport include both a local and non-local bend resistance.[8] The non-local bend resistance arises, first, because a junction filters the injected electrons, thus selectively populating the modes downstream from the junction. Then, the resistance caused by this non-equilibrium modal distribution at junctions further downstream is substantially different from that caused by an equilibrium distribution. The scale of the non-locality of the bend resistance is approximately the length to mix the different modes of the waveguide through elastic scattering.

The Hall resistance[10] of the straight-wire cross models does not show the experimentally oberved quenching when there are more than three modes open. If quenching were a phenomena intrinsic to quasi-one-dimensional ballistic structures, these straight-wire crosses would show quenching; therefore, the absence of quenching demonstrates that the geometry of the structure is crucial, that quenching is not intrinsic to quasi-one-dimensional structures (within the independent particle approximation). We find that structures which include a gradual widening of the wires near the junction region do show a supression of R_H over a wide range of fermi energies. Physically, the adiabatic widening leads to a selective population of the low-lying incoming modes which corresponds to a collimation of the electrons in the forward direction. The scattering of this non-equilibrium modal distribution by the junction produces the quenching.

An obvious implication of our explanation of the Hall resistance is that R_H should be sensitive to the junction geometry. Perhaps the best test of our theory would be to vary the change in width of the wire or the sharpness of the junction, perhaps using auxiliary gates; this should enable one to move smoothly from generic quenching behavior to a structure without quenching for $N_M \geq 3$. Also, if weakly coupled Hall probes were made to the current-carrying channel, the sawtooth behavior in Fig. 3b should be seen. More complicated cases have very recently been investigated experimentally by Chang and Chang[18] and Ford, et al.[19]. Chang and Chang showed that a uniform narrow wire is not necessary to produce generic quenching; instead a constriction on each lead of a wide cross structure (which by itself does not show quenching) is a necessary and sufficient condition for strong quenching, a result consistent with our model since the constriction introduces the necessary narrow region before the wide junction. In one relevant geometry studied by Ford, et al., a deflector is placed in the center of the junction. This sample showed no quenching at all, exactly as one would expect from our model since the deflector prevents forward transmission of the collimated beam. In another geometry, the lithographic mask includes a truncation of the square corner at a 45 degree angle much as in Fig. 6c. Ford, et al. suggested that electrons bent by the magnetic field should be reflected from the flattened region into the *wrong* probe, thus giving rise to a negative Hall resistance. We noted in connection with Fig. 6, that we do not find generic quenching for the simple structure in Fig. 6c and certainly not a strongly negative slope; however, if gradual grading is added prior to the 45 degree cuts in order to produce collimation, our preliminary results indicate that we do achieve negative slopes to the Hall resistance. This indicates that collimation is necessary in order to enhance the importance of the paths suggested by Ford, et al. and hence make their explanation of the negative Hall resistances applicable.

Since our explanation of the quenching is basically semiclassical, one should find generic quenching in wider wires if the junctions are widened proportionally in order to create the collimation effect. However it is important to remember that the controlling length-scale is the distance that the electrons travel before mixing the occupancy of the modes at the fermi level. This length has been found to be much shorter than the transport elastic mean-free-path in the case of the non-local bend resistance.[8] Our explanation points to the crucial role played by this mixing length-scale in determining the occurence of quenching.

Finally, we note that the selective population of modes in ballistic microstructures to which we appeal in explaining both the non-local bend resistance and the quenching of the Hall resistance has been used independently to explain two novel phenomena in the high field regime: the quantization of the Hall resistance at anomalous values[32] and the supression of the Shubnikov-deHaas oscillations[33] observed by the Delft/Philips collaboration. It appears that an emerging theme in transport in ballistic microstructures is the use of geometrical features to selectively populate quantum modes and hence to modify dramatically the transport coefficients.

ACKNOWLEDGEMENTS

We acknowledge valuable discussions with R. Behringer, M. Büttiker, A. M. Chang, Y. Imry, M. L. Roukes, A. Szafer, and G. Timp. The work at Yale was supported in part by NSF Grant No. DMR-8658135 and by the AT&T Foundation.

REFERENCES

1. For a review see H. Heirnrich, G. Bauer, and F. Kuchar, eds., **Physics and Technology of Submicron Structures** (Springer-Verlag, New York, 1988).
2. G. Timp, A. M. Chang, P. Mankiewich, R. Behringer, J. E. Cunningham, T. Y. Chang, and R. E. Howard, Phys. Rev. Lett. **59,** 732 (1987).
3. M. Büttiker, Phys. Rev. Lett. **57,** 1761 (1986).
4. R. Landauer, IBM J. Res. Develop. **1,** 233 (1957) and Z. Phys. B **68,** 217 (1987).
5. A. D. Stone and A. Szafer, IBM J. Res. Develop. **32,** 384 (1988).
6. H. U. Baranger and A. D. Stone, submitted to Phys. Rev. B.
7. P. A. Lee and D. S. Fisher, Phys. Rev. Lett. **47,** 882 (1981); H. U. Baranger, A. D. Stone, and D. P. DiVincenzo, Phys. Rev. B **37,** 6521 (1988).

8. G. Timp, H. U. Baranger, P. deVegvar, J. E. Cunningham, R. E. Howard, R. Behringer, and P. M. Mankiewich, Phys. Rev. Lett. **60,** 2081 (1988).
9. M. L. Roukes, A. Scherer, S. J. Allen Jr., H. G. Craighead, R. M. Ruthen, E. D. Beebe and J. P. Harbison, Phys. Rev. Lett. **59,** 3011 (1987).
10. H. U. Baranger and A. D. Stone, submitted to Phys. Rev. Lett.
11. The width in all of the straight wire cases, both with and without a magnetic field, is 21 sites. A hardwall boundary consists of simply truncating the lattice; for a harmonic cross-section, the single-site energies are varied within a structure of width 41. The graded-width wire varies from 21 to 41. To grade, we introduce a linear potential in the shaded region in the inset to Fig. 5b whose slope starts large and becomes zero near the junction. The B field is zero far from the junction and graded to the desired value within W of the junction; the non-uniformity of B does not affect the low field properties.
12. Y. Takagaki, K. Gamo, S. Namba, S. Ishida, S. Takaoka, K. Murase, K. Ishibashi, and Y. Aoyagi, Solid State Commun. **68,** 1051 (1988).
13. F. Lenz, J. T. Londergan, E. J. Moniz, R. Rosenfelder, M. Stingl, and K. Yazaki, Ann. Phys. **170,** 65 (1986).
14. F. M. Peeters, to be published in Superlattices and Microstructures and this proceedings.
15. R. L. Schult, D. G. Ravenhall, and H. W. Wyld, Phys. Rev. B **39,** 5476 (1989).
16. S. Das Sarma and F. Stern, Phys. Rev. B **32,** 8442 (1985) and F. F. Fang, T. P. Smith, and S. L. Wright, Surf. Sci. **196,** 1988 (1988).
17. C. J. B. Ford, T. J. Thornton, R. Newbury, M. Pepper, H. Ahmed D. C. Peacock, D. A. Ritchie, J. E. F. Frost, and G. A. C. Jones, Phys. Rev. B **38,** 8518 (1988).
18. A. M. Chang and T. Y. Chang, submitted to Phys. Rev. Lett..
19. C. J. B. Ford, S. Washburn, M. Büttiker, C. M. Knoedler, and J. M. Hong, submitted to Phys. Rev. Lett.
20. M. L. Roukes, private communication.
21. C. W. J. Beenakker and H. van Houten, Phys. Rev. Lett. **60,** 2406 (1988).
22. F. M. Peeters, Phys. Rev. Lett. **61,** 589 (1988).
23. D. G. Ravenhall, H. W. Wyld, and R. L. Schult, Phys. Rev. Lett. **62,** 1780 (1989).
24. G. Kirczenow, Phys. Rev. Lett. **62,** 1920 (1989) and submitted to Phys. Rev. B.
25. A. Kumar, S. E. Laux, and F. Stern, Appl. Phys. Lett. **54,** 1270 (1989).
26. J. H. Davies, Bull. Am. Phys. Soc. **34,** 589 (1989) and A. Kumar, S. E. Laux, and F. Stern, Bull. Am. Phys. Soc. **34,** 589 (1989).
27. C. W. J. Beenakker and H. van Houten, to be published in Phys. Rev. B.
28. A. Szafer and A. D. Stone, Phys. Rev. Lett. **62,** 300 (1989).
29. H. van Houten and C. W. J. Beenakker, to be published in *Nanostructure Physics and Fabrication,* edited by M. Reed (Academic Press, 1989).
30. Within the adiabatic approximation and in the absence of a magnetic field, paths involving multiple reflections from the confining potential contribute substantially to the transmission around a bend, thus making a simple estimate of the effect of the B field difficult. Such multiple-scattering paths contribute a much smaller fractional amount to the forward transmission.
31. A. Chang has pointed out that once the magnetic field is large enough to turn the collimated beam into one of the leads, the Hall resistance will be particularly large.
32. B. J. van Wees, E. M. M. Willems, C. J. P. M. Harmans, C. W. J. Beenakker, H. van Houten, J. G. Williamson, C. T. Foxon, and J. J. Harris, Phys. Rev. Lett. **62,** 1181 (1989).
33. B. J. van Wees, E. M. M. Willems, L. P. Kouwenhoven, C. J. P. M. Harmans, J. G. Williamson, C. T. Foxon, and J. J. Harris, Phys. Rev. B **39,** 8066 (1989).

ELECTRONIC PROPERTIES OF DOPING QUANTUM WIRES

Gerrit E.W. Bauer and Aart A. van Gorkum

Philips Research Laboratories
5600 JA Eindhoven, The Netherlands.

Self-consistent calculations of the electronic structure of Doping Quantum Wires (DQW's) are presented. DQW's in a GaAs host are characterized by many occupied subbands and extended electron distributions with enhanced mobilities and strong effects of magnetic fields. The electrons of a DQW in a Si matrix are found to be in the one-dimensional quantum limit. In spite of the strong confinement of the conduction electrons to the doping line charge, screening effects still render an improved mobility.

INTRODUCTION

Epitaxial growth utilizing lattice steps on tilted surfaces[1,2] is a promising alternative to lithographic methods to realize semiconductor structures in the nanometer regime.[3–5] Usually one tries to embed GaAs into AlGaAs-alloy to obtain one-dimensional confinement.[3–5] On the other hand research on the so-called δ-doping technique[6] has gained momentum in recent years because of promising possibilities for applications, especially in silicon devices.[7] One of us (A.v.G.) has recently proposed[8] combining the ideas of lattice step growth and δ-doping to create what we call Doping Quantum Wires (DQW).[9] Here we present self-consistent calculations of the electronic structure of an isolated DQW hosted by the archetypal semiconductors GaAs and Si. Our results should be a stimulus for crystal growers to fabricate these systems.

MODEL

We assume that the (donor) impurities in the DQW are distributed axially symmetrically about the wire axis parallel to the (110) crystal direction. Using the effective mass approximation the electronic structure is determined by local density-functional theory[10] and the exchange-correlation potential of Gunnarsson and Lundqvist.[11] We use the parameters $m_t = 0.19\,(0.067)$, $m_l = 0.916\,(0.067)$, $\kappa = 11.5\,(12.5)$ for Si (GaAs). The anisotropy of the Si

Science and Engineering of One- and Zero-Dimensional Semiconductors
Edited by S.P. Beaumont and C.M. Sotomajor Torres
Plenum Press, New York, 1990

effective mass normal to the wire is disregarded, as are small effects of anisotropy and valley degeneracy on the exchange-correlation potential. A homogeneous doping charge is assumed to have a Gaussian distribution normal to the wire with a full width at half maximum of 14Å. Densities and potentials are computed self-consistently up to a distance of 500 Å (Si) or 1000 Å (GaAs) from the wire axis. Because of the axial symmetry the subbands are labeled as nm, where $n - 1$ is the number of nodes and m the angular momentum in the

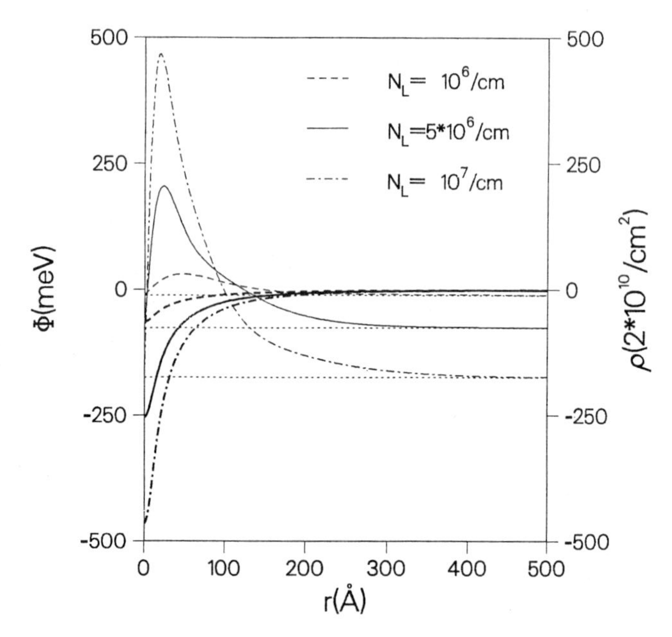

Fig. 1. Effective potential (bold curve) and free electron radial distribution of a GaAs doping quantum wire as a function of the total doping charge N_L. The zero's of the electron distributions coincide with the lowest subband edges (dotted lines). Fermi energies lie about 1-2 meV below the conduction band edge.

direction of the wire. The properties of DQW's depend strongly on the band structure of the host material as will be clear from the separate discussions for Si and GaAs.

GaAs

The results of the self-consistent calculations of a DQW in a GaAs host for three doping densities ($N_L = 10^6$, 5.10^6, and 10^7 cm^{-1}) are summarized in Fig. 1. Due to the small electron mass of GaAs many subbands (between 3 and 9) are occupied, the electron distribution is extended over several tenths of nanometers and the Fermi energy is close to the conduction band edge.

Disregarding the complexities of multi-subband transport theory the low-temperature mobility can be estimated semiclassically as

$$\mu_{QW}^{-1}(N_L) \simeq \int d\vec{\rho}\, \frac{n_{fe}(\vec{\rho})}{N_L}\, \mu_{3D}^{-1}(n_{fe}(\vec{\rho}), n_{imp}(\vec{\rho})),$$

where $\mu_{3D}^{-1}(n_{fe}, n_{imp})$ is the mobility of the bulk semiconductor with doping concentration n_{imp} and free electron concentration n_{fe}, $\vec{\rho}$ is the coordinate in the plane normal to the wire and N_L is the total doping (and free electron) concentration per unit of length. Calculating μ_{3D} with Thomas-Fermi screening in the Born approximation we find mobilities of about 1 $m^2/(Vs)$ which corresponds to an enhancement by a factor 2-4 compared to a bulk crystal homogenously doped by N_L^3. The above formula holds in the limit of high densities and broad impurity profiles and is used here beyond its formal validity. The competition between the spatial separation of electrons and donors and the reduced screening of the ionized impurity potential is believed to be real, however.

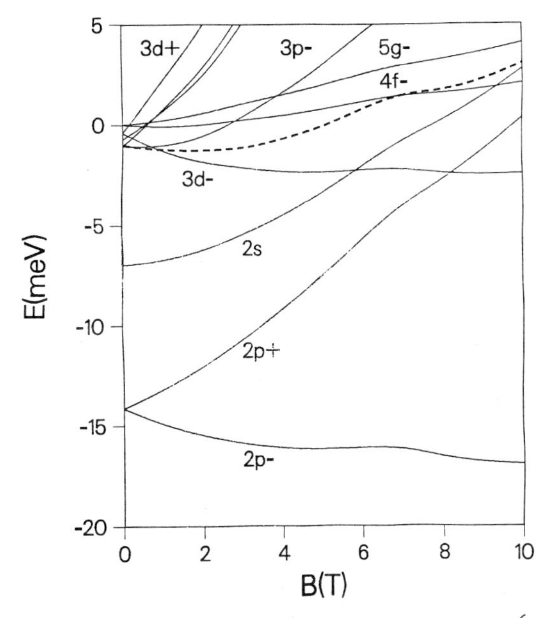

Fig. 2. Subband edges of a GaAs doping quantum wire ($N_L = 5.10^6/cm$) as a function of magnetic field parallel to the wire. The subbands are labeled by number of nodes ($= n - 1$) and angular momentum m. The dashed line indicates the Fermi energy.

The light effective mass is responsible for large effects of a magnetic field parallel to the wire axis. In Fig. 2 we observe a lifting of the orbital degeneracies already at weak magnetic fields. With increasing field excited subbands of a given symmetry are depopulated, but the lowest energy states with higher angular momentum become increasingly populated, a behaviour which is easily understood by the macroscopic degeneracy of Landau levels in the magnetic quantum limit. As observed in Fig. 3 the high momentum states can be described as semiclassical particles which revolve around the wire axis in a well-defined orbit. The present theoretical results should be tested by (polarized) infrared absorption or light scattering in high magnetic fields and magnetotransport experiments.

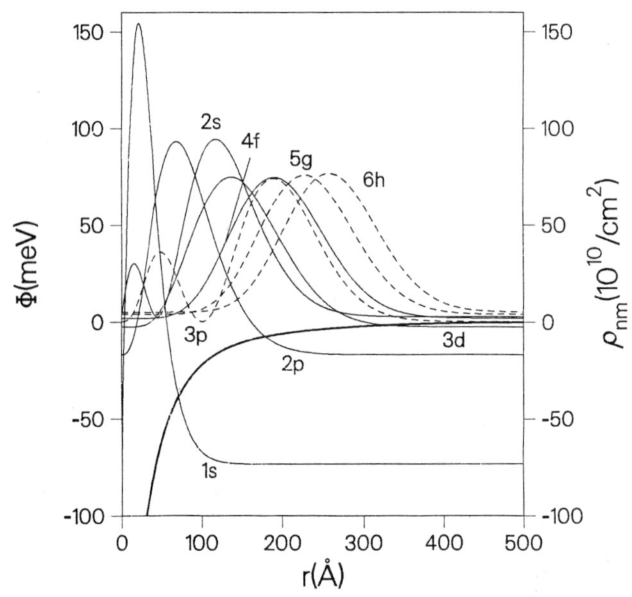

Fig. 3. Effective potential (bold curve) and normalized radial electron distribution of different subbands in a GaAs doping quantum wire ($N_L = 5.10^6$/cm) and a magnetic field of 10 T parallel to the wire. The subbands are labeled by number of nodes (+1) and angular momentum and the zero's of the electron distributions coincide with the subband energy. The distributions of unoccupied orbitals are plotted using dashed lines.

Si

For doping densities in the range of $10^6 - 10^7 cm^{-1}$ the first subbands of all six valleys turn out to be occupied in Si DQW's. The electron distribution is localized within less than 100 Å around the wire (Fig. 4) and the Fermi energies lie far below the conduction band edge. Exchange-correlation turns out to be important to arrive at these results.

It is well known that free carriers in a one-dimensional system will be localized at low temperatures. With increasing temperatures quantum interference effects will lose importance, however, and we may estimate mobilities in the Born approximation. From the strong binding of the conduction electrons to the doping line charge the effects of scattering from ionized random impurity potentials is more important in DQW's than in modulation-doped quantum wires discussed in Ref. 12 (see also Ref. 13). We estimate mobilities using form factors based on the self-consistent subband wave functions. The dielectric function is approximated by using the free-electron polarizability at zero wave vector. The mobilities depend strongly on the doping density, where the screening effect plays a decisive role.

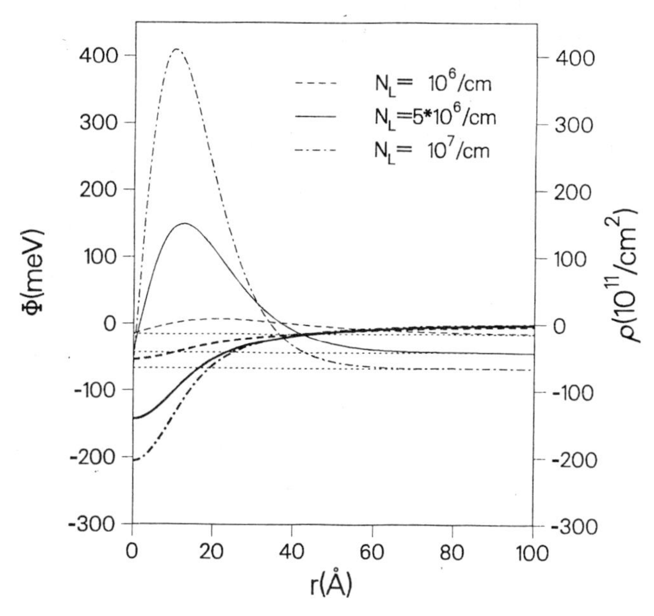

Fig. 4. Effective potential and free electron radial distribution of a Si < 110> δ-doping quantum wire as a function of the total doping charge N_L. The zero's of the electron distributions coincide with the lowest subband edges (dotted lines). Fermi energies are between 1-13 meV above the lowest subband edges for the present densities.

Extremely high mobilities calculated for lower densities should not be taken seriously since the underlying assumption that the system is metallic does not hold anymore. But also at a high density of 10^7 cm^{-1} the mobilities are found to be about 0.1 m^2/(Vs) which make DQW's very attractive for device applications. The single particle broadening amounts to a few eV's at higher densities, with significant contributions from forward scattering processes. The low Fermi energy means that the wire is difficult to ionize. Though negatively charged wires are only metastable a substantial amount of additional electrons can be bound with only small tunneling rates to the outside world.

ACKNOWLEDGEMENT

One of the authors (G.E.W.B.) would like to acknowledge support by Professor M.F.H. Schuurmans.

REFERENCES

1. P.M. Petroff, A.C. Gossard, and W.Wiegmann, Appl. Phys. Lett. 45:1071(1984).
2. T. Fukui and H. Saito, Appl. Phys. Lett. 50:824(1987).
3. M. Tanaka and H. Sakaki, Jap. J. Appl. Phys. 27:L2025(1988).
4. M. Tsuchiya, J.M. Gaines, R.H. Yan, R.J. Simes, P.O. Holtz, L.A. Coldren, and P.M. Petroff, Phys. Rev. Lett. 23:466(1989).
5. M. Tanaka and H. Sakaki, Appl. Phys. Lett. in press.
6. K. Ploog, J. Cryst. Growth 81:304(1987).
7. A.A. van Gorkum, K. Nakagawa, and Y. Shiraki, J. Cryst. Growth in press.
8. A.A. van Gorkum, patent filed March 1989.
9. A.A. van Gorkum and G.E.W. Bauer, to be published.
10. T. Ando, Z. Phys. B 26:751(1977).
11. O. Gunnarsson and B.I. Lundqvist, Phys. Rev. B 13:4274(1976).
12. H. Sakaki, Jpn. J. Appl. Phys. 19:L735(1980).
13. G. Fishman, Phys. Rev. B 34:2394(1986).

QUANTUM DOT RESONANT TUNNELING SPECTROSCOPY

M. A. Reed, J. H. Luscombe, J. N. Randall, W. R. Frensley, R. J. Aggarwal,* R. J. Matyi,§ T. M. Moore and A. E. Wetsel†

Central Research Laboratories
Texas Instruments Incorporated
Dallas, Texas 75265
U. S. A

ABSTRACT

The electronic transport through 3-dimensionally confined semiconductor quantum wells ("quantum dots") is investigated and analyzed. The spectra corresponds to resonant tunneling from laterally-confined emitter contact subbands through the discrete 3-dimensionally confined quantum dot states. Momentum non-conservation is observed in these structures.

INTRODUCTION

Carrier confinement to reduced dimensions has led to numerous important developments in basic semiconductor physics and device technology. Until recently, this confinement has only been realizable by an interface (such as Si/SiO_2 and GaAs/AlGaAs), and thus the confinement is in one dimension. Advances in microfabrication technology[1-3] now allow one to impose quantum confinement in additional dimensions, typically done by constricting or confining in lateral dimensions an existing 2D carrier system. Though the technology to create lateral confining potentials is considerably less advanced than the degree of control that exists in the vertical (epitaxial, or interface) dimension, remarkable progress has been made in elucidating the relevant physics of these systems.

Investigations into two-dimensionally confined "quantum wires"[4] were first attempted in metallic systems. However, subband spacings in metals are considerably smaller than in semiconductors for the same

* Department of Electrical Engineering and Computer Science, Massachusetts Institute of Technology, Cambridge, MA.
§ Department of Metallurgical and Mineral Engineering, University of Wisconsin at Madison, WI.
† Department of Physics, Harvard University, Cambridge, MA.

Science and Engineering of One- and Zero-Dimensional Semiconductors
Edited by S.P. Beaumont and C.M. Sotomajor Torres
Plenum Press, New York, 1990

dimensional size. The realization of semiconductor quantum wires[5,6] allowed the investigation of well-defined lateral subbands, and the creation of "electron waveguides".[7] These structures are excellent laboratories to study fundamentals of electronic transport,[8,9] and allow the investigation of the Landauer formalism in ballistic quantized point contacts,[10,11] non-locality,[12] and waveguide mode-mixing. Unfortunately, electronic transport through these structures can only be done near equilibrium (i.e., low voltages and temperatures) and do not shed light on more common non-equilibrium situations.

Recently, three-dimensionally confined "quantum dots" have been realized.[13] These structures are analogous to semiconductor atoms, with energy levels tunable by the confining potentials. These structures pose an experimental paradox distinct from the 2D and 1D structures mentioned above; to allow transport through the single electronic states the states cannot be totally isolated; i.e., the confining potential must be slightly "leaky", and thus the states are "quasi-bound". Additionally, contact to carrier resevoirs are non-trivial from an experimental and analysis viewpoint. We have adopted a configuration where the quasi-bound momentum component (and thus the resultant transport direction) is epitaxially defined in the form of a resonant tunneling structure, and additional confinement is fabrication-imposed. This configuration is distinct from all the above referenced quantum wire cases where the (unbound) current flow is along the interface; here it is through the interface. Because of this constraint, the system can be operated far from equilibrium, a significantly different case than the tunneling observed through discrete defect states.[14,15] Additionally, the structural experimental parameters are difficult to control in the defect tunneling configuration.

An alternative approach which circumvents the contact problem is to probe the density of states in isolated systems by capacitance measurements.[16] This near-equilibrium approach has the advantages of isolation, but lacks information of transport in and coupling to single electron states, since the 0D electron states are measured solely electrostatically. The advantage of studying a single dot spectrum is also sacrificed in capacitance measurements because of signal-to-noise limitations, and size fluctuations become an issue.

We present here a study of resonant tunneling through various quantum dot systems, and the bandstructure modeling necessary to understand the experimental electronic transport spectra.

QUANTUM DOT FABRICATION AND TUNNELING

Our approach used to produce quantum dot nanostructures suitable for electronic transport studies is to laterally confine resonant tunneling heterostructures. This approach embeds a quasi-bound quantum dot between two quantum wire contacts.

Figure 1 schematically illustrates the fabrication method used. An ensemble of AuGe/Ni/Au ohmic metallization dots (single or multiple dot regions) are defined by electron-beam lithography on the surface of the grown resonant tunneling structure. Creation of dots less than 500Å is possible, though we will show that the appropriate range for the typical epitaxial structure and process used is in the range 1000Å - 2500Å in diameter. A bi-layer polymethelmethacrylate (PMMA) resist and lift-off method is used. The metal dot ohmic contact serves as a self-aligned etch mask for highly anisotropic reactive ion etching (RIE) using BCl_3 as an etch gas. The resonant tunneling structure is etched through to the n^+ GaAs bottom contact, defining columns in the epitaxial structure (Figure 1(c)). A SEM of a collection of these etched structures is seen in Figure 2.

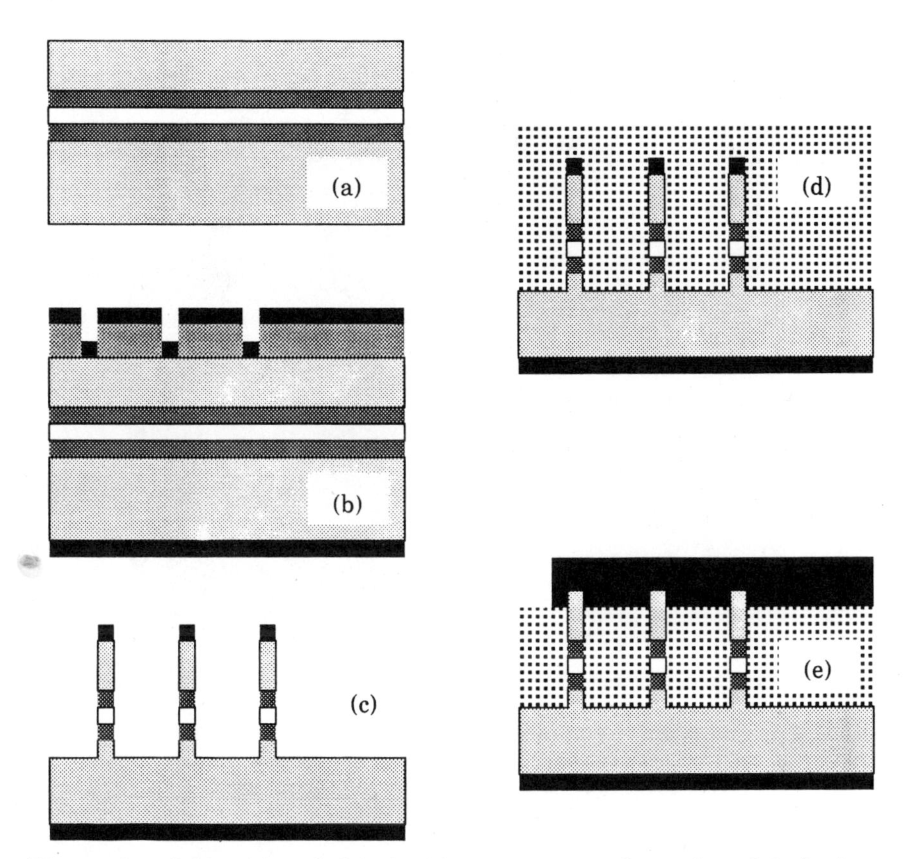

Figure 1. Schematic of fabrication sequence of quantum dot devices. (a) Resonant tunneling double barrier structure starting material, with n^+ GaAs contacts on top and bottom. (b) E-beam definition in PMMA of dots (singular or multiple; 3 shown here). Evaporation of AuGe/Ni/Au ohmic contacts. (c) Liftoff. BCl_3 reactive ion etch to define pillars with "dots" (white boxes). (d) Planarization with polyimide. (e) Etchback to ohmic top contacts. Evaporation of Au bonding pads.

To make contact to the tops of the columns, a planarizing and insulating polyimide is spun on the sample, then etched back by O_2 RIE to expose the metal contacts on the top of the columns. A gold contact pad was then evaporated over the top of the column(s). The bottom conductive substrate provids electrical continuity. Multiple columns can be connected in parallel for diagnostic purposes; however, all data discussed hereupon is for a single isolated column.

Figure 3 schematically illustrates the lateral (radial) potential of a column containing a quantum dot and the spectrum of 3-dimensionally confined electron states under zero and applied bias. A spectrum of discrete states would be expected to give rise to a series of resonances in transmitted current as each state is biased through the source contact. To observe lateral quantization of quantum well

Figure 2. Scanning electron micrograph of an array of anisotropically etched columns of predominantly GaAs which contain a quantum dot. The horizontal marker is 0.5 micrometer; the diameter of each of the columns is approximately 1000Å. The dark region on top of the column is the electron-beam defined Ohmic contact / etch mask. For transport measurements, only a single column in the field is defined by the lithography.

state(s), the physical size of the structure must be sufficiently small that quantization of the lateral momenta produces energy splittings > $3k_BT$. Concurrently, the lateral dimensions of the structure must be large enough such that pinch-off of the column by the depletion layers formed on the side walls of the GaAs column does not occur. Due to the Fermi level pinning of the exposed GaAs surface, the conduction band bends upward (with respect to the contacts Fermi level), and where it intersects the Fermi level determines in real space the edge of the central conduction pathcore. We can approximately express the radial potential $\Phi(r)$ in the column (for $(R-W) \leqq r \leqq R$), assumed axially symmetric, as

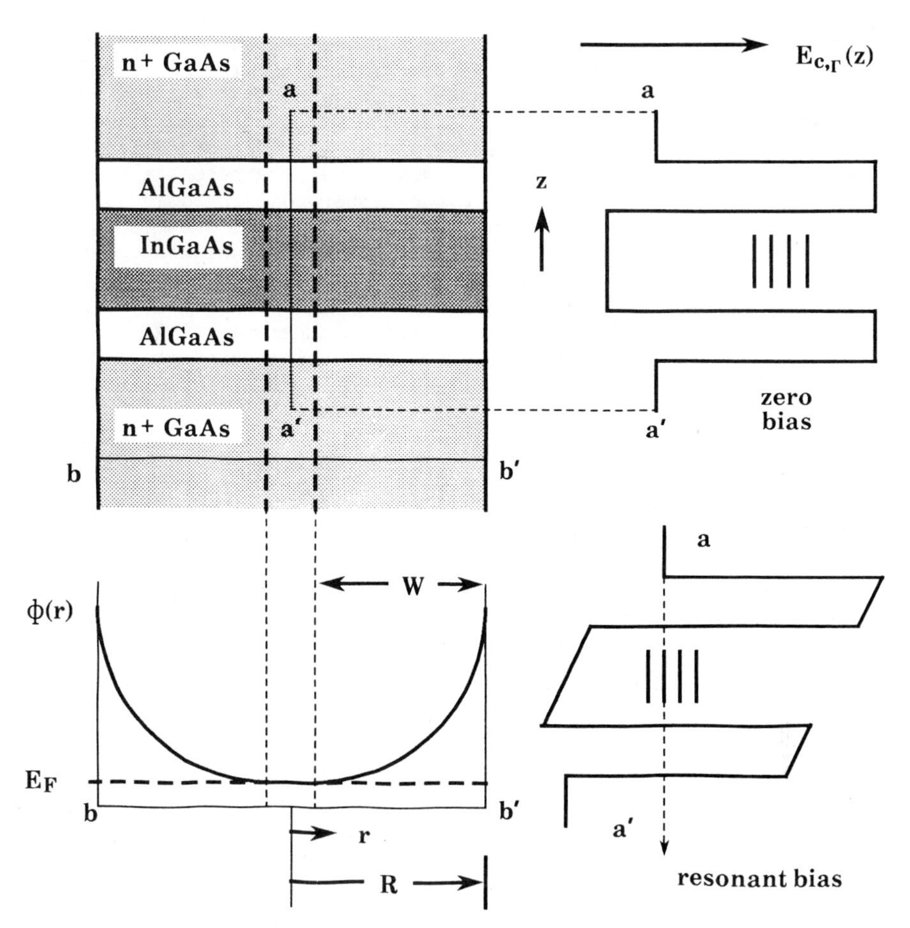

Figure 3. Schematic illustration of the vertical (a-a') and lateral (b-b') potential of a column containing a quantum dot, under zero and applied bias. Φ(r) is the (radial) potential, R is the physical radius of the column, r is the radial coordinate, W is the depletion depth, Φ_T is the height of the potential determined by the Fermi level (E_F) pinning, and $E_{c,\Gamma}$ is the Γ-point conduction band energy.

$$\Phi(r) = \Phi_T \left[1-(R-r)/W\right]^2$$

where r is the radial coordinate, R the physical radius of the column, W is the depletion depth, and Φ_T is the height of the potential determined by the Fermi level pinning. When the lateral dimension is reduced to 2W or less, the lateral potential becomes parabolic though conduction through the central conduction path core is pinched off. Observation of tunneling through the discrete states of a quantum dot necessitates fabrication within these design criteria.

Figure 4 shows the current-voltage-temperature characteristics of a quantum dot resonant tunneling structure successfully fabricated within these constraints. The structure lithographically is 1000Å in diameter and epitaxially is a n+ GaAs contact/AlGaAs barrier/InGaAs quantum well structure. At high temperature, the characteristic negative differential resistance of a double barrier resonant tunneling structure is observed. As the temperature is lowered, two effects occur. First, the overall impedence increases presumably due to the elimination of a thermally activated excess leakage current. Second, a series of peaks appears above and below the main negative differential resistance (NDR) peak. In the range of device bias 0.75V-0.9V, the peaks appear equally spaced with a splitting of approximately 50 mV. Another peak, presumably the ground state of the harmonic oscillator potential, occurs 80 mV below the equally spaced series.

The existence of the fine structure in the tunneling characteristics of this, and other, laterally confined resonant tunneling structures indicates the formation of laterally confined electronic states. However, a full indexing of the spectrum is needed to verify that the structure in the electrical characteristics is the discrete levels. To do this, a full 3D screening model of the quantum dot system is necessary.

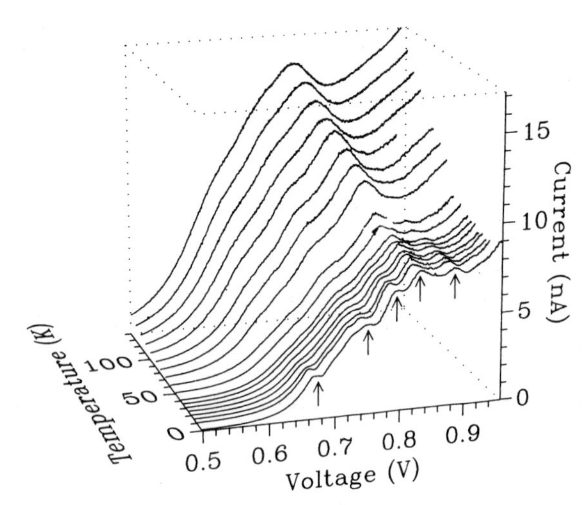

Figure 4. Current-voltage characteristics of a single quantum dot nanostructure as a function of temperature, indicating resonant tunneling through the discrete states of the quantum dot. The arrows indicate voltage peak positions of the discrete state tunneling for the T = 1.0°K curve.

1D RESONANT TUNNELING SPECTROSCOPY

Prior to understanding the detailed spectroscopy of a full 3-dimensionally confined system, let us first model and compare with experiment 1D modeling of resonant tunneling structures. Though resonant tunneling is qualitatively well understood, until now there lacks detailed quantitative spectroscopy of tunneling structures. We will use the epitaxial structure of the quantum dot presented previously to verify the modeling procedure and to illustrate the effect of the 3-D localization potential.

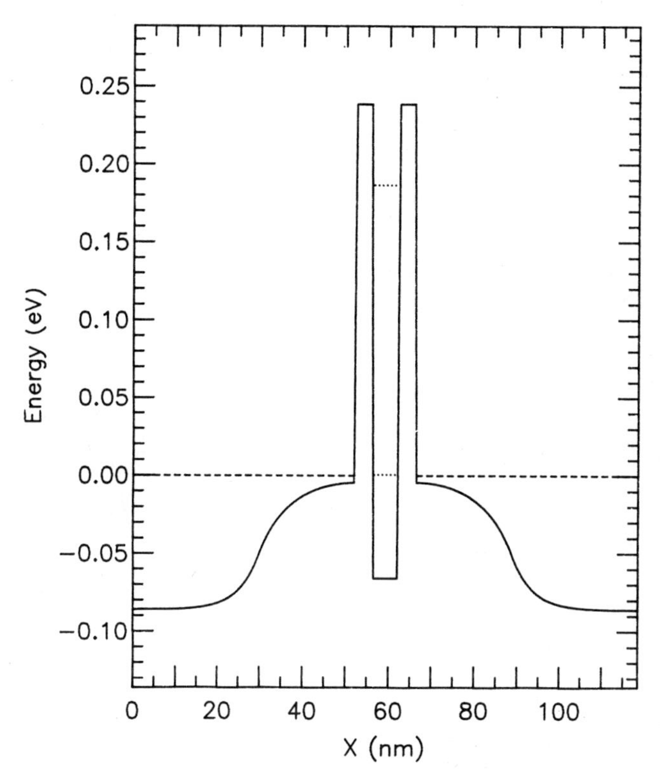

Figure 5. Self-consistent band diagram using Poisson's equation for the electrostatic potential of the quantum dot 1D epitaxial structure. The epitaxial dimension is denoted by x, in nanometers. The electrons in the contacts are treated in a finite-temperature Thomas-Fermi approximation. The simulation does not include current flow. The structure is a 40Å $Al_{.3}Ga_{.7}As$ barrier / 60Å $In_{.07}Ga_{.93}As$ quantum well / $Al_{.3}Ga_{.7}As$ barrier structure with contacts doped to $2x10^{18}$ cm^{-3}, at 77K. The Fermi level is denoted by a dashed line and the energies of the bound states are denoted by dotted lines.

Figure 5 shows a realistic conduction band profile of the quantum dot epitaxial structure at equilibrium. The model from which this Figure was obtained finds the self-consistent solution of Poisson's equations for the electrostatic potential. The electrons in the contacts are treated in a finite-temperature Thomas-Fermi approximation (i.e., these electrons are assumed to be in local equilibrium with the Fermi levels established by their respective contacts.) A result of this calculation, illustrated in the Figure, is that the band profile near the quantum well is significantly perturbed by the contact potential of the n^+-undoped junction. This shifts the resonant state upward (with respect to the n^+ GaAs Fermi level) from that expected from a naive flat-band picture.

Figure 6 shows the experimental current-voltage characteristics of a typical large areas mesa device of this epitaxial structure at 77K. The low bias peak, shown in Figure 6(a), is the resonance due to

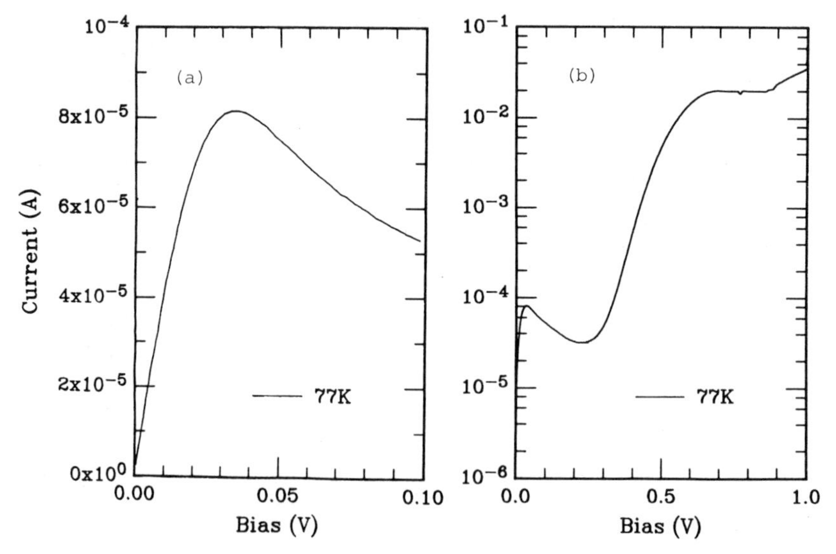

Figure 6. Current-voltage characteristics of the quantum dot epitaxial structure large mesa device, at T=77K. (a) Expanded low bias region showing resonant tunneling through the ground state of the quantum well. Note the large zero bias conductance. (b) Full scale, showing the n=1 (V=0.03) and n=2 (V~0.7)resonances.

tunneling through the ground state of the quantum well. Notice that the In content in the quantum well sufficiently lowers this quantum well state such that it lies below the Fermi level, but above the conduction band edge of the contact. The unique position of this state causes the observed large zero-bias conductivity of the structure and the low (30 mV) resonance, verified in the modeling as the voltage bias where the quantum well state falls below the emitter conduction band edge.

Nominal structural parameters were not sufficiently accurate to predict the correct resonant positions. It has been demonstrated that this method of modeling the resonant voltage peak positions yields greater accuracy for structural parameters than most present characterization techniques.[17] The nominal structural parameters were accurate to within their error bars; to fit the experimental resonant peak positions (Figure 6), it was necessary to assume a 60Å undoped $In_{0.07}Ga_{0.93}As$ quantum well instead of the nominal 50Å undoped $In_{0.08}Ga_{0.92}As$ quantum well. The high zero bias conductivity and low resonance imposes constraints on the model that allows for relatively

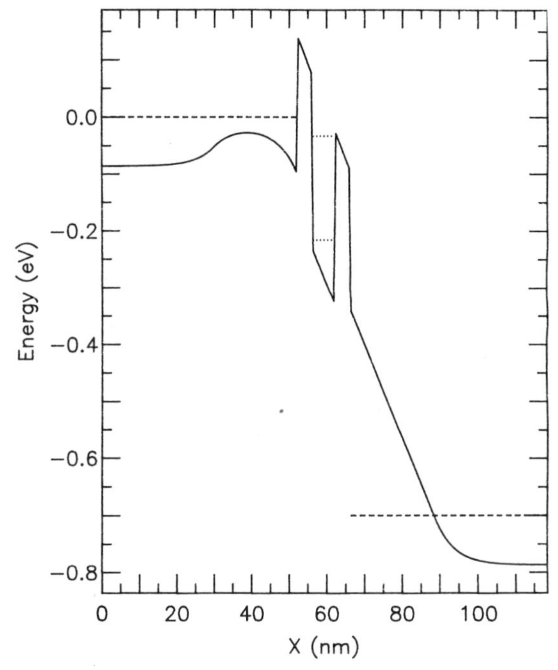

Figure 7. Self-consistent band diagram of the quantum dot 1D epitaxial structure at a bias of 0.7V and T=77K. The structure is a 40Å $Al_{.3}Ga_{.7}As$ barrier / 60Å $In_{.07}Ga_{.93}As$ quantum well / $Al_{.3}Ga_{.7}As$ barrier structure with contacts doped to $2x10^{18}$ cm^{-3}. The Fermi level is denoted by a dashed line and the energies of the bound (or virtual) states are denoted by dotted lines.

precise determination of the structural parameters. The only parameter which is not tightly constrained is the tunnel barrier height; however, the flat conductance region at approximately 0.7V due to the n=2 resonance implies that the n=2 state is becomming unbound at this bias due to the lowering of the collector-side barrier below the emitter Fermi level. Figure 7 shows this effect, which implies a Al content in the tunnel barriers of $\approx$30% instead of the nominal 25%.

Using these techniques, we can now attempt to understand the detailed spectroscopy of a full 3-dimensionally confined system. We have modeled the full screening potential of the quantum dot system taking into account the effects of lateral confinement. Cylindrical symmetry is assumed. The model self-consistently obtains the electrostatic potential in a zero-current theory from Poisson's equation utilizing a Thomas-Fermi approximation for the electron density. The solution of the electrostatic problem then provides the potential responsible for lateral quantization of electron states, which we obtain from the radial Schrodinger equation in cylindrical coordinates. The radial bound states in the contacts provide the minima of the emitter and collector subbands. Likewise the discrete quantum well levels, which in the absence of lateral confinement, would otherwise form a two-dimensional subband, are obtained from a solution of the radial Schrodinger equation. We shall first consider only the zero angular momentum ($l=0$) states.

The boundary conditions necessary for a solution to the quantum dot screening potential are considerably more complicated than for the 1D problem. At the center of the post ($r=0$), a simple Neumann condition of zero electric field was imposed. More involved is the question of the proper Dirichlet boundary condition to employ for the contact regions of this laterally-confined system. It is not enough to set the boundary potential in the degenerately-doped contacts to achieve charge neutrality, as one would have in a one-dimensional simulation or for bulk systems where surface effects are irrelevant. The restricted lateral extent of the quantum dot system, with the Fermi level pinning at the exposed outer lateral surface, implies a solution to the Poisson equation in the radial direction which is not a simple constant. Thus, to obtain a boundary condition in the contact regions for the full quantum dot system, we first do a 1D self consistent calculation for the radial direction, using the Laplacian for cylindrical coordinates. The boundary conditions for <u>this</u> calculation are again a zero field condition at the origin and another Neumann condition at the external radius set by an amount of surface charge necessary to support the value of the Fermi level pinning for $r=R$. To match up with the calculation for the full problem, it is assumed that there is negligible variation of the potential in the vertical direction in the vicinity of where the contact boundary conditions are to be imposed. The calculation is quite sensitive to the boundary condition specified at the outer lateral surface. We have employed a Neumann condition where the slope is determined by the surface charge. We allow this quantity to vary in the vertical direction. Our model assumes, to a first approximation, a constant density of surface states per unit area, independent of the material composition or doping level. We assume however that these states are occupied according to a Fermi-Dirac distribution, with the value of the Fermi level pinning acting as a local chemical potential. This rudimentary model of the surface charge distribution effectively "pins" the computed potential at the external lateral surface to the desired Fermi level pinning value. The calculation itself adjusts the occupation of surface states to self-consistently achieve a constant surface potential in the vertical direction (for zero bias) independent of material or doping level variations.

The equilibrium solution to the 3D screening problem using the quantum dot epitaxial structure and the measured physical radius of the column is displayed in Figure 8. The electron potential energy surface is plotted as a function of radius (R) and epitaxial (z) dimensions. The radial extent is 0-500Å and the vertical length is approximately 2000Å, centered about the double barriers. The energy scale is defined relative to the Fermi energy, thus the potential at the external radius equals 0.7V. The contours in the contact regions

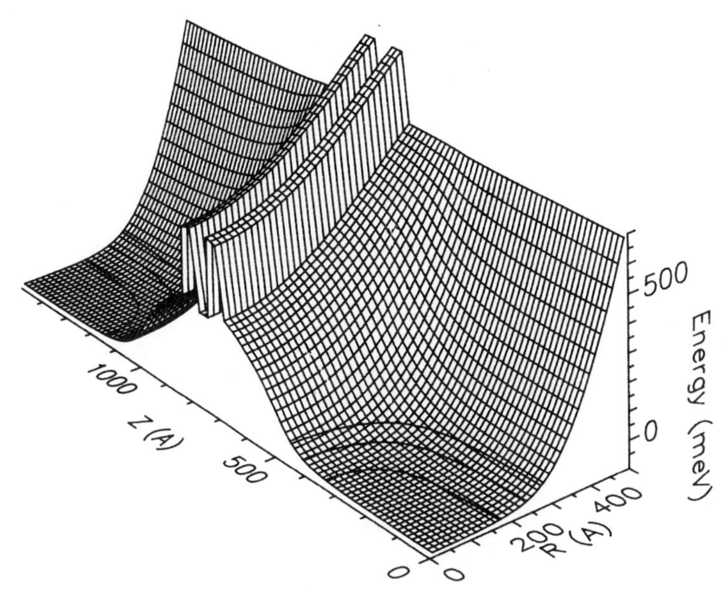

Figure 8. Self-consistent 3D band diagram of the quantum dot detailed previously (Figure 4), at equilibrium. The electron potential energy surface is plotted as a function of radius (R) and epitaxial (z) dimensions. The contours in the contact regions are the occupied laterally-defined subbands. For clarity, the quantum dot energy levels are not drawn.

are the occupied laterally-defined subbands that lie below the Fermi level. For this specific case of radial dimension and contact doping level, three contact subbands are occupied. The subband energies are determined by solving the radial Schroedinger equation. For clarity, the quantum dot energy levels are not drawn in this Figure.

The equilibrium solution demonstrates strong depletion in the region of the quantum dot due to the radial depletion, exacerbated by the z-dependent doping profile. This is demonstrated in Figure 9, which is just the r=0 contour of Figure 8. The contact subbands are denoted by dashed lines below the dotted Fermi level (E=0). It is clear that the lateral depletion has a dominant effect on lifting the double barrier structure significantly above the level previously determined only by the z-doping profile. The quantum dot states determined by solving the radial Schroedinger equation are shown as dashed lines between the barriers. These are the quantum dot states arising from the previous quantum well ground state (n_z=1); the excited state (n_z=2) quantum dot state are virtual. Previous mis-identification of the quantum dot resonances at with the n_z=2 resonance[13] was due to a fortuitous coincidence of the dot resonant voltages (~0.7V; see Figure 4) and the size of the quantum dot internal depletion barrier ($V_{1st\ resonance}$ ~ 2x(300 meV)).

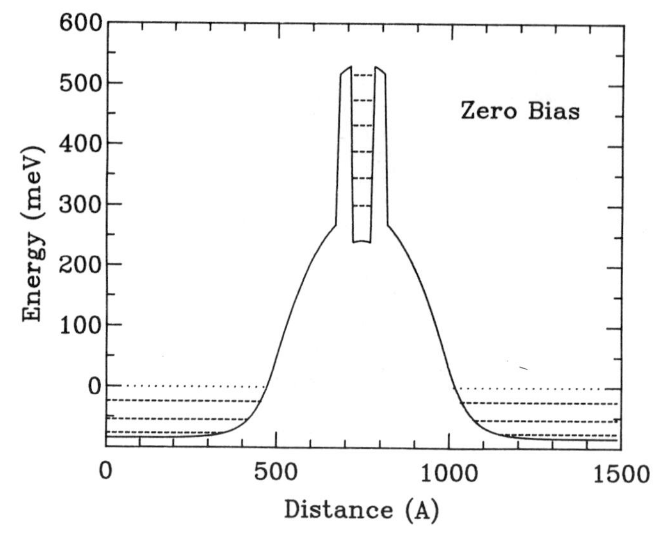

Figure 9. Radial slice (r=0) of the self-consistent 3D band diagram of the quantum dot (Figure 8) at equilibrium. The large internal depletion barrier due to the lateral confinement and the epitaxial doping profile is evident.

It has been suggested[18] that the observed quantum dot spectrum can be explained as resonances when the quantum dot states are biased through the emitter subband states with increasing device bias. To determine if this mechanism quantitatively explains the spectrum, we solve the 3D self-consistent screening quantum dot model at applied bias, to determine the variation of the emitter and quantum dot energy levels with applied voltage. Figure 10 shows the electron potential energy surface at finite applied voltage. For clarity, the subbands and levels are not shown.

Figure 11 shows the crossings of the emitter subband levels (n') with the quantum dot levels (n) as a function of applied bias. The parameters of the quantum dot model were the same as detailed above except the width of the undoped spacer layer. The present 3D model could only accomodate "box-like" (sharp) doping profiles, instead of

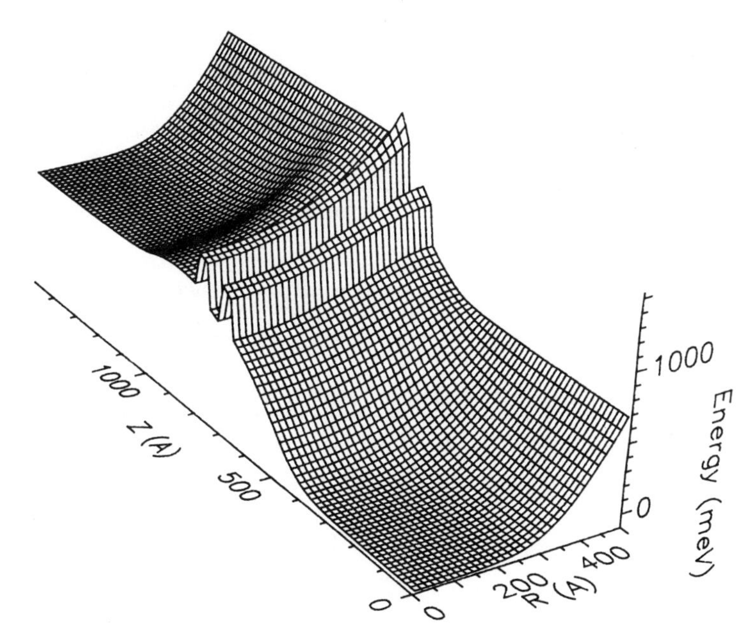

Figure 10. Self-consistent 3D band diagram of the quantum dot detailed previously at V=0.835V. For clarity, the contact subbands and quantum dot energy levels are not drawn.

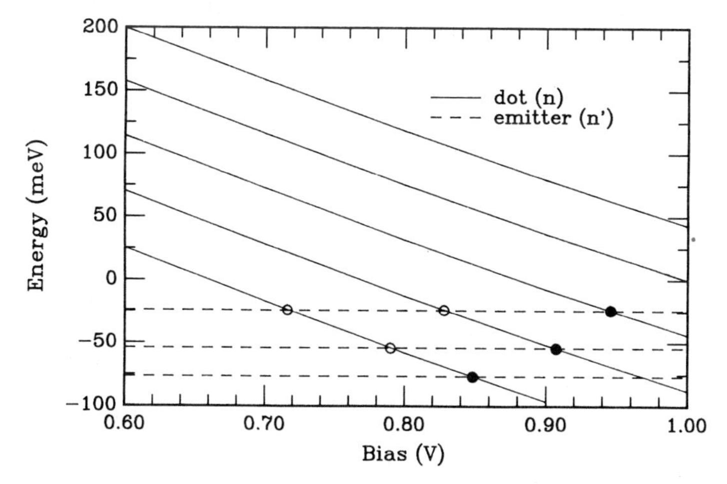

Figure 11. Emitter subband levels (n') and the quantum dot levels (n) as a function of applied bias. The circles denote the crossings; solid for momentum-conserving (n=n') transitions and open for momentum-nonconserving (n≠n') transitions.

the graded doping profile of the epitaxial structure. It was found that the absolute voltage values of the crossings were very sensitive to spacer thickness in the box-like profile model, but that the relative spectral spacing was invarient over a wide (~.1V) range; i.e., a change of spacer thickness "translates" the spectrum along the voltage axis. The spacer thickness is thus used as a fitting parameter; indeed, not only are the structural characterization or nominal growth parameters insufficiently accurate to yield this information, but the statistical fluctuations of dopants on this scale becomes important. For the spatial region included in the 3D model, there are less than 800 dopants.

Figure 12 shows the crossings of the emitter subbands with the quantum dot states, transposed onto the 1.0°K I-V characteristic of the quantum dot, with a spacer width of 177Å and with the initial and final state index numbers labeled (n'-n). There is general agreement between the experimental and predicted peak voltage positions,

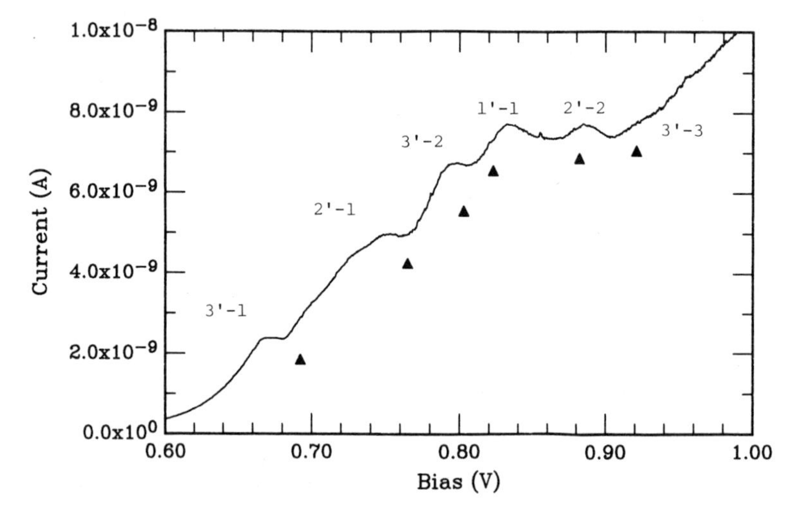

Figure 12. I-V characteristic at T=1.0°K of the quantum dot discussed previously, with predicted resonant peak positions and initial and final state index numbers (n'-n).

especially the anomolously large splitting of the first resonance. This can be seen as a consequence of the subband-level crossing mechanism, when more than one lateral subband is below the Fermi level. The experimental peaks differ from the experimental peak positions by at most 15 mV, which corresponds to approximately 5 meV in energy. This is in good agreement, considering the approximations of zero-current , homogeneous dopant distributions, and perfect radial symmetry. It should also be noted that the experimental measurement is current, which implies that an integration over the density of emitter states should be done for a strict comparison. It is possible that peaks may be shifted in voltage or even "washed out" when this is correctly done; however, the qualitative and quantitative agreement of the peak positions suggests this may not be a significant effect.

An additional corroboration of this spectroscopy and the peak indexing is found in the temperature dependence of the quantum dot peaks. Figure 13 shows the spectrum of the quantum dot at T=1.0°K and 50°K. The three lowest voltage peaks disappear by 50°K, and the spectrum is dominated by the single 1'-1 transition. This is expected, since when the subband spacing is less than $3k_BT$, thermal smearing will destroy the well defined subband structure. As can be seen in Figure 11, this occurs at approximately 7 meV, or 80°K, in reasonable agreement with the observed temperature dependence. In this high temperature limit, the structure emulates an unconfined 1D resonant tunneling diode, with the resonance determined by the 1'-1 crossing.

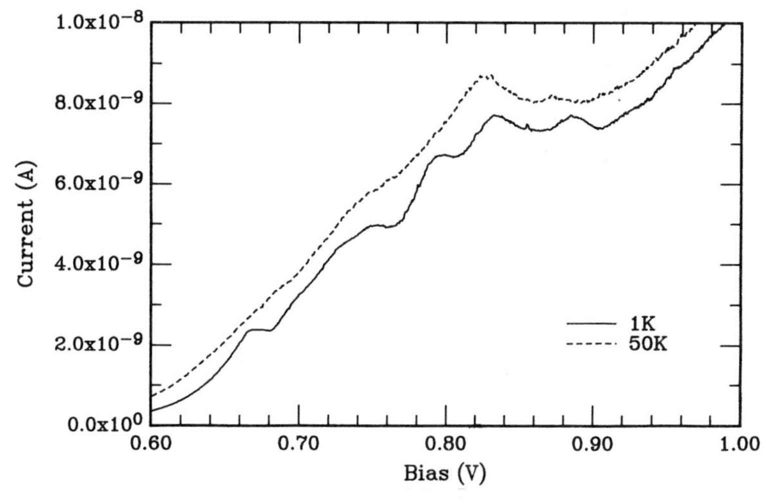

Figure 13. Current-voltage characteristics at T=1.0°K and 50°K for the quantum dot.

Finally, the predicted 3'-3 transition appears to be absent in the spectrum, except for a very weak structure at .92V-.93V. However, this is not unexpected since the collector barrier becomes sufficiently low that the state becomes virtual, similar to that discussed earlier (see Figure 7). Additionally, this implies that the lower bias resonances are due to states localized in the quantum dot and not due to the density of states in the collector contact.

The preceeding calculations are for angular momenta (l) equal to zero. Higher angular momentum states can be calculated, and effectively split the spectra into lx(number of n'-n crossings). Such extra structure does not seem evident in our experimental data, though sharper peak shapes are very desirable. Preliminary magnetic field studies up to 9.0T show no obvious Zeeman split discrete level peaks, which would be observable by 2.0T if higher angular momenta states were occupied.

SUMMARY

Quantum dot spectra are explained well by the transitions between laterally-confined emitter subbands and quantum dot discrete levels. We have indexed the transitions, and can thus determine selection rules for the transitions. The observation of the momentum-nonconserving transitions ($n' \neq n$) show that n is not a conserved quantity in a quantum dot system. The "hourglass" topography of the electron energy surface may be responsible for this (absence of a) selection rule. Fabrication improvements may allow us to create structures that critically explore and test these observations.

ACKNOWLEDGEMENTS

We wish to thank B. Bate for constant encouragement and support, and R. K. Aldert, E. D. Pijan, D. A. Schultz, P. F. Stickney, and J. R. Thomason for technical assistance. This work was sponsored by the Office of Naval Research, the Army Research Office, and the Air Force Wright Avionics Laboratory.

REFERENCES

1. A. N. Broer, W. W. Molzen, J. J. Cuomo, and N. D. Wittels, Appl. Phys. Lett. 29, 596 (1976).
2. R. E. Howard, P. F. Liao, W. J. Skocpol, L. D. Jackel, and H. G. Craighead, Science 221, 117 (1983).
3. H. G. Craighead, J. Appl. Phys. 55, 4430 (1984).
4. R. A. Webb, S. Washburn, C. P. Umbach, and R. B. Laibowitz, Phys. Rev. Lett. 54, 2696 (1985).
5. H. van Houten, B. J. van Wees, M. G. J. Heijman, and J. P. Andre, Appl. Phys. Lett. 49, 1781 (1986).
6. K. F. Berggren, T. J. Thornton, D. J. Newson, and M. Pepper, Phys. Rev. Lett. 57, 1769 (1986).
7. G. Timp, A. M. Chang, P. Mankiewich, R. Behringer, J. E. Cunningham, T. Y. Chang, and R. E. Howard, Phys. Rev. Lett. 59, 732 (1987).
8. R. Landauer, IBM J. Res. Dev. 1, 223 (1957).
9. M. Büttiker, Phys. Rev. Lett. 57, 1761 (1986).
10. B. J. van Wees, H. van Houten, C. W. J.Beenakker, J. G. Williamson, L. P. Kouwenhoven, D. van der Marel, and C. T. Foxon, Phys. Rev. Lett. 60, 848 (1988).
11. D. A. Wharam, T. J. Thornton, R. Newbury, M. Pepper, H. Ahmed, J. E. F. Frost, D. G. Hasko, D. C. Peacock, D. A. Ritchie, and G. A. C. Jones, J. Phys. C 21, L209 (1988).
12. G. Timp, H. U. Baranger, P. deVegvar, J. E. Cunningham, R. E. Howard, R. Beringer, and P. M. Mankiewich, Phys. Rev. Lett. 60, 2081 (1988).
13. M. A. Reed, J. N. Randall, R. J. Aggarwal, R. J. Matyi, T. M. Moore, and A. E. Wetsel, Phys. Rev. Lett. 60, 535 (1988).
14. A. B. Fowler, G. L. Timp, J. J. Wainer, and R. A. Webb, Phys. Rev. Lett. 57, 138 (1986).
15. T. E. Kopley, P. L. McEuen, and R. G. Wheeler, Phys. Rev. Lett. 61, 1654 (1988).
16. T. P. Smith III, K. Y. Lee, C. M. Knoedler, J. M. Hong, and D. P. Kern, Phys. Rev. B 38, 2172 (1988).
17. M. A. Reed, W. R. Frensley, W. M. Duncan, R. J. Matyi, A. C. Seabaugh, and H.-L. Tsai, to be published 27 March 1989 Appl. Phys. Lett.
18. G. W. Bryant, Phys. Rev. B 39, 3145 (1989).

FREQUENCY-DEPENDENT TRANSPORT IN QUANTUM WIRES

Bernhard Kramer and Jan Mašek*

Physikalisch-Technische Bundesanstalt, Bundesallee 100,
3300 Braunschweig, F.R. Germany
*Institute of Physics, Academy of Science, Na Slovance 2,
180 40 Prague 8, CSSR

INTRODUCTION

Coherent quantum transport phenomena have been the subject of many theoretical and experimental studies since the discovery of quantum interference oscillations in the magnetoresistance of thin metallic cylinders (Altshuler et al., 1981; Sharvin and Sharvin, 1981). Universal reproducible stochastic fluctuations in the magnetoresistance of small metallic systems (Washburn and Webb, 1986) have been attributed to the interference of elastically scattered electrons at randomly distributed impurities. In quasi one-dimensional inversion layers in MOSFETs, similar fluctuations have been observed when changing the gate voltage at low temperatures (Fowler et al., 1982). They are due to localisation of electron states. Most recently, a new quantisation phenomenon was discovered in the conductance of geometrically constricted inversion layers of high-quality GaAlAs heterostructures (van Wees et al, 1988; Wharam et al., 1988). It is believed that in these systems, at very low temperatures (< 1K) the electrons behave coherently and are not scattered elastically within distances of several μm, due to the absence of phase-randomising scattering events and of impurities. The quantisation of the conductance can then be interpreted as a consequence of the size quantisation of the electronic energy levels (Imry, 1986; Sharvin, 1965; Johnston and Schweitzer, 1988; Kramer and Mašek, 1988; Isawa, 1988; Kawabata, 1989).

All of the above-mentioned phenomena were observed in the dc transport. There have been almost no frequency-dependent transport investigations in the "mesoscopic" regime (which will be defined in brief) although one might expect interesting effects to occur upon changing the electric field as a function of time. Theoretical investigations have been restricted to the frequency dependence of the conductivity of disordered systems (Albers and Gubernatis, 1978; Saso, 1984). Only recently, time-dependent quantum coherence phenomena have been predicted theoretically in the region of weak localisation (Altshuler et al., 1987, 1988). The frequency dependence of the conductance fluctuations in one-dimensional disordered systems has been investigated numerically (Mašek and Kramer, 1988).

It is the main purpose of this article to review the work we have been doing during the past year (Mašek and Kramer, 1989) on frequency-dependent transport phenomena in quasi one-dimensional systems of finite width that are at most, weakly disordered within a

Science and Engineering of One- and Zero-Dimensional Semiconductors
Edited by S.P. Beaumont and C.M. Sotomajor Torres
Plenum Press, New York, 1990

finite interval and in which phase coherence is present or at most, only weakly destroyed. The frequency-dependent electric field is restricted to a finite interval, and it is constant, although the latter is not really necessary for the explanation of some of the effects described below, as we shall see. The model we use is qualitatively sketched in Fig. 1.

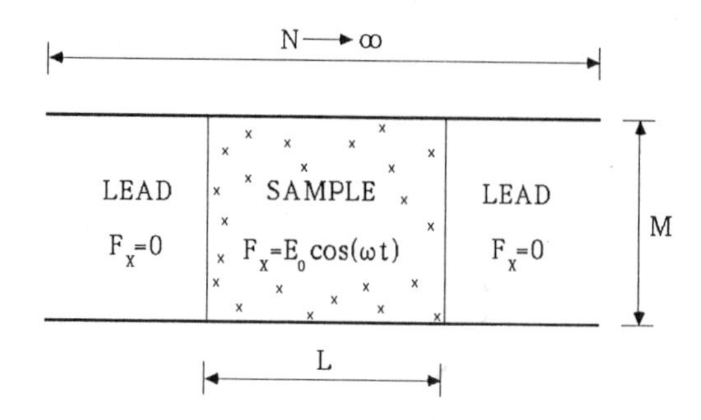

Fig. 1. Model for the conductance of mesoscopic systems

In the next section we shall introduce a classification of the transport regimes. The various characteristic lengths competing with each other at low temperatures are briefly discussed: then the quantum mechanical linear response theory of the frequency-dependent conductance of (almost) ballistically-behaving, non-interacting electrons in quasi one-dimensional systems is reviewed. It is shown that in an ideal system the dc conductance is quantised in units of e^2/h independently of the spatial variation of the electric field. The dependence on the width of the system, the Fermi energy, and the magnetic field is discussed and compared with experiments.

Some phenomena which have not yet been observed experimentally are predicted: the frequency-dependent conductance is shown to exhibit oscillations which provide insight into the spatial variation of the electric field. The quantized dc conductance in the presence of weak disorder exhibits sharp anti-resonances just at the steps. The breakdown of the quantisation when increasing the disorder is shown to coincide with the onset of conductance fluctuations of constant magnitude.

A simple model for incorporating phase-breaking processes is proposed and explicitly solved. The frequency-dependent conductance is shown to contain information about the phase-breaking length. Possible experimental verification of the predicted phenomena is briefly discussed.

CLASSIFICATION OF TRANSPORT REGIMES

Several length scales are of importance for a description of the low temperature transport. At zero temperature these are the *Fermi wavelength* $\lambda_F = 2\pi/k_F$, and the *elastic mean free* path l given by $l^{-1} = \text{const}\ \overline{v^2}\ k_F^{d-3}$ for a weak random (impurity) potential v. It corresponds to the mean distance between two elastic scattering events. If the disorder becomes larger, l decreases and the Born approximation will no longer be valid. When $l \lesssim \lambda_F$, the transport must be described quantum mechanically. The transport properties will then be governed by a *localisation length* $\lambda > l$ which describes the asymptotic exponential decay of the wave functions in the random potential. At finite temperature, inelastic, phase-randomising, scattering events are important. The corresponding *phase coherence length* L_φ plays a key role in the understanding of transport properties of metals. It is given by the diffusion length between two successive phase-breaking scatterings

$$L_\varphi = (D\,\tau_\varphi)^{1/2}$$

where τ_φ is the phase-breaking time. It increases with decreasing temperature according to $\tau_\varphi \sim T^{-p}$. p is of the order of unity and depends on the scattering mechanism (Schmid, 1985; Anderson et al., 1979; Lee and Ramakrishnan, 1985; Altshuler and Aronov, 1985). For instance, if electron-electron collisions dominate, we have p = 1 (Altshuler and Aronov, 1985), and for electron-phonon scattering p = 3...4 (Schmid, 1985) depending on the temperature range. L_φ must be larger than λ or even larger than the system size L in order to make quantum effects observable in the transport properties. In the regime of coherent quantum transport, the charge carriers are almost exclusively influenced by randomness and not by the energy and phase-randomising processes (Table 1).

Table 1. Classification of the regimes of transport in disordered systems. L: system diameter; L_φ: phase-breaking length; l: elastic mean free path, λ: localisation length; λ_F: Fermi wavelength. *Italics denote the regimes treated in this paper.* Effects that have not yet been observed experimentally are marked with an asterisk*

length scales	transport regimes		effects in resistance
$l < \lambda << \lambda_F < L_\varphi << L$	hopping		"$T^{-1/4}$ – law"
$\lambda_F << l < L_\varphi << \lambda < L$	weak localisation	incoherent	log T behavior
$\lambda_F < L_\varphi < l << \lambda < L$	diffusive		T^p behavior
$\lambda_F < L_\varphi < L << l < \lambda$	*incoherent ballistic*		*T–dependent oscillations**
$\lambda_F < L << L_\varphi < l < \lambda$	ballistic		*quantisation*
			*oscillations in ac resistance**
$\lambda_F << l < L < \lambda < L_\varphi$	quasi-ballistic	coherent	*quantum interference*
$l < \lambda < \lambda_F << L < L_\varphi$	strong localisation		exponential increase*

In the ballistic regime the above relation between the phase-breaking length and the phase-breaking time is certainly not valid. Here, L_φ is probably a lower limit of the true phase-breaking length.

FREQUENCY-DEPENDENT CONDUCTANCE

There are various formulations of the transport theory of small systems (an overview of the present status is given in the IBM J. Research and Development **32**, No. 3, 1988). We use the linear response theory since it is capable of treating the frequency dependence. For simplicity we consider for the moment the model shown in Fig. 1. A monochromatic, spatially homogeneous electric field $F = (F_x, 0)$, $F_x = F_0\, e^{i\omega t}$, is applied within a finite portion of the length L to a wire of the length N ($\rightarrow \infty$), in the x direction and of the finite cross-section A (perpendicular to the x direction) described by a Hamiltonian H. The conductance $\Gamma(\omega)$ can then be defined by considering the energy absorption rate $P \equiv \Gamma(\omega)\, F^2_{rms}\, L^2$ where F_{rms} is the root meat square of F in the limit $F_{rms} \rightarrow 0$.

From the linear response theory one obtains (Fisher and Lee, 1981)

$$\Gamma(\omega) = \frac{h}{2L^2} \int dE \frac{f(E) - f(E+\hbar\omega)}{\hbar\omega} \times$$
$$\times \sum_{a,b} |J_{a,b}|^2 \, \delta(E - E_a) \, \delta(E + \hbar\omega - E_b) \ . \qquad (1)$$

f(E) is the Fermi distribution function.The current matrix elements J_{ab} are given by

$$J_{ab} = \int_L dx \int_A d^{d-1}r \, \frac{e\hbar}{2im} \left[\frac{\partial \psi_a^*}{\partial x} \psi_b - \psi_a^* \frac{\partial \psi_b}{\partial x} \right] . \qquad (2)$$

$\psi_{a,b}$ and $E_{a,b}$ are the eigenfunctions and energies of H, respectively; d is the dimensionality. This definition should work perfectly well for $\omega \neq 0$. The electric field may, for instance, be introduced by shining electromagnetic radiation onto a piece of the length L of the entire wire at least in a *gedanken experiment.* The absorbed energy may be measured without referring to chemical potential differences. The conductivity may be defined as

$$\sigma(\omega) = \Gamma(\omega) \, L/A \qquad (3)$$

Formally, one may rewrite eq. (1) in terms of Green's functions $G^{\pm}(E) = (E \pm i\varepsilon - H)^{-1}$

$$\Gamma(\omega) = \int dE \, \frac{f(E) - f(E+\hbar\omega)}{\hbar\omega} \, \Gamma(E,\omega) \qquad (4)$$

with

$$\Gamma(E,\omega) = -\frac{\hbar}{4\pi L^2} \mathrm{Tr} \left\{ J^{(L)} \Delta G(E) \, J^{(L)} \Delta G(E+\hbar\omega) \right\} . \qquad (5)$$

Here, $\Delta G = G^+ - G^-$, and $J^{(L)}$ is the operator of the current, spatially averaged over the part of the system where $F \neq 0$

$$J^{(L)} = \frac{e\hbar}{2im} \int_L dx \int_A d^{d-1}r \int d^d r' \, \delta(r-r') \times$$
$$\times \left[\frac{\partial}{\partial x} |r\rangle\langle r'| - |r\rangle \frac{\partial}{\partial x'} \langle r'| \right] . \qquad (6)$$

This form of $\Gamma(\omega)$ is especially useful for analytical (Lee et al., 1987; Altshuler et al., 1986) and numerical purposes (Mašek and Kramer, 1989; Szafer and Stone, 1989)

FREE ELECTRONS IN A QUASI ONE-DIMENSIONAL SYSTEM

It is instructive to evaluate eq. (1) for free electrons

$$H = \frac{p^2}{2m} \qquad (7)$$

confined within a strip of width M with periodic boundary conditions in the x direction. The wave functions are then of the form

$$\psi_{k\mu}(x,y) = N^{-1/2}\,\varphi_\mu(y)\,e^{ikx} \tag{8}$$

with $\langle\varphi_\mu|\varphi_{\mu'}\rangle = \delta_{\mu\mu'}$. The corresponding energies are, as a result of transversal quantization,

$$E_\mu(k) = \hbar^2 k^2 / 2m + E_\mu \; . \tag{9}$$

Inserting these into eq. (1), at T = 0 one easily obtains

$$\Gamma(\omega) = \frac{e^2}{h} \sum_{\mu=1}^{\mu(E_F)} \frac{\sin^2 \frac{\omega L}{2v_\mu}}{\left(\frac{\omega L}{2v_\mu}\right)^2} \; . \tag{10}$$

$v_\mu = \sqrt{2(E_F - E_\mu)/m}$ is the Fermi velocity in the μ-th sub-band. $\mu(E_F)$ is the index of the uppermost occupied sub-band, i.e. $E_F > E_{\mu(E_F)}$.

This result demonstrates two remarkable facts (Fig. 2). First, in the limit of zero frequency,

$$\Gamma(0) = \frac{e^2}{h}\,\mu(E_F) - o\left(\frac{\omega L}{2v_\mu}\right)^2 \tag{11}$$

is quantised in integer units of e^2/h, the integer $\mu(E_F)$ being generally given by the number of "open" transport channels. This is the number of intersections of the energy dispersion $E_\mu(k)$ with E_F divided by two. It reduces to the number of occupied sub-bands only in some simple cases, such as the parabolic bands eq. (9). Secondly, as a function of $\omega L/2v_\mu$ the channel contributions to $\Gamma(\omega)$ oscillate. The zeros are given by $\omega L/2v_\mu = n\pi$ ($n = 1, 2, 3, \ldots$). If the distance $2\pi\, v_\mu/\omega$ which can be travelled by an electron with the velocity v_μ within the period $2\pi/\omega$ of the oscillating electric field equals L/n, on average no current will flow in the system! This can be interpreted as an interference phenomenon due to the coherence between the wave functions $e^{\pm ikx}$ of the electron and the hole.

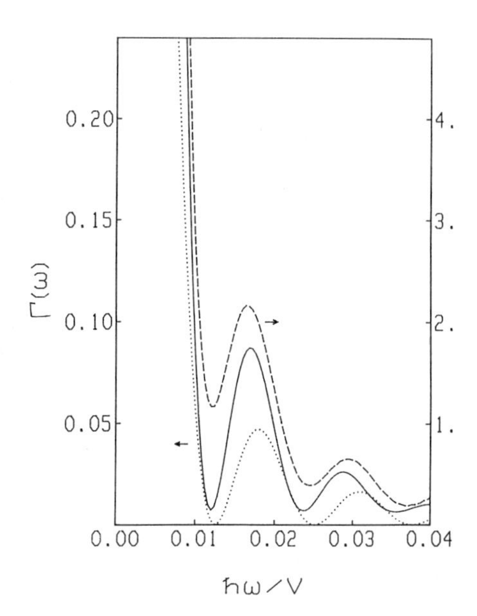

Fig. 2. Frequency-dependent conductance $\Gamma(\omega)$ in units of e^2/h for confined electrons in a tight binding system of the length L = 1000 (see section on lattice electrons). The dotted curve ($M = 1$, $E_F = 0$) corresponds to the one-channel case, the dashed curve ($M = 100$, $E_F = 0$) corresponds to 100 open channels, the full curve ($M = 3$, $E_F = V$) corresponds to two channels

The oscillations of $\Gamma(\omega)$ provide information on v_F, i.e. the electron density and on L, the length of the portion of the wire with $F \neq 0$, i.e. the spatial behavior of the electric field. We will discuss this point in more detail in a later section.

CONFINED LATTICE ELECTRONS

The influence of a periodic potential, a random potential and a magnetic field can be studied by starting from the tight-binding Hamiltonian

$$H = \sum_{j=-N/2}^{N/2} \left\{ \sum_{m=1}^{M} |jm> \varepsilon_{jm} < jm| + \sum_{m=1}^{M-1} \left(|jm><jm+1| + |jm+1><jm| \right) \right\} +$$

$$+ V \sum_{j=-N/2}^{N/2-1} \left\{ \sum_{m=1}^{M} \left(|jm><j+1\ m|\, e^{i\alpha m} + |j+1\ m><jm|\, e^{-i\alpha m} \right) \right\} . \tag{12}$$

Here, $\{|jm>\}$, ε_{jm} are an orthonormal set of lattice site states (square lattice, lattice constant a = 1), and the corresponding site energies, respectively. They represent a potential energy which may also be taken at random according to some distribution function $P(\varepsilon)$ within a finite part of the entire system, j = 1 ... L, for instance, in order to account for randomness. The hopping matrix elements in the y direction are taken as unity, in the x direction they are V. The phase factors $\exp\{\pm i\alpha m\}$ included in the last terms describe the effect of a perpendicular magnetic field (Luttinger, 1951). $\alpha = eB/\hbar$ is the number of flux quanta in the unit cell multiplied by 2π. We again take periodic boundary conditions in the x direction.

The operator of the average current is now

$$J^{(L)} = \frac{e}{ih} \sum_{j=2}^{L} \sum_{m=1}^{M} \left\{ |jm> v_m <j-1\ m| - |j-1\ m> v_m^* <jm| \right\} \tag{13}$$

with $v_m = V \exp\{-i\alpha m\}$.

For the ordered case ($\varepsilon_{jm} = 0$) the conductance may again be explicitly evaluated (Kramer and Mašek , 1988). Due to the choice of boundary conditions and the gauge of the vector potential, the eigenstates of H may be written as

$$|\psi_{k\mu}> = \sum_{m=1}^{M} c_{m\mu}(k) \sum_{j=-N/2}^{+N/2} e^{ikj} |jm> \tag{14}$$

where $-\pi < k = 2\pi n/N \leq \pi$, and $\mu = 1 \ldots M$. The coefficients $c_{m\mu}(k)$ obey

$$c_{m+1\mu}(k) + c_{m-1\mu}(k) + 2V \cos(k+\alpha m)\, c_{m\mu}(k) = E_{k\mu}\, c_{m\mu}(k) . \tag{15}$$

For $\alpha = 0$ (vanishing magnetic field) the electronic motion in the x and y directions decouples, $c_{m\mu}(k) \equiv c_{m\mu}$, and therefore

$$E_{k\mu} = 2V \cos k + E_{\mu} . \tag{16}$$

E_μ, now denoting the centers of the 1d sub-bands, are given by the Schrödinger equation of an isolated layer containing M lattice sites. Examples of the density of states (DOS) are shown in Fig. 3. With increasing M the band structure near the band edges becomes more

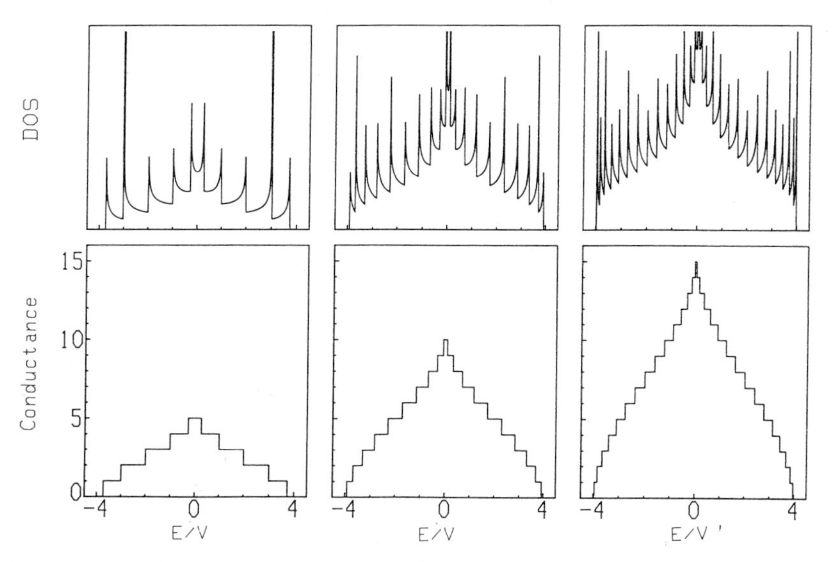

Fig. 3. Density of states in arbitrary units and dc-conductance in units of e^2/h as a function of the Fermi energy for M = 5, 10, 15 (from the left to the right).

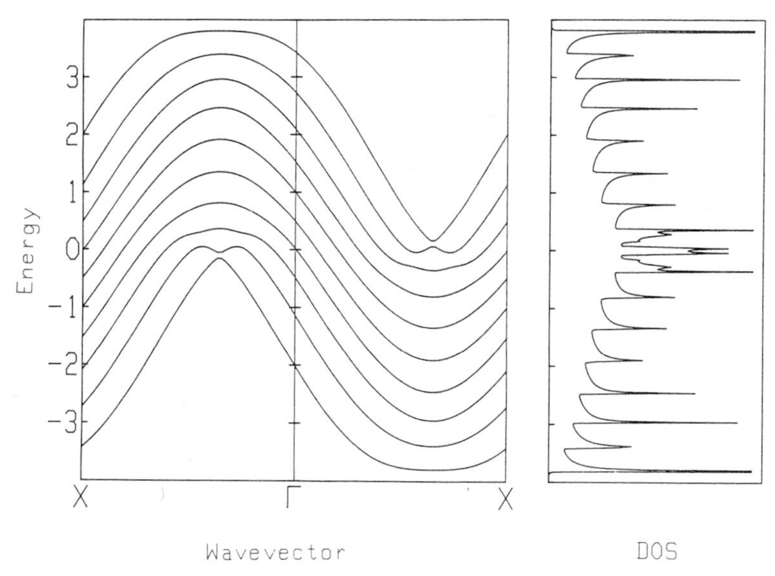

Fig. 4. Magnetic band structure $E_{k\mu}$ in units of V = 1 (left) and the density of states (in arbitrary units) for a finite magnetic field ($\alpha = 0.2$). The width is M = 10.

and more similar to the one obtained in effective mass approximation. The striking feature of the DOS is the 1d van Hove singularities at the edges of the sub-bands. For $\alpha \neq 0$ the band structure becomes more complicated, as shown in Fig. 4. Additional singularities appear in the centre of the band due to the interplay between the spatial periodicity and the magnetic field.

QUANTISATION OF DC CONDUCTANCE

It can be shown that the conductance may again be written as a sum of independent contributions $\Gamma_\mu(E,\omega)$ of transport channels.

$$\Gamma(\omega) = \sum_{\mu=1}^{M} \int dE \, \frac{f(E) - f(E+\hbar\omega)}{\hbar\omega} \, \Gamma_\mu(E,\omega) \ . \tag{17}$$

$\Gamma_\mu(E,\omega)$ may be evaluated analytically in a closed form which is similiar to eq. (10) for $\alpha = 0$ (Mašek and Kramer, 1989). For $\alpha \neq 0$ we have not yet found a closed representation of the channel contributions $\Gamma_\mu(\omega)$.

Figure 5 shows a selection of results for $\alpha = 0$. It is obvious that, with excellent accuracy,

$$\Gamma(L, M, \omega, E_F, \alpha = 0) = \Gamma(L \cdot \omega, M, E_F) \tag{18}$$

and that the oscillations do exist even for large M. The latter is due to the fact that the smallest sub-band Fermi velocities determine the position of the minima in Γ at low frequencies.

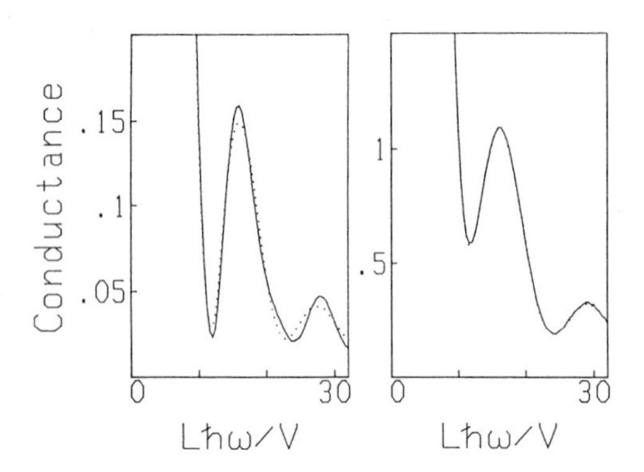

Fig. 5. Scaling behavior of the dimensionless conductance for various widths M of the sample and positions of the Fermi level M = 5, E_F = 0 (left); M = 50, E_F = 0 (right); full lines: results for L = 100; dots: results for L = 1000.

For $\omega \to 0$ it is easy to see that

$$\Gamma(L, M, E_F, 0, \alpha) = \Gamma(M, E_F, 0, \alpha) = \frac{e^2}{h} \mu(M, E_F, \alpha) \tag{19}$$

where $\mu(M, E_F, \alpha)$ is again given by half the number of intersections $E^\alpha_{\mu k} = E_F$, and defines the number of open transport channels.

The quantisation of the conductance is thus independent of L, and is observable as a function of E_F (electron density), width M, and the magnetic field α (Fig. 6).

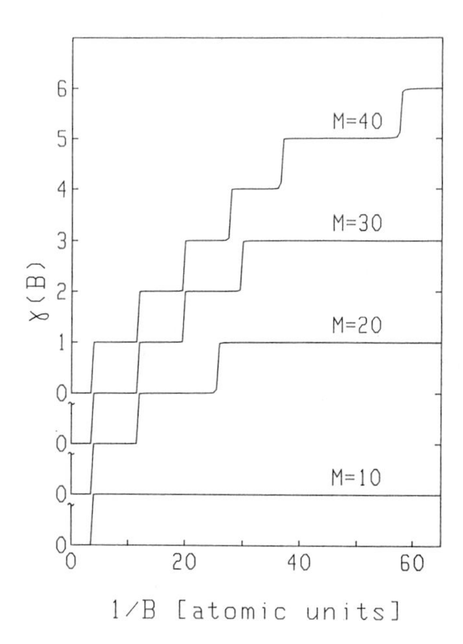

Fig. 6. Dimensionless conductance as a function of the inverse of the magnetic field in atomic units for several sample widths M.

INFLUENCE OF DISORDER

The influence of a random potential on the quantisation for $\omega \to 0$ and the oscillatory behavior may be studied by taking ε_{jm} independently at random, and distributed according to a box distribution function of width W. $\Gamma(\omega)$ will then depend on the configuration $\{\varepsilon_{jm}\}$. One has to consider the average of the conductance.

$$\overline{\Gamma(\omega)} = \int \dots \int \prod_{jm} d\,\varepsilon_{jm}\, P(\varepsilon_{jm})\, \Gamma\left(\omega \left\{\varepsilon_{jm}\right\}\right) . \tag{20}$$

For a specific configuration one may compute the conductance by using a recursive method described elsewhere (MacKinnon, 1985; Mašek and Kramer, 1989). The average of $\Gamma(0)$ over many realisations of the set of site energies is shown in Fig. 7 for two different L. The average conductance is still quantised in the presence of a moderate random potential. However, the "unit" is now smaller than e^2/h.

Figure 8 shows the dependence of the step height $\Delta\Gamma(0)$ on the magnitude of the random potential. $\Delta\Gamma(0)$ decreases monotonically with W. The fluctuations of $\Gamma(0)$, $\Delta = (\overline{\Gamma(0)^2} - \overline{\Gamma(0)}^2)^{1/2}$ in the statistical ensemble are also shown. $\Delta(\omega)$ increases monotonically for small W and saturates for $W \gtrsim V$ at the value of $\Delta_c \approx 0.33\ e^2/h$. At this value of the disorder, the height of the steps also becomes smaller than Δ. The average conductance is no longer quantised.

Throughout the paper we have omitted the spin degeneracy of the electrons. Taking into account the spin factor of 2 we have $\Delta_c = 0.66\ e^2/h$ which is close to the value obtained by the perturbation theory for the universal conductance fluctuations in 2d metallic systems $\Delta_c = 0.862\ e^2/h$ (Lee et al., 1987).

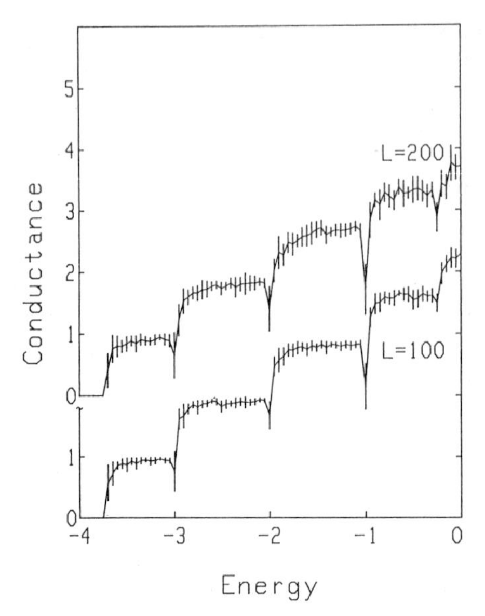

Fig. 7. Dimensionless conductance as a function of the Fermi energy for weak disorder (W = 0.2) for two lengths of the sample, L. The width is M = 15. The root mean square deviations from the average conductance are shown as vertical bars.

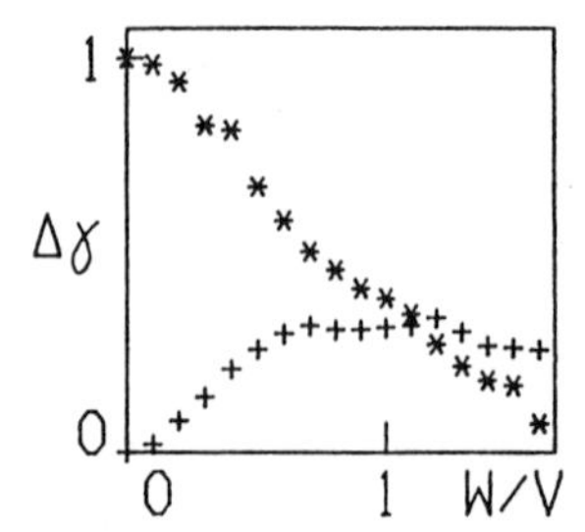

Fig. 8. The dependence of the step height $\Delta\gamma = \langle \gamma(-1.5) \rangle - \langle \gamma(-2.5) \rangle$ (*), and of the r.m.s. deviation of the conductance (+) on the strength of the disorder W/V for L = 100.

A remarkable feature of the average conductance in the presence of moderate disorder is the existence of sharp anti resonances at the band edges (Fig. 7). They are caused by disorder-induced mixing of the sub-bands and are *not* due solely to a density of states effect (Mašek et al. 1989). The conducting states of the lower sub-band are mixed with the localised states just below the band edge of the higher sub-band. This hybridisation introduces additional scattering events and strongly decreases the conductance.

INHOMOGENEOUS ELECTRIC FIELD

Eq. (1) for the conductance may be generalised to the case of a spatially inhomogeneous monochromatic electric field (Velicky et al., 1989). The starting point is the linear relationship between induced current and applied electric field F(x)

$$j(x) = \int \sigma(x, x'; \omega)\, F(x')\, dx' \quad , \tag{21}$$

where

$$\sigma(xx'; \omega) = e^2 \int dE \, \frac{f(E) - f(E+\hbar\omega)}{\omega} \, \mathrm{Tr}\Big(\mathbf{j}(x)\ \delta(E-\mathbf{H})\ \mathbf{j}(x')\ \delta(E+\hbar\omega-\mathbf{H})\Big) \ . \tag{22}$$

We shall assume the electric field to be vanishing sufficiently rapidly outside the interval $L/2 < x < L/2$. The current density operator is (cf. eq. (2)) $\mathbf{j}(x) = 1/2m\, [\mathbf{p}, \delta(x-\mathbf{x})]$.

By using the sub-band representation one can easily deduce that σ may be decomposed into independent non-local contributions. For $\hbar\omega/(E_F - E_\mu) << 1$ they may be explicitly evaluated

$$\sigma_\mu(xx'; \omega) = \frac{e^2}{h} \{\cos(\omega x/v_\mu)\cos(\omega x'/v_\mu) + \sin(\omega x/v_\mu)\sin(\omega x'/v_\mu)\} \; . \qquad (23)$$

From the power absorbed by the electrons

$$P(\omega) = \frac{1}{2}\int dx \int dx' \; F(x)\, \sigma(xx'; \omega)\, F(x') \qquad (24)$$

we obtain

$$\Gamma(\omega) \equiv P(\omega) \Big/ \frac{1}{2} U^2 = \frac{e^2}{h} \sum_{\mu=1}^{\mu(E_F)} L(\omega/v_\mu) \Big/ L(0) \; . \qquad (25)$$

for the conductance. Here the Fourier-transformed autocorrelation function of the electric field

$$L(k) = \int dx \int dx' \; E(x+x')\, E(x')\, e^{ikx} \qquad (26)$$

was introduced, and

$$L(0) = \int E(x+x')\, E(x')\, dx\, dx' = U^2 \qquad (27)$$

is nothing but the square of the voltage drop across the sample of the length L. *Thus the frequency-dependent conductance essentially measures the autocorrelation function of the electric field when $T = 0$. It is only for $\omega \rightarrow 0$ that the field distribution drops out of the conductance.*

A few model calculations illustrate this result (Fig. 9). The simplest case is a constant field within the interval [–L/2, L/2]. The autocorrelation function of F is then a triangle (model I). This of course gives the previously obtained result of eq. (10). The second example shows the effect of smoothing the shape of F(x) (model II). Here, the box function is broadened with a Gaussian. The correlation function is much less influenced, and in the spectrum $\Gamma(\omega)$ the zeros survive. This is due to the convolution theorem. If the width of the Gaussian becomes comparable with L, the zeros become important (modell III). If the field was divided among two intervals separated by the zero field region, the correlation function would have three maxima, the central one corresponding to the correlation within each of the non-zero field intervals. A "double slit diffraction pattern" of $\Gamma(\omega)$ would result. Relaxation of the phase coherence between the two non-zero field regions would yield a conductance that is one half of the one corresponding to the model I. This is nothing other but the classic result that two (incoherent) resistances in series are simply added.

The fact that the ac conductance is given by the Fourier-transformed autocorrelation function presents the experimental possibility of gaining some insight into the spatial behavior of the field without introducing additional voltage probes which are known to influence the transport behavior in the coherent regime (Washburn and Webb, 1986).

INCOHERENT TRANSPORT

In this section we shall describe the case in which $L_\varphi < L$ but still $L_\varphi > M$ (Kramer and Mašek, 1989). Phase coherence is then relaxed in the x direction. We assume that it will still be possible to superpose the total conductance from independent channel contributions and restrict the following considerations to the case where there is a single channel case.

Let us consider a set of phase-randomising scattering processes (Fig. 10) taking place at sites $r_1 \ldots r_N$ so that $-L/2 = x_0 < x_1, < x_2, < \ldots < x_N < L/2 = x_{N+1}$. The scatterings define in-

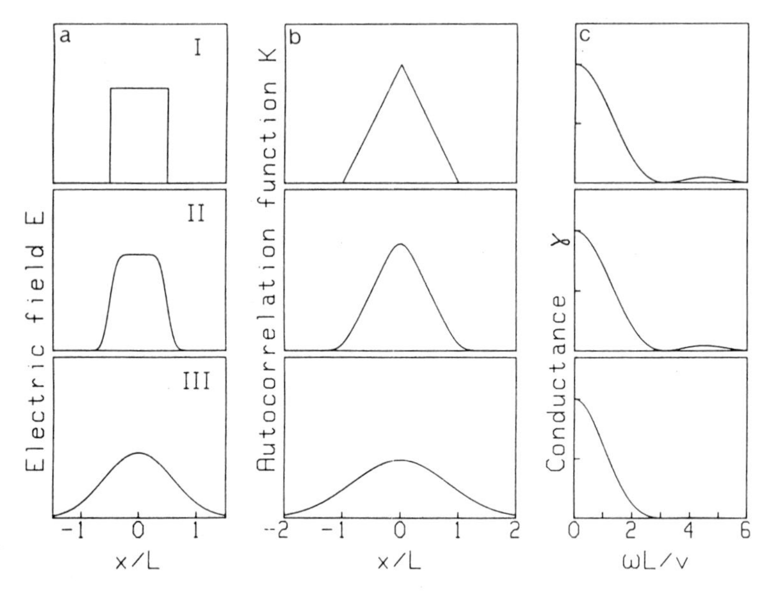

Fig. 9. Electric field distribution E(x) (a), its autocorrelation function K(x) (b), and the corresponding frequency-dependent 1d conductance γ in units of e^2/h (c) for the models I, II and III defined in the text.

tervals $\Delta_j = y_{j+1} - x_j$ which have the distribution $P(\Delta)$. Furthermore, we assume that the effect of the scatterings is solely to randomise the phases of the states

$$\psi(r) = e^{i\phi_j}\chi(r) \quad , \quad x_j < x < x_{j+1} \tag{28}$$

with random phases ϕ_j. $\chi(r)$ is the wave function in the absence of phase randomisation. The wave function $\psi(r)$ cannot be defined around the points r_j where the single particle picture breaks down, but the contribution of these infinitesimal intervals to the matric elements is negligible. On this assumption, the evaluation of the integral in eq. (2) is straightforward. The square of the average current is given by

$$\left|\int_0^L dx\, J_{ab}(x)\right|^2 = \sum_{jj'=0}^{N} e^{i(\Delta\phi_j^{ab}-\Delta\phi_{j'}^{ab})} \int_{x_j}^{x_{j+1}} dx\, J_{ab}(x) \int_{x_{j'}}^{x_{j'+1}} dx'\, J^*_{ab}(x') \ . \tag{29}$$

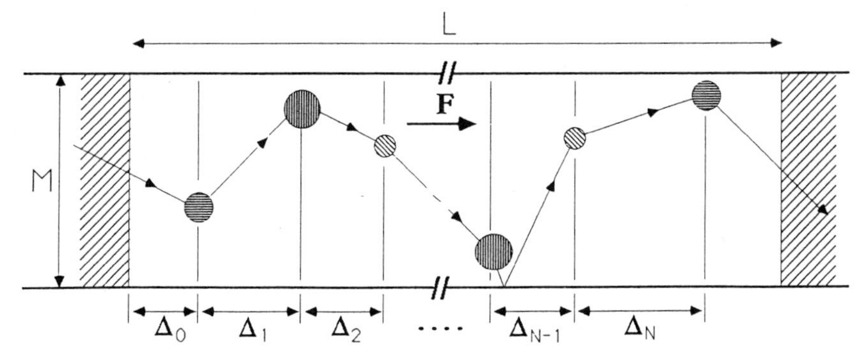

Fig. 10. Model of phase-breaking scattering processes. M: system width; L: system length; Δj: distance between two scattering events parallel to the electric field **F**.

Averaging with respect to the phase differences yields

$$\Gamma(\omega) = \frac{e^2}{h} \sum_{j=0}^{N} \left[\frac{\sin \omega\Delta_j/2v_F}{\omega L / 2v_F}\right]^2 . \tag{30}$$

Finally, we average over the number of the scatterings, and the positions where they take place. Averaging over the distribution of Δ, P(Δ) results in the conductance being closely related to the Fourier transform of P(Δ),

$$\frac{d}{d\omega}\left(\omega^2 \Gamma(\omega)\right) = 2\frac{v_F}{L}\int_0^\infty d\Delta\, P(\Delta) \sin(\omega\Delta / v_F) . \tag{31}$$

P(Δ) may be evaluated for independent scattering events. The probability that N scatterings take place within the "sample", L, is then given by the Poisson distribution

$$p_N(L) = \frac{1}{N!}\left(\frac{L}{L_\varphi}\right)^N e^{-L/L_\varphi} \tag{32}$$

The distribution function $p_N(\Delta, L)$ of the distances Δ on the condition that N scattering events take place within L is then given by

$$p_N(\Delta,L) = N\frac{(L-\Delta)^{N-1}}{L^N}\left(\theta(\Delta)-\theta(\Delta-L)\right) \tag{33}$$

for $N \geq 1$. θ(x) is the step function. The total distribution function P(Δ,L) is obtained as a weighted sum of all $p_N(\Delta, L)$.

$$P(\Delta,L) = \frac{1}{L_\varphi} e^{-\Delta/L_\varphi}\left\{\theta(\Delta)-\theta(\Delta-L)\right\} + e^{-L/L_\varphi}\,\delta(\Delta-L) . \tag{34}$$

Correspondingly, when eq. (31) is used, the conductance is

$$\Gamma(\omega) = \frac{e^2}{h} \frac{L_\varphi}{L} \left(\frac{v_F}{\omega L_\varphi}\right)^2 \left\{ \ln\left[1 + \frac{b^2}{a^2}\right] + 2\,\mathrm{Re}\left[E_1(a+ib) - E_1(a+i0)\right]\right\}$$

$$+ e^{-L/L_\varphi} \left(\frac{\sin \omega L/2v_F}{\omega L/2v_F}\right)^2 \tag{35}$$

with $a = L/L_\varphi$ and $b = L\omega/v_F$, $E_1(z)$ is the exponential integral (Abramowitz and Stegun, 1965).

The result of eq. (35) is plotted in Fig. 11 for several ratios L_φ/L. The purely ballistic limit is recovered for $L_\varphi \gg L$. On the other hand, it does not contain the Drude limit since the scatterings are assumed to be completely independent.

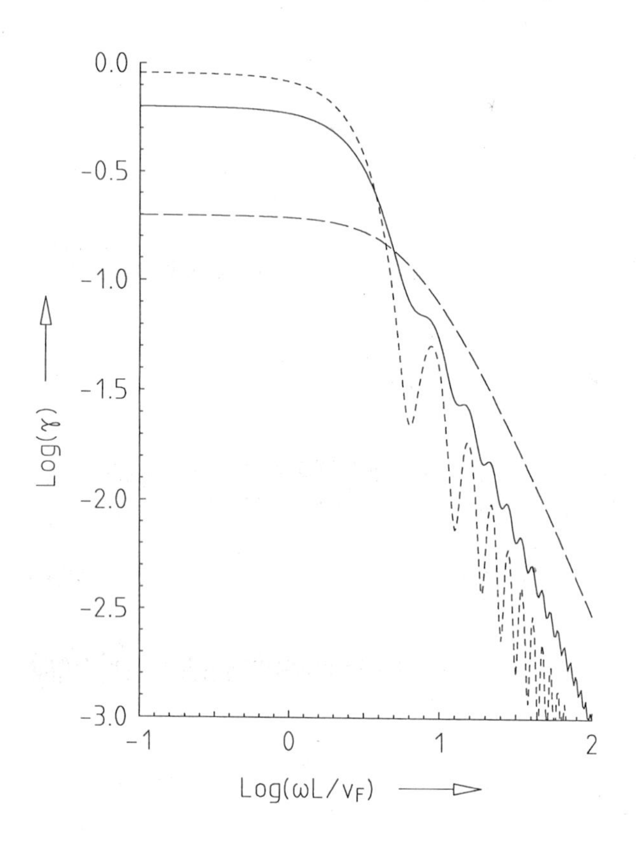

Fig. 11. Frequency-dependent dimensionless conductance γ in atomic units of e^2/h as a function of $\omega L/v_F$ in the case of independent scattering events for $L_\varphi = L/5$ (dashed line), $L_\varphi = L$ (full line) and $L_\varphi = 5L$ (dotted line)

CONCLUSIONS

We have presented a theory of the quantum transport in thin wires based on linear response. A number of phenomena have been discussed which are caused by the coherence of the quantum mechanical motion of the particles. The simplest of these is the quantisation of the dc conductance in units of e^2/h in ballistic systems. This phenomenon has been observed in 2d point contacts. We have shown that the ac conductance of ballistic systems should exhibit quantum oscillations which are a direct consequence of the coherent motion of the electron and the hole. This phenomenon is very similar to the interference of optical waves at slits, and has not yet been experimentally observed.

We have demonstrated that the ac conductance contains information about the spatial distribution of the electric field. This opens up possibilities of measuring the spatial dependence of the field without introducing additional voltage probes that are known to change the transport behavior of coherent systems.

The ac conductance can also be used to obtain information about the phase coherence length. We have seen that the frequency dependence measures the Fourier-transformed distribution function of L_φ. If the phase-breaking scattering events are statistically independent, the ac conductance may be evaluated in a closed form. The quantum interference oscillations are destroyed with increasing frequency of phase-breaking events. However, in order to obtain the classical Drude dependence, a certain degree of correlation between the phase-breaking events must be introduced (Kramer and Masek, 1989).

The oscillations in $\Gamma(\omega)$ should be experimentally accessible. Since the positions of the minima are given by $n\pi = L\omega/2v_\mu$, $n = 1, 2, 3, \ldots$, $v_\mu \sim \sqrt{E_F - E_\mu}$, one had only to match the difference between the Fermi energy, i.e. the electron density, and the position of the sub-band edge E_μ to a value which ensures that the position of the first minimum ($n = 1$) is within the GHz regime. This can be achieved in principle by adjusting the split gate voltage and/or the magnetic field to the onset of a plateau.

The second phenomenon which we have predicted concerns the sharp (anti) resonant structure in the dc conductance in the presence of disorder. It should be possible to observe this effect, which is due to electron localisation, by slightly distorting a GaAlAs sample with impurities.

Besides giving information about the internal properties of the system, the experimental verification of these new interference phenomena would also be very important for confirmation of the underlying linear transport theory.

REFERENCES

Abramowitz, M., Stegun, I.A., 965, Handbook of Mathematical Functions (Dover Publications, New York)

Albers, R.C., Gubernatis, J.E., 1978, Phys. Rev. B, **17**, 4487

Altshuler, B.L., Kravtsov, V.E., Lerner, Z.V., 1986, Sov. Phys. - JETP Letters **43**, 441; Zh. Eksp. Teor. Fiz. 91, 2276

Altshuler, B.L., Kravtsov, V.E., Lerner, I.V., 1987, Zh. Eksp. Teor. Fiz. **94,** 258

Altshuler, B.L., Kravtsov, V.E., Lerner, I.V., 1988, Springer Proc. Phys. **28**, 300

Altshuler, B.L., Aronov, A., Spivak, B.Z., 1981, Sov. Phys. JETP Letters **33**, 94

Altshuler, B.L., Aronov, A., 1985, in: "Electron-Electron Interactions in Disordered Systems", A.L. Efros, M. Pollak, eds., Elsevier, New York

Anderson, P.W., Abrahams, E., Ramakrishnan, T.V., 1979, Phys. Rev. Letters **43,** 718

Fisher, D.S., Lee, P.A., 1981, Phys. Rev. B **23**, 6851

Fowler, A.B., Hartstein, A., Webb, R.A., 1982, Phys. Rev. Letters **48,** 196

Imry, Y., 1986, Physics of Mesoscopic Systems, in: "Directions in Condensed Matter Physics", G. Grinstein and G. Mazenko, eds., 101, World Scientific, Singapore

Isawa, Y., 1988, J. Phys. Soc. Japan **57**, 3457

Johnston, R., Schweitzer, L., 1988, J. Phys. C **21**, L861

Kawabata, A., preprint, to be published

Kramer, B., Mašek, J., 1988, J. Phys. C **21**, L1147

Kramer, B., Mašek, J., 1989, to be published

Lee, P.A., Ramakrishnan, T.V., 1985, Rev. Mod. Phys. **57**, 287

Lee, P.A., Stone, D., Fukuyama, H., 1987, Phys. Rev. B **35**, 1039

Luttinger, J.M., 1951, Phys. Rev. B **4**, 814

MacKinnon, A., 1985, Z. Phys. B **59,** 385

Mašek, J., Kramer, B., 1988, Sol. St. Commun. **68**, 611

Mašek, J., Kramer, B., 1989, Z. Phys. B**75**, 37
Mašek, J., Lipavsky, P., Kramer, B., 1989, J. Phys. C, in press
Saso, T., 1984, J. Phys. C **17**, 2905
Schmid, A., 1985, Springer Ser. Sol. St. Sci. **61**, 212
Sharvin, Y.V., 1965, Zh. Eksp. Teor. Fiz. **48**, 984
Sharvin, D.Yu., Sharvin, V.Yu., 1981, Sov. Phys. JETP Letters **34**, 272
Szafer, A., Stone, A.D., 1989, Phys. Rev. Letters **62**, 300
Velicky, B., Mašek, J., Kramer, B., 1989, to be published
Washburn, S., Webb, R.A., 1986, Adv. Phys. **35**, 375
Wees, van, B.J., Houten, van, H., Beenakker, C.W.J., Williamson, J.G., Kouwenhoven, L.P., Marel, van der, D., and Foxon, C.T., 1988, Phys. Rev. Letters **60**, 848
Wharam, D.A., Thornton, T.J., Newbury, R., Pepper, M., Ahmed, H., Frost, J.E.F., Hasko, D.G., Peacock , D.C., Ritchie, D.A., and Jones, G.A.C., 1988, J. Phys. C **21**, L209

TRANSPORT THROUGH ZERO - DIMENSIONAL STATES IN A QUANTUM DOT

L.P. Kouwenhoven, B.J. van Wees, and C.J.P.M. Harmans

Faculty of Applied Physics, Delft University of Technology
P.O.Box 5046, 2600 GA Delft, The Netherlands

J.G. Williamson

Philips Research Laboratories, 5600 JA Eindhoven, The Netherlands

1. INTRODUCTION

The importance of magnetic edge channels for transport in the quantum Hall regime is no longer a question of doubt. This simplifies considerably the basic description of the transport properties of a two dimensional electron gas (2DEG) in a high magnetic field.[1] The right- and left-moving electrons, which carry the net current, are located at opposite boundaries of the sample and travel in truly one-dimensional edge channels. Under usual circumstances all occupied edge channels contribute equally to the conductance, which results in the quantization of the Hall resistance. However, it was recently shown that edge channels can be selectively populated or detected, when quantum point contacts (QPC's) are used as current or voltage probes respectively.[2,3] An anomalous integer quantum Hall effect was observed, determined by the number of transmitted edge channels through the QPC probes, instead of the number of occupied edge channels in the bulk 2DEG. It was also concluded that the one dimensional transport through magnetic edge channels can occur adiabatically (i.e. with conservation of quantum-channel-number and velocity direction) even on macroscopic length scales much larger than the (zero magnetic field) transport mean free path.[3,4]
We have used these properties of edge channels in combination with the selective transmission properties of QPC's for the construction of a one dimensional interferometer.[5] For this the edge channels are led through a quantum dot via two QPC's. Closed loops of edge channels are formed in the dot, resulting in zero dimensional states due to the finite circumference of the dot. The zero dimensional states show up as pronounced oscillations in the conductance as the magnetic field is varied, with maxima occurring whenever the energy of a zero dimensional state coincides with the Fermi energy. The experimental results are in good agreement with theory and confirm the edge channel description of confined electron transport in a quantizing magnetic field.

2. DEVICE DESCRIPTION

Figure 1a shows schematically the Hall-bar geometry of our device, which is defined in the 2DEG of a high mobility GaAs-$Al_{0.33}Ga_{0.67}As$ heterostructure. The electron density of the 2DEG is 2.3 10^{15} m^{-2} and the transport mean free path is 9 μm. On top of the heterostructure two pairs (A and B) of metallic

Science and Engineering of One- and Zero-Dimensional Semiconductors
Edited by S.P. Beaumont and C.M. Sotomajor Torres
Plenum Press, New York, 1990

gates are fabricated by combined optical and electron-beam lithography and standard lift-off techniques. Application of a negative voltage of -0.2 V on the gates depletes the electron gas underneath them. This creates a disc in the 2DEG with a diameter of 1.5 μm. The narrow channel between pairs A and B is already pinched-off at this gate voltage. Two 300 nm wide QPC's form the connection between the wide 2DEG regions and the disc. A further reduction of the gate voltage V_g to a larger negative value, increases the depletion region around the gates. At the QPC's the depletion regions overlap and form a saddle shaped potential barrier. The barrier height increases with lower gate voltages until the QPC's are pinched-off at -1 V. The separate operation of gate pair A or B also enables to measure the (quantized) transport through the individual QPC's, and compare them to the conductance of the complete device. A comparison of the mean free path (9 μm) to the device dimensions (1.5 μm) indicates that the transport through the quantum dot will be ballistic. The measurements are performed at 20 mK with current biasing (0.5 nA) between contacts I_1 and I_2. The direction of the magnetic field is such that voltage contact V_1 is in electrochemical equilibium with I_1, and correspondingly V_2 with I_2. In this configuration a so called two-terminal conductance is measured, defined as the ratio between the net current through the device and the difference in electrochemical potential between current source and drain.

3. TRANSPORT IN A QUANTIZING MAGNETIC FIELD

The relevant electron states for transport are only those at the Fermi energy E_F. In a high quantizing magnetic field their motion is determined by the guiding energy: [6]

$$E_G = E_F - (n-1/2)\hbar\omega_c \pm g\mu_B B \tag{1}$$

Equation 1 implies that the current carrying electrons follow (in absence of scattering) equipotential lines V(x,y) determined by the condition $E_G = -eV(x,y)$. In the bulk of the 2DEG the electrostatic potential V(x,y) will be nominally flat and it will rise at the boundary of the sample. From $E_G = -eV(x,y)$ it follows that the current carrying electrons with different Landau level index n follow different equipotential lines, but are all located at the sample boundary.[7] Due to their spatial location, the Landau level intersections with the Fermi energy are often called magnetic edge channels. It can be shown that the transport through these edge channels is one dimensional, including the equal contribution of e^2/h per spin-split edge channel to the conductance.[8]

Using their controllable barrier height E_B, QPC's can be used as selective edge channel transmitters.[2-4] Those edge channels for which $E_G < E_B$ will be reflected by a QPC and those with $E_G > E_B$ can pass

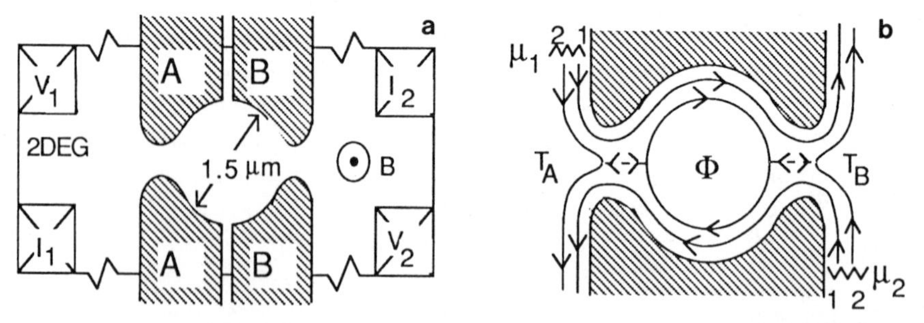

Figure 1. (a) Schematic sample lay-out, showing current (I_1,I_2) and voltage contacts (V_1,V_2) and the two pairs A and B of metallic gates, which define the quantum dot in the 2DEG of a Hall-bar. (b) Location of edge channels along the device boundary, illustrating the partial transmission of the second edge channel. 0D-states are formed in the closed edge channel loop in the dot. (from ref.5)

through the QPC. If N denotes the number of fully transmitted channels and T the partial transmitted upper edge channel, the two-terminal conductance G_{2t} of a single QPC is given by:

$$G_{2t} = \frac{e^2}{h}(N + T) \qquad (2)$$

With $E_G = E_G\,(B)$ and $E_B = E_B\,(V_g)$, it follows that the number of transmitted channels can be changed by varying the magnetic field or the gate voltage. Conductance quantization occurs in those intervals for B and V_g where $T = 0$.

4. EDGE CHANNELS IN A QUANTUM DOT

The fact that edge channels follow equipotential lines along the sample boundary, makes it possible to lead the edge channels along all kinds of geometrical configurations. Recently, it was shown that at sufficiently high magnetic fields this can be done without the appearance of inter channel transitions.[2-4] In figure 1b we have schematically shown the locations of the lowest two occupied edge channels for the quantum dot geometry. The number of occupied edge channels in the bulk 2DEG may be larger than the two considered in figure 1b. The rest of the bulk edge channels are fully reflected by the QPC's and thus do not contribute to the conductance. The edge channels are fed by reservoirs at electrochemical potentials μ_1 at the left and μ_2 at the right of the confined region. In figure 1b the lowest edge channel is fully transmitted through the device and contributes the quantized amount e^2/h to the conductance. The upper channel is only partially transmitted by a fraction T_A through QPC A and T_B through QPC B. Inside the dot this edge channel forms a closed loop. The finite circumference of the dot constitutes a second confinement to the one dimensional (1D) edge channel loop. A second quantum number m (besides the Landau level index n) arise from the rotational symmetry in the dot. Electrons within a single edge channel, but with different quantum number m now follow different equipotential lines. With r_m the radius of the m-th state, the difference in location between two consecutive 0D-states is expressed by the Bohr-Sommerfeld quantization rule: [8,9]

$$\pi B\,(r^2_{m+1} - r^2_m) = \phi_o \qquad \text{with: } \phi_o = h/e \text{ the flux quantum} \qquad (3)$$

Equation 3 implies that two consecutive states enclose a difference of ϕ_o in magnetic flux. While n indexes the 1D-edge channels, the second quantum number m denotes the zero dimensional (0D) states formed within an edge channel. The 0D-states can be viewed as constructive interference of electron waves circulating along the circumference of the quantum dot. Constructive interference occurs whenever the enclosed flux equals an integer number of flux quanta.

5. TRANSPORT THROUGH ZERO DIMENSIONAL STATES

The controllable coupling to the quantum dot provided by the QPC's, enables to make a connection to the 0D-states for transport measurements. The fully transmitted edge channels contribute e^2/h to the conductance G_D of the dot. Including the partial transmission T_D of the upper edge channel, G_D is given by:

$$G_D = \frac{e^2}{h}(N + T_D) \qquad (4)$$

T_D will depend on the transmissions T_A and T_B of the individual QPC's as well as the interference effect inbetween the two barriers. T_D can be easily calculated [9] by considering an incoming wave Ψ_{in} from the left of the dot. The right- and left- moving waves Ψ_R and Ψ_L in the dot are mutually connected by: $\Psi_R = \sqrt{T_A}\cdot\Psi_{in} + \sqrt{(1-T_A)}\cdot\Psi_L$ and $\Psi_L = \sqrt{(1-T_B)}\cdot\Psi_R\cdot\exp(i\upsilon)$, when both are evaluated at QPC A. υ is the phase acquired by a wave in one revolution around the dot. With $\Psi_{out} = \sqrt{T_B}\cdot\Psi_R$ for the outgoing wave at the right side of the dot it follows for $T_D = |\Psi_{out}|^2 / |\Psi_{in}|^2$:

$$T_D = \frac{T_A\, T_B}{1 - 2\sqrt{(1-T_A)(1-T_B)}\cos\upsilon + (1-T_A)(1-T_B)} \tag{5}$$

Related to the enclosed Aharonov-Bohm flux $\phi = B{\cdot}A$, with A the enclosed area, υ is given by $\upsilon = 2\pi BA/\phi_o$. It can be seen from equation 5 that constructive interference occurs (when T_D is maximal), whenever the enclosed flux equals an integer number of flux quanta. If the enclosed flux is varied, T_D and thus the conductance G_D will oscillate, with the amplitude being determined by the individual transmissions T_A and T_B. For nearly zero transmissions T_A and T_B (i.e. for very weak coupling), discrete 0D-states are formed in the dot. They show up as very narrow (Lorentzian) peaks in the conductance whenever their energy coincides with the Fermi energy. Equation 5 further shows that if one of the transmissions T_A or T_B is equal to zero, there is no longer a coupling for transport to the 0D-states resulting in a quantized conductance G_D. Note that equation 5 is exactly the formula for a 1-dimensional interferometer.[10] While in our case the phase is determined by the Aharonov-Bohm flux, equation 5 also holds for a cavity inbetween two barriers, where the cavity length times the longitudinal wave vector determines the phase υ.

6. RESULTS

The measurements we present in this paper are all performed at a fixed gate voltage V_g = -0.35 V and a varying magnetic field B. In figure 2 the two-terminal magneto conductance G_D of the dot is plotted. G_D shows quantized (spin-split) plateaus above 1T at integer multiples of e^2/h. Note that despite the two-terminal measurement no Shubnikov-de Haas resistance oscillations originating from the wide regions of 2DEG are seen, superposed on the quantized plateaus. This is due to the non-equilibrium population between the transmitted and reflected edge channels by the QPC's, which is studied in detail in ref.4. The plateaus indicate one dimensional transport ($T_D = 0$) through the dot, whereby edge channels are either fully transmitted or completely reflected by the QPC's. At the transitions between the plateaus ($T_D \neq 0$) we have the situation of figure 1b, where transport through 0D-states is expected.

To show this we have measured, on an expanded scale, the transition from the second to the third plateau, where the lowest two channels are fully transmitted (N=2) and the third channel is partially transmitted ($T_D \neq 0$). First we have plotted in figure 3a and 3b the conductances G_A and G_B of the individual QPC's to enable a comparison with the conductance of the complete device. The conductance

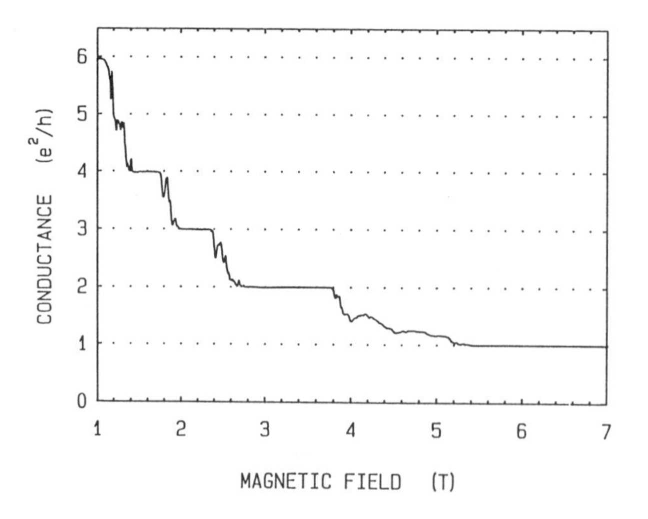

Figure 2. Conductance G_D of the quantum dot as a function of magnetic field. The quantized (spin-split) plateaus indicate 1D-transport, while at the transitions transport through 0D-states is expected.

G_A and G_B are obtained with zero voltage on gate pair B and A respectively. The increasing magnetic field gradually reduces the transmissions T_A and T_B of the third edge channel from 1 to 0 (see figure 3a and 3b). The irregular structure can be attributed to random interferences within the QPC's itself. Note that van Loosdrecht et al.[11] have reported regular Aharonov-Bohm oscillations in the conductance of a single QPC. In figure 3c the conductance of the complete device is plotted, measured with both gate pairs in operation. Superposed on top of the gradual transition, large oscillations are seen in the region where G_A and G_B are not quantized. The amplitude modulation of the oscillations is up to 40% of the conductance contribution e^2/h of a single channel. The fact that the oscillations do not exceed $3e^2/h$ nor drop below $2e^2/h$, illustrates that the conductance modulation originates from the third edge channel only. According to equation 5 the envelope of the oscillations is determined by the transmissions T_A and T_B. We will show below that the measured envelope is in good agreement with equation 5, when energy averaging of the finite temperature is taken into account.

Figure 4a shows the oscillations on an even more expanded scale illustrating their regularity. The period B_o of the oscillations slowly varies from $B_o = 2.5$ mT at $B = 2.5$ T to $B_o = 2.8$ mT at $B = 2.7$ T. For very weak coupling to the dot, the oscillations appear as narrow discrete peaks, as can be seen in figure 4b. The transmission peaks arise whenever the energy of the 0D-states coincide with the Fermi energy resulting in a resonant transmission.

The 0D-states belonging to other edge channels have also been observed. In figure 4c the oscillations are shown belonging to the second edge channel. Their period ($B_o = 5.3$ mT at $B = 5.1$ T) differs from the period of the oscillations of the third edge channel discussed above. Also oscillations from the fourth ($B_o = 2.1$ mT at $B = 1.85$ T) and fifth edge channel ($B_o = 1.4$ mT at $B = 1.25$ T) have been observed. The origin for the difference in period for different edge channels will be discussed below. The observation of distinct periods for different transitions, indicates that the conductance modulation originates from single edge channels only.

The energy separation between consecutive 0D-states can be estimated from the dependence of the oscillations on energy averaging. The oscillations disappear for temperatures above 200 mK and for voltages across the sample above 40 μV, which both lead to an energy separation of about 40 μeV.

It is interesting to note that regular oscillations have also been observed at fixed magnetic field and varying gate voltage. For $B = 2.5$ T the observed period was 1 mV, if V_g was varied on both gate pairs. When V_g was kept fixed (-0.35 V) on one gate pair and varied on the other, the observed period was 2

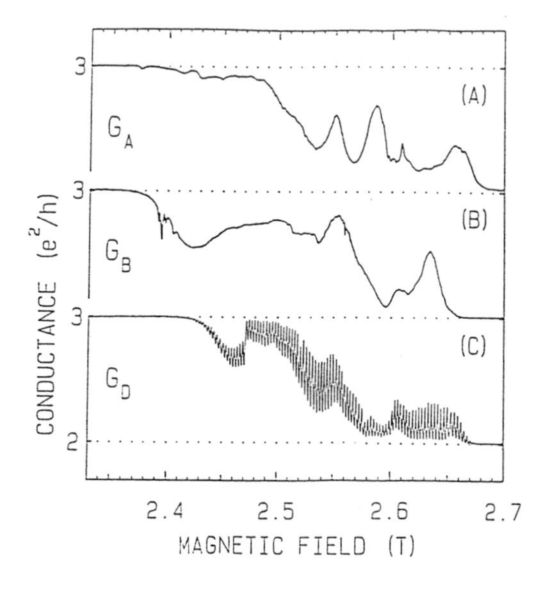

Figure 3. (a) Conductance G_A of QPC A. (b) Conductance G_B of QPC B. (c) Conductance G_D of the quantum dot showing enlarged the transition from $3e^2/h$ to $2e^2/h$. Large oscillations up to 40% of e^2/h are seen in the region where G_A and G_B are not quantized. (from ref.5)

mV. Hence it can be concluded that a change in gate voltage varies the enclosed area by the edge channels. Thus our device also provides an electrostatic control of the resonant transmission through 0D-states.

7. DISCUSSION

Equation 5 shows that the envelope of the oscillations is determined by the coupling to the quantum dot through the transmissions T_A and T_B. Only for very weak and equal coupling ($T_A = T_B \approx 0$) the amplitude modulation can approach 100% of e^2/h. To illustrate this we have calculated the envelope function from the measured conductances G_A and G_B (see figure 3a and 3b) of the individual QPC's. The outer curves in figure 5 are calculated for $\cos\upsilon = 1$ (upper curve) and $\cos\upsilon = -1$ (lower curve) substituted in equation 5. This envelope function would be the amplitude modulation at zero temperature. The conductance G_D calculated from the measured G_A and G_B and taken temperature averaging into account, is also shown in figure 5. A fixed period of 3 mT is chosen for the oscillations. At finite temperature the conductance G_D (T) is given by $G_D(T) = \int G_D(E)\cdot[\partial f(E,T)/\partial E]\cdot dE$ in which f(E,T) is the Fermi distribution function and G_D (E) is the energy dependent conductance at zero temperature. The latter can be obtained from equation 4 and 5 by noting that a change of 2π in the phase υ corresponds to a change in energy of 40 μeV. The calculation is done for an effective temperature of 30 mK, which is the sample temperature (20 mK) plus a contribution from the finite voltage ($\approx 6\ \mu$V) across the sample. Comparison of figure 5 with figure 3c shows a good agreement of the measured with the calculated modulation of the oscillations. Also the shape of the oscillations, which is rounded-off for strong coupling and peaked for weak coupling, appear the same in the measurement as in the calculation.

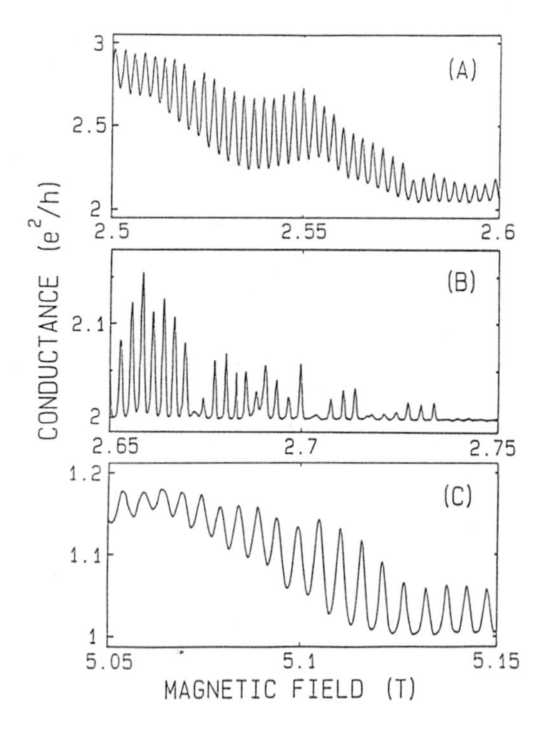

Figure 4. (a) Conductance G_D, enlarged from figure 3c, showing the oscillations (period = 2.5 mT) belonging to the third edge channel. (b) Resonant transmission through 0D-states. In this region the conductance of the third channel is almost zero, except when the energy of a 0D-state coincides with the Fermi energy. (c) Conductance G_D showing the oscillations (period = 5.3 mT) belonging to the second edge channel. (from ref.5)

The conductance oscillations, described in this paper are reminiscent to the Aharonov-Bohm effect observed in small metal[12] or semiconductor rings[13]. However, in these systems the electrons are already confined in the ring in the absence of a field. The conductance of such rings oscillate as a function of B, with period ϕ_o/A (A is the fixed enclosed area by the ring) even if the wires are not one dimensional. In fact this Aharonov-Bohm effect quenches for high magnetic fields when twice the cyclotron radius $l_c = [\hbar \cdot (2n+1)/eB]^{1/2}$ (denoting the wave function width) becomes smaller than the wire width.[13] In our device the current is carried by 1-dimensional edge channels, which are only formed when a sufficiently high magnetic field is applied. A variation of the magnetic field changes the location of these edge channels (equation 1). The change in enclosed flux $\Delta\phi$ resulting from the change in field ΔB can be written as:

$$\Delta\phi = \Delta\,(B\pi r^2) = \pi r^2\,\Delta B + B\,2\pi r\,\Delta r = (\pi r^2 + \frac{B 2\pi r}{eE}\frac{\partial E_G}{\partial B})\,\Delta B \qquad (6)$$

The change in edge channel radius is given by $\Delta r = \Delta E_G/(eE)$, in which E is the radial electric field at the location of the edge channel. Evaluation of equation 6 with r = 750 nm, B = 2.5 T, and rough estimate $E = 3 \cdot 10^4$ V/m [14] shows that the second term (which is negative) can be of the same order as the first one. Therefore the observed period $B_o = \phi_o \Delta B/\Delta\phi$ is not simply related to the enclosed area by the edge channel, but depends on the magnetic field and the shape of the electrostatic potential in which the electrons are confined. From equation 6 it follows that the difference in quantum number n for the different edge channels leads to a different period in magnetic field. The fact that a single, well defined period is observed for each transition, clearly indicates that this oscillation originates from a single edge channel. Note that this latter conclusion supplies strong evidence that in the ballistic regime the net current is completely carried by edge channels.

In summary we have combined the one dimensional transport properties of edge channels with the selective transmission capability of quantum point contacts to study the transport properties of a quantum dot. Zero dimensional states are formed in the dot whenever edge channels are confined in closed loops. The zero dimensional states show up as pronounced oscillations in the conductance when the enclosed flux is varied. The experimental results are in good agreement with theory and confirm the elementary description of confined electron transport in a quantizing magnetic field.

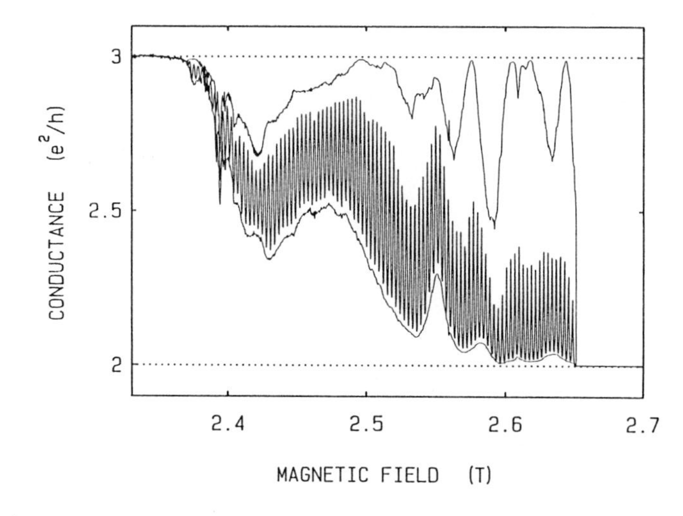

Figure 5. Calculation of the envelope function (outer curves) from the measured conductances G_A and G_B (see figure 3a and 3b) and of the conductance G_D from equation 5 for which an energy averaging is taken into account of 30 mK.

ACKNOWLEDGEMENT

We thank C.W.J. Beenakker, L.W. Molenkamp and A.A.M. Staring for valuable discussions, M.E.I. Broekaart, S. Phelp and C.E. Timmering at the Philips Mask Centre, C.T. Foxon, J.J. Harris and the Delft Centre for Submicron Technology for their contribution in the fabrication of the devices, and the Stichting F.O.M. for financial support.

REFERENCES

1. M. Büttiker, Phys. Rev. B 38, 9375 (1988).
2. B.J. van Wees, E.M.M. Willems, C.J.P.M. Harmans, C.W.J. Beenakker, H. van Houten, J.G. Williamson, C.T. Foxon, and J.J. Harris, Phys. Rev. Lett. 62, 1181 (1989).
3. S. Komiyama, H. Hirai, S. Sasa, and S. Hiyamizu, preprint.
4. B.J. van Wees, E.M.M. Willems, L.P. Kouwenhoven, C.J.P.M. Harmans, J.G. Williamson, C.T. Foxon, and J.J. Harris, Phys. Rev. B 62, xxxx (1989).
5. The main results described in this paper have been submitted to Phys. Rev. Lett: B.J. van Wees, L.P. Kouwenhoven, C.J.P.M. Harmans, J.G. Williamson, C.E. Timmering, M.E.I. Broekaart, C.T. Foxon, and J.J. Harris.
6. *The Quantum Hall Effect*, edited by R.E. Prange and S.M. Girvin (Springer-Verlag, New York, 1987).
7. We ignore scattering potentials and screening effects, which we consider to be irrelevant in our ballistic device. A self-consistent calculation of confining potential and electron occupation is given by: U. Wulf, V. Gudmundsson, and R.R. Gerhardts, Phys. Rev. B 38, 4218 (1988).
8. B.I. Halperin, Phys. Rev. B 25, 2185 (1982).
9. A theoretical study of a 2D dot to which narrow leads are attached is given by:U. Sivan, Y. Imry, and C. Hartzstein, preprint. U. Sivan, and Y. Imry, Phys. Rev. Lett. 61, 1001 (1988).
10. M. Büttiker, preprint.
11. P.H.M. van Loosdrecht, C.W.J. Beenakker, H. van Houten, J.G. Williamson, B.J. van Wees, J.E. Mooij, C.T. Foxon, and J.J. Harris, Phys.Rev. B 38, 10162 (1988).
12. R.A. Webb, S. Washburn, C.P. Umbach, R.B. Laibowitz, Phys. Rev. Lett. 54, 2696 (1985).
13. G. Timp, A.M. Chang, J.E. Cunningham, T.Y. Chang, P.Mankiewich, R. Behringer, and R.E. Howard, Phys.Rev. Lett. 58, 2814 (1987); C.J.B. Ford, T.J. Thronton, R. Newbury, M. Pepper, H. Ahmed, C.T. Foxon, J.J. Harris, and C. Roberts, J. Phys. C, Solid State Phys. L325 (1988).
14. At the 2DEG boundary the electrostatic potential changes by an amount E_F/e (≈9mV) in a depletion region which is about 300 nm wide. This gives a typical field strength $E \approx 3 \cdot 10^4$ V/m.

LATERAL SURFACE SUPERLATTICES AND QUASI-ONE-DIMENSIONAL STRUCTURES IN GaAs

D. A. Antoniadis, K. Ismail, and Henry I. Smith

Massachusetts Institute of Technology
Cambridge, Massachusetts 02139

ABSTRACT

Two types of field-effect transistors on a modulation-doped GaAs/GaAlAs heterostructure will be discussed. The first type has a gate electrode that consists of a 0.2 μm-period Ti/Au grid which presents a tunable, two-dimensional periodic potential modulation (lateral surface-superlattice, LSSL) to electrons traveling from source to drain. The other type consists of 100 parallel GaAs/GaAlAs channels, each about 40 nm-wide where the electron density is controlled by backgate bias or illumination. Conductance measurements at 4.2 K exhibit clearly a superlattice effect (i.e., coherent back-diffraction) and provide evidence of sequential resonant tunneling in the LSSL-type device. Similar measurements show clear evidence of quasi-one-dimensional (Q1D) density of states and corresponding electron mobility modulation in the Q1D type devices.

INTRODUCTION

In 1970 Esaki and Tsu[1] proposed incorporating an artificial superlattice into a semiconductor in pursuit of a novel negative differential resistance (NDR) device. This was experimentally verified by Esaki and Chang[2] in a grown GaAs/AlGaAs superlattice. With the advances in epitaxial growth techniques, such layers have been grown with periodicities down to a few monolayers.

However, despite the precise control of layer thickness, these structures have three major shortcomings: first, although the electron can see a superlattice in one direction, it behaves as a free electron in the two other directions (the plane of the growth). This prevents the formation of true minigaps, as pointed out by Kroemer.[3] Second, the strength of the periodic modulation, or equivalently the size of the minigaps, which depends on the material composition, cannot be controlled after the growth. Third, the mobility in such structures is relatively low, as compared to that in modulation-doped structures.

Those three shortcomings can be partially or completly overcome in a lateral surface superlattice (LSSL), independently proposed by Sakaki et al.[4] and Bate[5], and implemented by Warren et al.[6] in a Si/SiO_2 double

Science and Engineering of One- and Zero-Dimensional Semiconductors
Edited by S.P. Beaumont and C.M. Sotomajor Torres
Plenum Press, New York, 1990

gate field-effect transistor. In such a structure the periodic modulation is induced by a surface field-effect, and thus its strength is tunable. Moreover, if the electrons are confined to two dimensions, as is the case in Si/SiO_2 and modulation-doped GaAs/AlGaAs heterostructures, the degrees of freedom for scattering into free electron states are reduced to one in the case of a grating, and to zero in the case of a grid structure. Furthermore, if the GaAs/AlGaAs modulation-doped material system is used, a very high electron mobility and a long mean-free-path can be obtained.

With the ability to control the strength of the periodic potential at the two-dimensional electron-gas (2DEG) plane by means of the field-effect of a gate it is possible to study the behavior of electronic systems not only in a scattering potential but in a confining potential as well. For example, a grating-gate surface structure can produce a superlattice potential at moderate bias near threshold[7] and multiple parallel quasi one dimensional wires at biases below threshold[8]. Similarly a grid-gate can produce both a 2-D lateral superlattice or zero-dimensional quantum wells.

The confinement of electron transport to two dimensions, as is the case in Si/SiO_2 MOSFETs and GaAs/AlGaAs MODFETs, has been studied extensively and exploited in device applications. Quantum-wires and quantum-boxes can be constructed by further restricting the electron motion in one or both lateral directions, respectively. The confinement can be achieved by etching techniques,[9-13] by selective ion implantation[14-15] by using the field-effect to create Q1D conducting channels[16-18] or by pinching-off a two-dimensional (2D) conducting channel using a split-gate.[19-20] Study of electron transport in quantum wires has already provided evidence of localization[9-11,15,20] and quantum interference effects.[5,6]

Experimental evidence of Q1D electronic subbands in the Si/SiO_2 system has been obtained by studying transport in multiple parallel wires[16] and, more recently, in single wires[17] defined by a dual-stacked-gate configuration. To our knowledge, evidence in the GaAs/AlGaAs system has been provided by experiments with optical excitations,[18,21] with magnetoresistance,[12] with capacitance,[22] and more recently with conductance[8].

A dramatic increase in mobility for electrons confined to the lowest subband in a quantum wire has been discussed in literature (e.g., Sakaki [23]) This is because the probability of elastic scattering depends on the density of accessible final states, which in a one-dimensional conductor decreases with increasing Fermi energy. As will be seen, this work provides the first experimental evidence confirming those predictions.

In what follows, device fabrication and transport measurements at 4.2 K and 77 K are discussed for two types of devices fabricated in MBE grown AlGaAs/GaAs heterostructures: (a) Grid-gate 2-D superlattice MODFETs, and (b) multiple parallel quantum wire (MPQW) channel MODFETs.

DEVICE FABRICATION

Schematic top views and cross sections of the LSSL and the MPQW MODFETs are shown in Figs. 1 and 2, respectively. The GaAs/AlGaAs layers were prepared by molecular beam epitaxy. After growing a thick, undoped buffer layer on top of a semi-insulating substrate, an undoped 7.5 nm-thick AlGaAs spacer and a 42 nm-thick Si-doped (1×10^{18} cm^{-3}) AlGaAs layer were grown. The Al content was 28%. Finally, a 20 nm, doped GaAs cap was grown on top. In such a structure electrons from the doped AlGaAs layer diffuse to, and get confined at, the interface with the undoped GaAs layer due to the bandgap difference between the two materials.

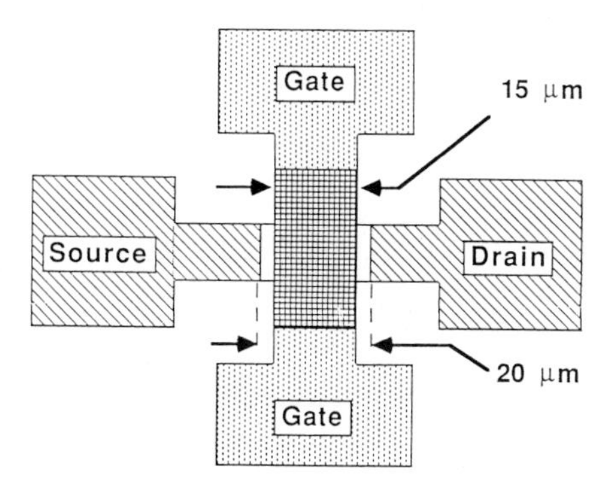

Fig. 1. Schematic top view of a LSSL-MODFET. The gate length is 15 μm, as indicated, and the gate width is 20 μm, defined by the mesa etching.

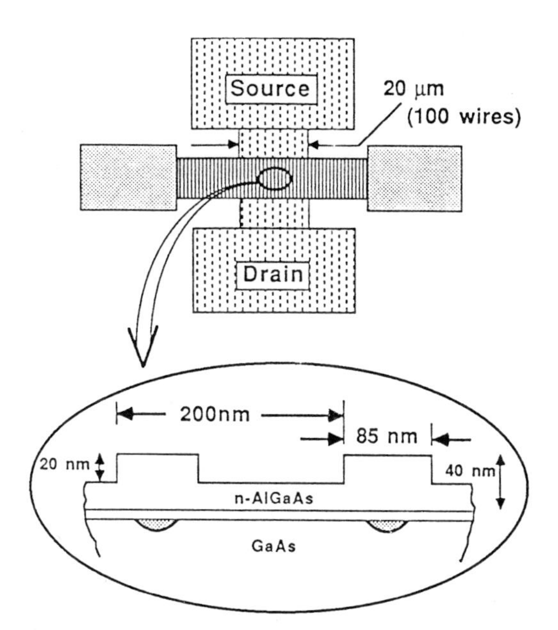

Fig. 2. Schematic top view and magnified cross section of the multiple parallel quantum wires. The gate length is 5 μm and the gate width is 20 μm, defined by the mesa etching. The Ohmic contacts to the left and right are test contacts to examine transport perpendicular to the wires.

Schematic top views and cross sections of the LSSL and the MPQW MODFETs are shown in Figs. 1 and 2, respectively. The GaAs/AlGaAs layers were prepared by molecular beam epitaxy. After growing a thick, undoped buffer layer on top of a semi-insulating substrate, an undoped 7.5 nm-thick AlGaAs spacer and a 42 nm-thick Si-doped (1×10^{18} cm^{-3}) AlGaAs layer were grown. The Al content was 28%. Finally, a 20 nm, doped GaAs cap was grown on top. In such a structure electrons from the doped

AlGaAs layer diffuse to, and get confined at, the interface with the undoped GaAs layer due to the bandgap difference between the two materials.

The process flows for the grid-gate and MPQW devices are very similar. In fact our early grid-gate devices were fabricated simultaneously with the MPQW devices by splitting the grid-gate formation step into two perpendicular grating formations[24]. However, better yields result from direct grid lithography and this is the process that resulted into the devices reported here.

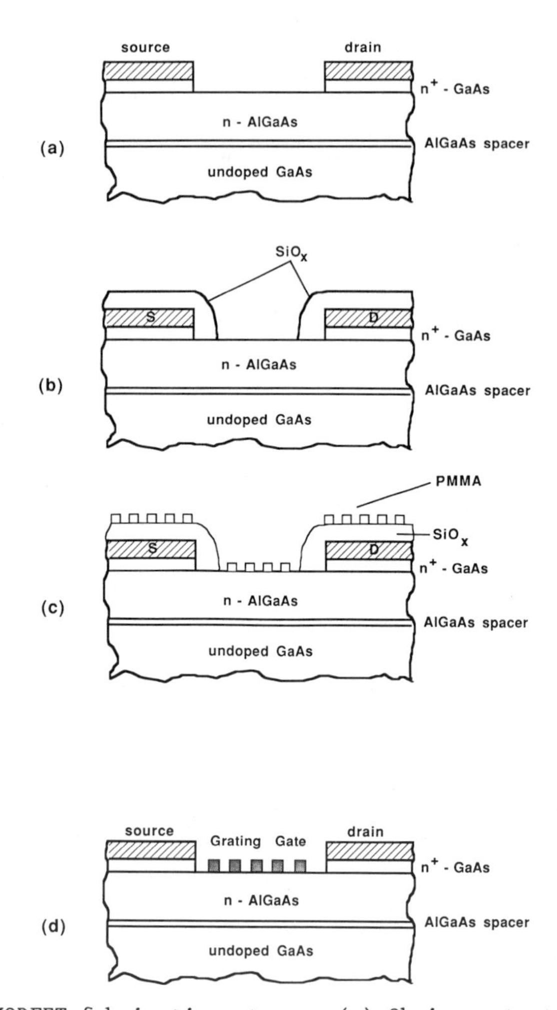

Fig. 3. LSSL MODFET fabrication steps: (a) Ohmic contact evaporation and liftoff, (b) sacrificial oxide layer evaporation and liftoff, (c) combined deep-UV and x-ray exposure of PMMA, (d) Schottky metal evaporation, liftoff, and etching of residual lines and sacrificial oxide. The mesa isolation and contact reinforcement steps are not shown.

The processing steps are summarized in Fig. 3. Starting with an optical lithography step to define the Ohmic contact pads, Ni/Au/Ge/Ni/Au was e-beam evaporated, defined by lift-off and sintered at 450°C for 2 min. Then the GaAs cap layer was etched away everywhere except underneath the source and drain, where it helps the formation of good Ohmic contacts. Another lithography step was then used to expose photoresist everywhere except in the active channel between source and drain. SiO_x

was e-beam evaporated and lifted-off in the channel area. This SiO_x is only a sacrificial layer that is etched away at the end of the process.

Since it is essential in the LSSL device that the fine-period lines of the gate be electrically connected to the coarse gate pads, both were fabricated using a single liftoff step. Polymethylmetacrylate (PMMA) was spun on, at a thickness of 0.25 μm. The gate pads were defined by deep-UV lithography using a Xe-Hg short arc lamp. The intensity was 15 mW/cm^2 centered at a wavelength of 220 nm.

After the deep-UV exposure, and prior to the x-ray lithography step, the sample was baked at 110°C for 15 min to allow the PMMA to reflow slightly. Our experience has shown that without this baking step the PMMA lines (defined by the subsequent x-ray lithography step) near the large deep-UV-exposed area are weakened due to the standing wave effect of the 220 nm Xe-Hg line. X-ray lithography was then used to define a 200-nm-period grid or grating in PMMA for grid-gate LSSL and MPQW devices respectively. Exposure was with the C_k line (λ = 4.5 nm) emitted from a graphite target bombarded by a 5 kV electron beam. The masks consisted of a 1 μm-thick polyimide membrane with a 200-nm-period grid or grating (100 nm nominal linewidth) of 0.1 μm-thick gold absorber. The grating was produced by holographic lithography followed by metal shadowing, oxygen RIE, and liftoff. The grid mask was produced by two x-ray exposures of a grating at 90° rotation. The working masks were held electrostatically in intimate contact with the substrate.

After the x-ray exposure, the sample was developed in a 3:2 mixture of isopropyl alcohol (IPA) and methylisobutylketone (MIBK). The deep-UV and the x-ray exposures develop simultaneously. This was followed by evaporation and liftoff of 20 nm Ti and 15 nm Au.

In the active channel the Ti/Au lines form the Schottky grid-gate of the LSSL devices. In the MPQW device the grating lines extend from source to drain contact and serve as a mask in the subsequent etching of 20 nm of AlGaAs by ion milling with argon at an energy of 500 eV. Subsequent process flow is the same for both device types: using an HF solution, the sacrificial SiO_x was etched and the residual metal lines on it were lifted-off. The devices were then isolated by a deep mesa etch in a 3:1:15 $NH_4OH:H_2O_2:H_2O$ solution. Finally, a 0.3 μm Ni/Au reinforcement layer was evaporated and defined by liftoff on all contact pads.

The LSSL devices had a fixed channel width of 20 μm and three different channel lengths, 5,10 and 20 μm. The corresponding gate lengths were 3,6 and 15 μm, respectively, symmetrically positioned in the channel. The MPQW devices had about 100 parallel, 100-nm-wide lines, and lengths of either 5 or 10 μm. The actual conducting linewidth is much smaller than the delineated linewidth. The method for estimating the electrical wire width is described later on.

ELECTRON TRANSPORT IN GRID-GATE MODFETs

The measurements that are discussed here were carried out in devices fabricated as described above in GaAs/AlGaAs heterostructure samples with Hall mobility of 300,000 cm^2/V.s at 4.2 K, and a sheet carrier concentration of 2 x 10^{11} cm^{-2}, as reported in Ref. 25. The device configuration is shown in Fig. 1. We performed two different types of measurements.

The first was drain-source current I_{DS} as a function of gate bias V_{GS}, for a fixed drain-source bias V_{DS}. Figure 4 shows a family of such curves for three different drain bias conditions. As V_{GS} is swept, the charge concentration is increased, or equivalently, the Fermi energy is raised, passing through the superlattice-induced minibands and minigaps. Whenever the Fermi energy corresponds to the energy range of a minigap, the current should drop to zero, assuming a ballistic motion at absolute zero temperature. In practice, the energy distribution is broadened due to the finite temperature, by about kT (0.36 meV at 4.2 K), and due to elastic scattering by about $h/2\pi\tau_{el}$ (0.05 meV at 4.2 K, where τ_{el} is the elastic scattering time).

In our previous experiments on grating-gate devices[7] the electrons could scatter into free electron states perpendicular to the superlattice axis. However, since the mean-free-path was longer than the grating periodicity, weak structure in the I_{DS} vs. V_{GS} curve, and corresponding transconductance oscillations, could still be observed. In the present experiment with a grid there is a periodic potential in each of the two orthogonal directions, and thus true minigaps are possible. For the current to drop, the size of the minigaps should be larger than the cumulative energy broadening effects, described above. From previous calculations [26], the size of the minigaps was estimated to be around 1.0 meV.

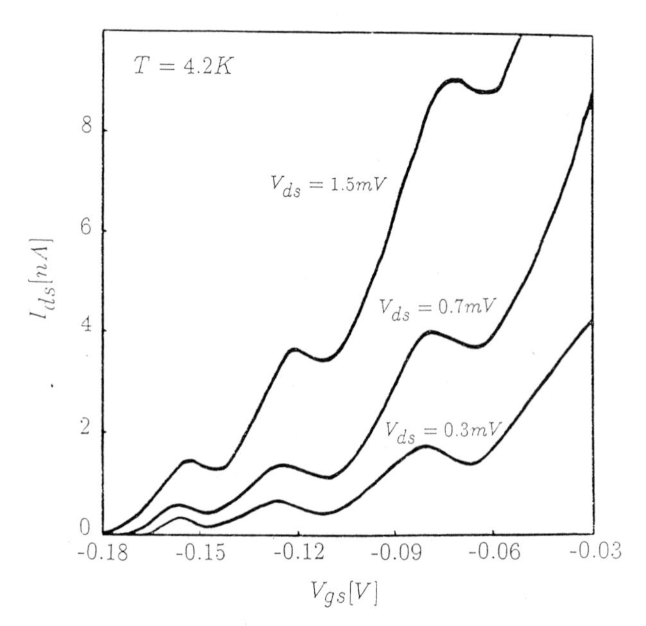

Fig. 4. I_{DS} as a function of V_{GS} for three different V_{DS} values.

In Fig. 4 we can see clear drops in I_{DS} as a function of V_{GS}. The number of experimentally detectable minigaps, i.e., the number of current drops, was consistent with the theoretical predictions, ranging between three and five. The spacing between the current values at which the drops were observed consistently increased, as expected for a parabolic conduction band. It is also interesting to note that the current drops became weaker for higher V_{DS} and vanished completely at around 15 mV. Assuming an inelastic (coherence) length of 1.5 μm, the channel can be divided into 10 coherent regions along the source-drain axis. At V_{DS} = 15 mV, and assuming a uniform field distribution, the voltage drop along any coherent region is about 1.5 mV, slightly larger than the estimated miniband separation.

We believe that this is the first clear observation of minigaps in a grid-gate superlattice. This is significant particularly keeping in mind that the period (0.2 μm) is more than one order of magnitude longer than in grown superlattices. Currently, we are working on combining a semi-

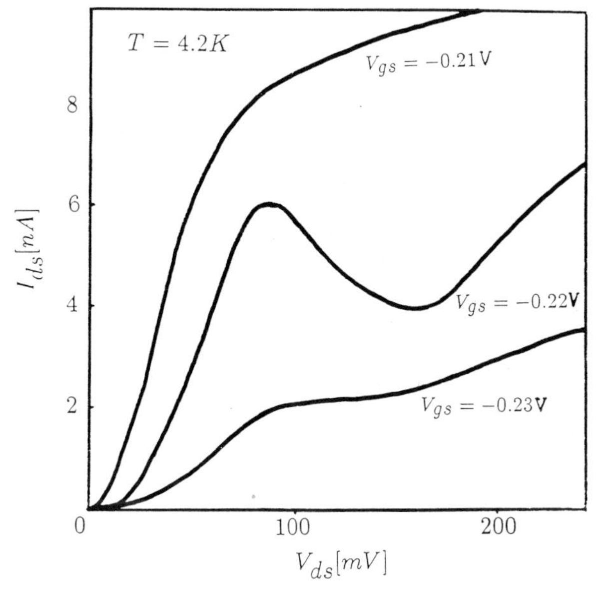

Fig. 5. I_{DS} as a function of V_{DS} for three different V_{GS} values. Only at V_{GS} = -0.22 V is a clear NDR observed.

classical 2-dimensional simulation with a quantum mechanical calculation, to allow us to quantitatvely correlate the experimental results with the theoretical predictions.

In the second type of experiment I_{DS} was measured while sweeping V_{DS}, with V_{GS} fixed to values corresponding to weak conduction. This is very similar to the experiments performed on grown superlattices with thick barriers.[27] Figure 5 shows I_{DS} as a function of V_{DS} at three different V_{DS} values. A negative differential resistance (NDR) is obtained at V_{DS} around 90 mV and V_{GS} = -0.22. Out of a total of 12 grid-gate devices, 10 showed the NDR region, within 20 mV of the same V_{DS}. Two devices showed even a second NDR at around 200 mV. The NDR effect was very sensitive to the gate bias condition. As shown in Fig. 5, at slightly higher or lower gate bias the effect almost vanished.

Figure 6 shows I_{DS} vs. V_{DS} at both 4.2 K and 77 K; at the higher temperature the NDR has become a plateau. In earlier grating-gate devices at 4.2 K, structure was observed only in the differential conductance and not in the current directly.

The cause of the NDR is not understood. However, due to the low V_{DS} at which it is observed, an explanation based on the Gunn effect, or on real space transfer, can be excluded. If the NDR were a result of inter-band scattering, one would expect to see the same effect in continuous-gate devices, which we did not. Also, it would be very difficult to explain the extreme sensitivity of the NDR feature to gate bias. This gate sensitivity leads us to speculate that the NDR may be a manifestation of sequential resonant tunneling across the periodic potential barriers in some part of the channel. A similar effect has been observed

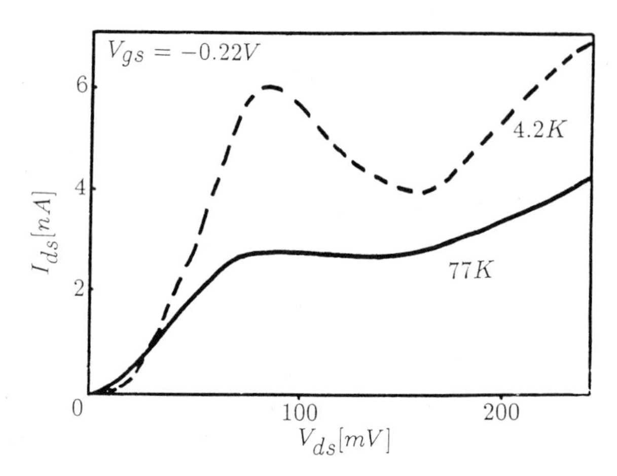

Fig. 6. I_{DS} as a function of V_{DS} at 4.2 K and 77 K for V_{GS} = -0.22 V.

earlier in grown superlattices.[27] If so, this would be the first observation of resonant tunneling in a planar structure, where the barrier height is modulated by an external bias. Due to the relatively wide gate lines (60-100 nm) in our devices, one might expect a vanishingly small tunneling probability through the field-induced barriers. However, as the gate bias is raised the Fermi energy is brought very close to the tops of the barriers. In addition, the barriers in our case are not rectangular, but rather rounded off and narrower close to the top, which was observed in a semiclassical 2-dimensional numerical simulation.[11] Both effects may add to increase the tunneling probability.

MPWQ CHANNEL MODFETs

The MPQW devices were fabricated in modulation doped GaAs/AlGaAs samples similar to the grid-gate MODFETs but with Hall mobility of 250,000 $cm^2/V \cdot s$ at 4.2 K. Two means of modulating the charge concentration (Fermi energy) in the wires were utilized: applying bias to the substrate or changing the light intensity using an LED. To improve the effectiveness of the backgate bias the samples were thinned to about 5 μm. Since the samples did not exhibit persistent photoconductivity, the conductance response to changing the light intensity was instantaneous. The device configuration is shown in Fig. 2. The measurements were performed at a drain-source bias (V_{DS}) lower than 5 mV to avoid heating the electrons. Figure 7 shows the drain-source current (I_{DS}) as a function of substrate bias (V_{SUB}) (solid curve) at V_{DS} = 1 mV. The resulting structure in current was completely reproducible after temperature cycling. The same type of structure was observed when sweeping the LED intensity at low V_{SUB}. In fact, the current plateaus in both cases corresponded within 10% to the same values of current. No measurable substrate current could be detected throughout the range of measurements. Performing the same experiment on a 2D device having the same total width (20 μm) and length (5 μm), the current increased monotonically with either substrate bias or LED intensity. In Fig. 8 we show the derivative of the current with respect to the substrate bias. For the MPQW at 4.2 K, strong oscillations and even a negative value can be clearly seen. Even at 77 K there is still evidence of the same effect, however, much weaker. This we believe is a strong evidence for the presence of Q1D subbands in our MPQW.

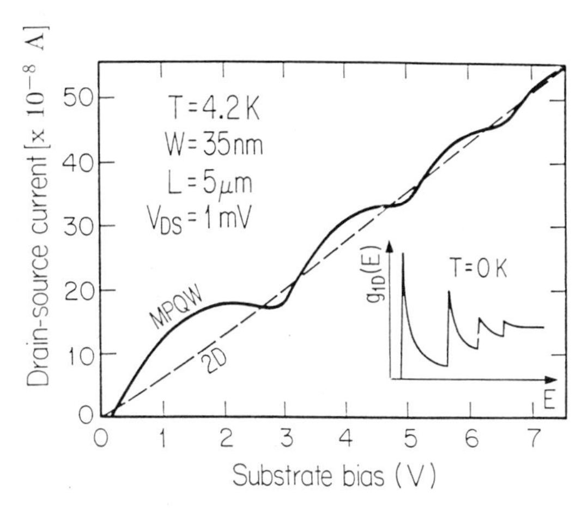

Fig. 7. I_{DS} as a function of substrate bias (V_{SUB} for the multiple parallel quantum wires (solid line) and scaled down 2D channel (dashed line) (V_{DS} = 1 mV, L_{DS} = 5 μm). The inset shows the theoretical density of states in a Q1D wire at 0 K.

To evaluate the results more quantitatively, we had to start by estimating the width of the conducting wires. We based our estimate on the assumption that the conductivity of the MPQW at high values of V_{SUB}, where oscillations are no longer observed, is the same as that of a 2D gas. On this basis, the effective channel area of the MPQW was 5.7 times less than that of the 2D device. In other words, out of the 20 μm total width, about 3.5 μm were conducting. This divided by the number of periods (100) yields an individual wire width of 35 nm at high V_{GS}. The same scaling was repeated at 77 K yielding a scaling factor of 5.6. In Fig. 7 the scaled 2D current is shown in comparison to that of the MPQW at 4.2 K, and in Fig. 8 the scaled transconductance of the 2D device is compared to that of the MPQW at both 4.2 K and 77 K.

The reason behind scaling the 2D current to match the Q1D current at high V_{SUB} is to make sure that the presence of the subbands does not affect the scaling factor. In fact, it was only around V_{SUB} = 6.8 V, that the conductance perpendicular to the wires started being measurable. It was about 500 times smaller than the parallel conductance. This indicates the transition from a Q1D to a 2D behavior.

To support this scaling principle further, we performed a 2D numerical simulation of the structure. To take the presence of surface states into account, the Fermi level at the surface of the AlGaAs layer was assumed to be pinned at 0.8 V below the conduction band. The results indicated a potential well 36 mV deep, with a width ranging between 25 and 40 nm, consistent with our scaling experiments. A rough calculation of the position of the energy levels relative to the bottom of the well gave the following: E_0 = 5 meV, E_1 = 15 meV, E_2 = 25 meV, and E_3 = 33 meV.

Whenever the Fermi energy is raised such that populating the next subband is allowed, intersubband scattering starts taking place, result-

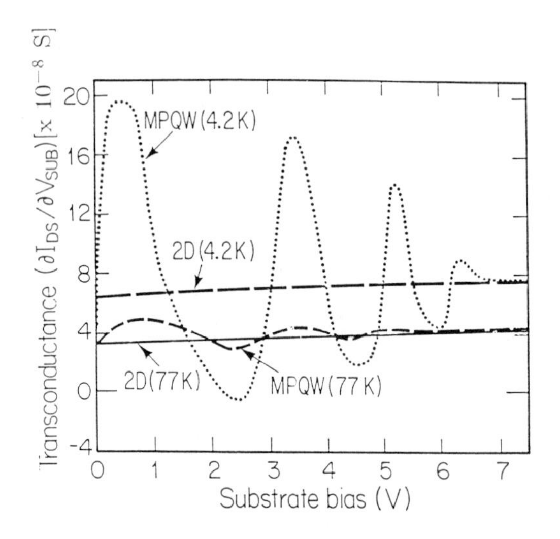

Fig. 8. Tranconductance ($\partial I_{DS}/\partial V_{SUB}$) for the multiple parallel quantum wires at 4.2 K and 77 K, and for the scaled 2D channel at 4.2 K and 77 K, as a function of substrate bias (V_{SUB}).

ing in a drop in the mobility. One thus expects, from the calculated number of bands, to see the current drop, or level-three times (corresponding to the second, third and fourth level), in agreement with the results in Fig. 7. Moreover, since the effect was observed at 77 K, the energy level separation must be larger or at least comparable to kT, which is 6 meV at that temperature. The calculated subband separation (10 meV) is consistent with this observation.

It is interesting to note, since the charge density in both the 2D case and the MPQW is affected the same way by the substrate bias, that the increase in current in the MPQW relative to the 2D case, at some particular bias points, must be related to an increase in the mobility. Similarly, a relative decrease in current at some other points indicates a reduction in the mobility. This is explainable in terms of the difference between the density of states in 1D and 2D. In Q1D wires the density of states (shown schematically in the inset of Fig. 6 at 0 K) reflects the probability of scattering. The higher the Fermi energy is raised in a particular band the less the number of available empty states, and hence the less the probability of elastic scattering. Once the Fermi energy matches the next subband there is a high probability of energy conserving interband scattering, due to the availability of empty states in the new subband, which can cause negative differential mobility. It is thus not surprising that the transconductance in Fig. 8, for both the MPQW and the 2D cases, reflects the true shape of the corresponding density of states, after adding thermal broadening.[17] A maximum increase in mobility by a factor of 2.9, with respect to the 2D mobility, is observed in the first subband. On the other hand, a negative differential mobility is observed at the onset of populating the second subband. This result is the first experimental confirmation of the previous theoretical predictions.[23]

ACKNOWLEDGEMENTS

This work was sponsored by the Air Force Office of Scientific Research under contracts AFOSR-85-0376C and AFOSR-88-0304. The Joint Services Electronics Program, under contract DAAL-03-89-C-0001, provided partial support of the Submicron Structures Laboratory, where the devices were fabricated. The authors would like to thank P. Bagwell for providing his theoretical calculations, and T. Orlando and D. Tsui for interesting discussions. They also acknowledge the help of W. Chu, A. Yen, J. Carter and T. McClure at various stages of device fabriction, and C. Fonstad and the MBE group at MIT for their help with sample growth.

REFERENCES

1. L. Esaki and R. Tsu, *IBM J. Res. Dev.* **14**, 61 (1970).

2. L. Esaki and L.L. Chang, *Phys. Rev. Lett.* **33**, 495 (1974).

3. H. Kroemer, *Phys. Rev.* **33**, 495 (1974).

4. H. Sakaki, K. Wagatsuma, J. Hamasaki and S. Saito, *Thin Solid Films* **36**, 497 (1976).

5. R.T. Bate, *Bull. Am. Phys. Soc.* **22**, 407 (1977).

6. A.C. Warren, D.A. Antoniadis, H.I. Smith and J. Melngailis, *IEEE Elect. Dev. Lett.* **6**, 294 (1985).

7. K. Ismail, W. Chu, D.A. Antoniadis and H.I. Smith *Appl. Phys. Lett.* **52**, 1072 (1988).

8. K. Ismail, D.A. Antoniadis and H.I. Smith, *Appl. Phys. Lett.*, **54**, 1130 (1989).

9. W.J. Skocpol, L.D. Jackel, E.L. Hu, R.E. Howard and L.A. Fetter, *Phys.Rev. Lett.* **49**, 951 (1982).

10. R.G. Wheeler, K.K. Choi, A. Goel, R. Wisnieff, and D.E. Prober, *Phys.Rev. Lett.* **49**, 1674 (1982).

11. H. van Houten, B.J. van Wees, M.G.J. Heijiman, and J.P. Andre *Appl. Phys. Lett.* **49**, 1781 (1986).

12. T. Demel, D. Heitmann, P. Grambow, and K. Ploog, *Phys. Rev. B*, **38**, 12732 (1988).

13. G. Timp, A.M. Chang, J.E. Cunningham, T.Y. Chang Mankiewich, R. Behringer and R.E. Howard, *Phys. Rev. Lett.* **58**, 2814 (1987).

14. C.P. Umbach, S. Washburn, R.B. Laibowitz, and R.A. Webb, *Phys. Rev. B* **30**, 4048 (1984).

15. T. Hiramoto, K. Hirakawa, and T. Ikoma, *J. Vac. Sci. and Tech. B* **6**, 1014 (1988).

16. A.C. Warren, D.A. Antoniadis, and H.I. Smith, *Phys. Rev. Lett.* **56**, 1858 (1986).

17. J.H.F. Scott, M.A. Kastner, D.A. Antoniadis, H.I. Smith, and S. Field, *J. Vac. Sci. and Tech.*, B6(6), 1841 (1988).

18. W. Hansen, M. Horst, J.P. Kotthaus, U. Merkt, Ch. Sikorski, and K. Ploog, *Phys. Rev. Lett.* **58**, 2586 (1987).

19. T.J. Thornton, M. Pepper, H. Ahmed, D. Andrews, and G.J. Davies, *Phys. Rev. Lett.* **56**, 1198 (1986).

20. H.Z. Zheng, H.P. Wei, and D.C. Tsui, *Phys. Rev. B* **34**, 5635 (1986).

21. J. Cibert, P.M. Petroff, G.J. Dolan, D.J. Werder, S.J. Pearton, A.C. Gossard and H.H. English, *Superlattices and Microstructures* **3**, 35 (1987).

22. T.P. Smith, H. Arnot, J.M. Hong, C.M. Knoedler, S.E. Laux, and H. Schmid, *Phys. Rev. Lett.* **59**, 2802 (1987).

23. H. Sakaki, Jpn. *J. Appl Phys.* **19**, L735 (1980).

24. K. Ismail, W. Chu, D.A. Antoniadis, and H.I. Smith, *J. Vac. Sci. Technol.* **B6(6)**, 1824 (1988).

25. K. Ismail, W. Chu, A. Yen, D.A. Antoniadis and H.I. Smith, *Appl. Phys. Letts.*, 54, 460 (1989).

26. P. Bagwell, Masters Thesis, MIT, (1987).

27. F. Capasso, K. Mohammed, and A.Y. Cho, *Appl. Phys. Lett.*, 48, 478 (1986).

MAGNETOTRANSPORT AND INFRARED RESONANCES IN LATERALLY PERIODIC NANOSTRUCTURES

Jörg P. Kotthaus and Ulrich Merkt

Institut für Angewandte Physik, Universität Hamburg
Jungiusstrasse 11, D-2000 Hamburg 36, F.R. Germany

1. INTRODUCTION

Quantum confinement of electrons at a semiconductor interface results in quasi-two-dimensional (2D) electron systems in which the electronic motion is unbound within the interface plane but quantized perpendicularly to it. The unique electronic properties of such 2D electron systems have been widely investigated for more than two decades. Refined lithographic technologies now make it possible to further restrict the electronic motion at semiconductor interfaces in the lateral directions to dimensions in the range around 100 nm which are becoming comparable to the Fermi wavelength of electrons in 2D electron systems in semiconductor heterostructures.[1] At low temperatures and for high quality device structures such as GaAs-AlGaAs heterojunctions other relevant electronic length scales such as the elastic mean free path and the phase coherence length can easily exceed 1 μm and thus can be much larger than lateral confinement dimensions W. Under these conditions lateral quantization phenomena have recently become observable in a variety of semiconductor nanostructures using various confinement schemes.

Here we want to summarize recent investigations carried out in our laboratory on laterally periodic field effect devices with periodicities in the range between 200 nm and 500 nm.[2-8] In these devices electrostatic confinement via field effect is used to create either 2D systems with strong periodic modulation of the electron density, or arrays of 1D electron inversion channels, or even arrays of effectively 0D quantum dots. Using holographic lithography such structures have been fabricated on GaAs, InSb, and Si homogeneously on comparatively large areas of typically 10 mm^2 and thus can be used for investigations with infrared transmission spectroscopy as well as static or quasi-static transport experiments.[9]

In the following we will discuss experimental results obtained on two particular laterally periodic nanostructures, namely a GaAs-AlGaAs

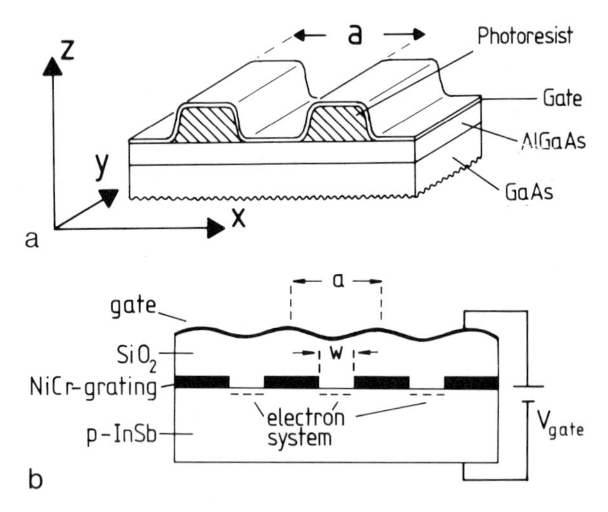

Fig. 1. Schematic cross section through laterally periodic field effect devices. (a) GaAs-AlGaAs heterojunction with a periodically microstructured gate and (b) InSb-MOS-structure with imbedded Schottky grating.

heterojunction with a periodically structured gate and an InSb metal-oxide-semiconductor (MOS) device with a Schottky grating imbedded in the gate oxide. Schematic cross sections through these structures are given in Fig. 1(a,b). In fabricating the GaAs device in Fig. 1(a) we start with a standard modulation doped heterojunction grown by molecular beam epitaxy which contains a single 2D electron system at the GaAs-AlGaAs interface which typically has a 4.2 K mobility of 300,000 cm^2/Vs and an electron density of $3x10^{11}$ cm^{-2}. On top of the heterojunction we holographically define a photoresist grating which then is covered with a semitransparent layer of NiCr serving as a gate. Application of a negative gate bias V_g causes depletion of electrons at the heterojunction interface primarily in those regions where the gate is closest to that interface. Increasing negative gate bias thus results in an increasing periodic modulation of the inversion electron density N_s until at voltage $V_g=V_d$ the electron system is transformed into a series of narrow parallel inversion channels. With further negative bias $V_g<V_d$ the linear electron density N_ℓ and width W of these inversion channels are further reduced until the connection to source and drain contacts are pinched and the electron number in a given channel becomes fixed. By variation of the gate voltage the GaAs structure thus can be tuned from a 2D electron system to a 1D electron system and thus is particularly suited to study the 2D to 1D transition as discussed in the following section. The same device is also well suited to study electronic phenomena in 2D electron systems with moderate periodic density modulation. In section 5 we will discuss novel magnetoresistance oscillations recently observed in such a system[7,10,11] and will show that these reflect lateral superlattice effects in a magnetic field.

In the InSb device sketched in Fig. 1(b) a NiCr Schottky grating is prepared on the free surface of InSb and serves to pin the Fermi level within the band gap of InSb. On top of this a SiO_2 layer is deposited by plasma-enhanced chemical vapor deposition and covered with a semitransparent NiCr gate. Application of a positive gate bias results in formation of narrow electron inversion channels in the region between the Schottky grating. This device is a relatively simple, voltage-tunable 1D system as demonstrated in section 3. If a cross grating is fabricated instead of the linear Schottky grating the structure is similarly suited to induce arrays of quantum dots below the open spaces of the Schottky grating.[8] Cross gratings can be prepared using double exposure in the holographic lithoraphy when the resist covered sample is rotated by an angle of 90° before the second exposure. Wires and dots in the InSb device can be charged without direct contacts to the inversion electrons since the InSb substrate has a finite resistivity even at liquid helium temperatures. Initial results obtained for quantum dots are presented in section 4.

2. FROM TWO- TO ONE-DIMENSIONAL BEHAVIOR IN GaAs HETEROJUNCTIONS

The gated heterojunction device sketched in Fig. 1(a) is well suited to study the transition from 2D to 1D electronic behavior since such a transition can be induced by solely changing the gate bias. At voltage V_d the differential capacitance of the device decreases substantially reflecting the transition from a density-modulated 2D electron system to an array of parallel inversion wires. One way to establish lateral quantization in such inversion wires as their width W becomes comparable to the Fermi wavelength of the inversion electrons is to study Shubnikov-de Haas (SdH) oscillations of the magnetoresistance in a magnetic field perpendicular to the interface plane. For a homogeneous 2D electron system these are known to be periodic in 1/B and the periodicity is a measure of electron density N_s. For a 2D system with modulated electron density they still remain periodic in 1/B but now reflect an average electron density $\bar{N}_s$. When lateral quantization sets in there are only a finite number of 1D subbands occupied already at magnetic field B=0. With increasing magnetic field these 1D subbands increase their energetical spacing and become depopulated.[12] This gives rise to SdH-like oscillations which are no longer periodic in 1/B. This is illustrated in Fig. 2(a) where for a device as in Fig. 1(a) the quantum index of the magnetoresistance maxima is plotted versus its position in reciprocal magnetic field at gate voltages just above and below V_d=-0.5 V. The deviation from 1/B periodicity is clearly visible at low magnetic fields and can be well described by a parabolic confinement model as entered in Fig. 2(a).

Though the confining potential is only expected to be parabolic at very low densities N_ℓ and should become more square-well like at higher ones,[13] we believe the parabolic model to yield reasonably accurate values for the density N_ℓ, the 1D subband spacing $\hbar\Omega$, and the electronic width W_F at the Fermi energy. In Fig. 2(b) the 1D subband spacing extracted from the analysis of the magnetoresistance data in Fig. 2(a) assuming parabolic confinement is displayed as squares versus gate voltage. Lateral quantization becomes already visible just above gate voltages V_d and for $V_g \leq V_d$ yields subband spacings of about 2 meV in this particular device.

Another way to study the transition from 2D to 1D electronic behavior is to investigate the infrared electronic excitations in a transmission experiment.[2,4,8] For laterally periodic structures with periods which are much smaller than the infrared wavelengths one can

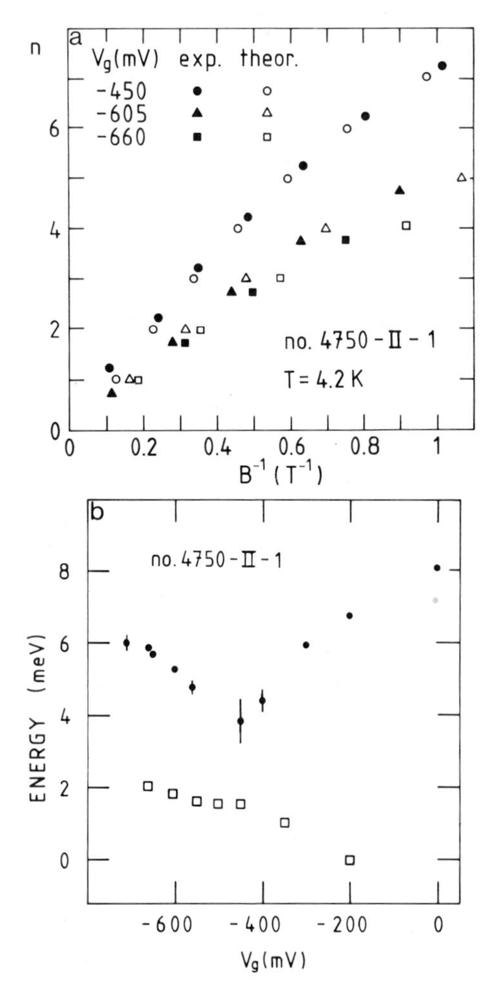

Fig. 2. Evaluation of magnetoresistance oscillations in a GaAs-AlGaAs heterojunction with a laterally periodic gate of period a=400 nm. (a) index n of the maxima in dR/dV_g vs reciprocal magnetic field as observed (filled symbols) and fitted (open symbols) with a parabolic confining potential at various gate voltages V_g. (b) Energy of the 1D subband spacing derived from the magnetoresistance oscillations (open squares) in comparison to energy of infrared resonances at B=0 (filled circles) vs V_g. From Ref. 5.

observe both q=0 modes driven by the spatially uniform component of the infrared field as well as modes with in-plane wave vector $q=(2\pi/a)n$ ($n=1,2,...$) excited by the spatially modulated components of the infrared field. For a 2D electron system at magnetic field B=0, the q=0 component gives rise to Drude absorption whereas the $q\neq0$ components excite 2D plasmons with well defined wave vector. Such 2D plasmons excited at wave vectors $q=2\pi/a$ are clearly visible in the $V_g=0$ spectrum of Fig. 3. There the relative change in transmission $-\Delta T/T=-(T(V_g)-T(V_t))/T(V_t)$ with $V_t=-1$ V is displayed versus infrared frequency. The Drude contribution to $-\Delta T/T$ is not visible at frequencies above 10 cm^{-1} for high mobility heterojunctions such as the one here. With decreasing V_g, i.e., decreasing $\bar{N}_s$, the 2D plasmon resonance with frequency $\omega_p\propto(\bar{N}_s)^{1/2}$ initially shifts as expected to lower frequency. Around voltage $V_g=V_d$ the resonance changes its character and suddenly starts to increase in frequency with decreasing V_g. Also, the signal strength at $V_g\leq V_d$ is much stronger than at $V_g=0$. The latter reflects that the resonance now becomes excited by the uniform (q=0) component of the infrared field. In fact, the distinction between q=0 and $q=2\pi/a$ modes becomes unnecessary once the electron system develops into a set of parallel wires of period a. The changed character of the resonance and the observation that it increases in frequency with decreasing $\bar{N}_s$ makes it natural to associate this resonance with a transition between 1D subbands. However, as also known from 2D intersubband resonance, such an infrared induced intersubband transition is usually shifted to energies above the subband spacing by collective effects.

The 1D intersubband energy may be written as $\hbar\omega=\hbar(\Omega^2+\omega_d^2)^{1/2}$, where $\hbar\Omega$ is the 1D subband spacing and $\hbar\omega_d$ denotes collective contributions. Such collective effects which in simplest approximation are associated with classical depolarization[2,4,5] are naively expected to become less important with decreasing carrier density and to vanish as $\omega_d\propto(\bar{N}_s)^{1/2}$. Insofar it is surprising that the comparison of 1D subband spacings and the infrared resonance energies as displayed in Fig. 2(b) for a particular sample suggests that collective effects here are most important at the lowest carrier density. This comparison also shows that the size of the collective effects in such GaAs-AlGaAs heterojunctions is so large that a reliable determination of 1D subband spacings from infrared spectra alone becomes virtually impossible. Though the quantitative description of such collective effects is improving[14,15] it still is not possible to correctly predict the collective shifts in a particular experimental situation.

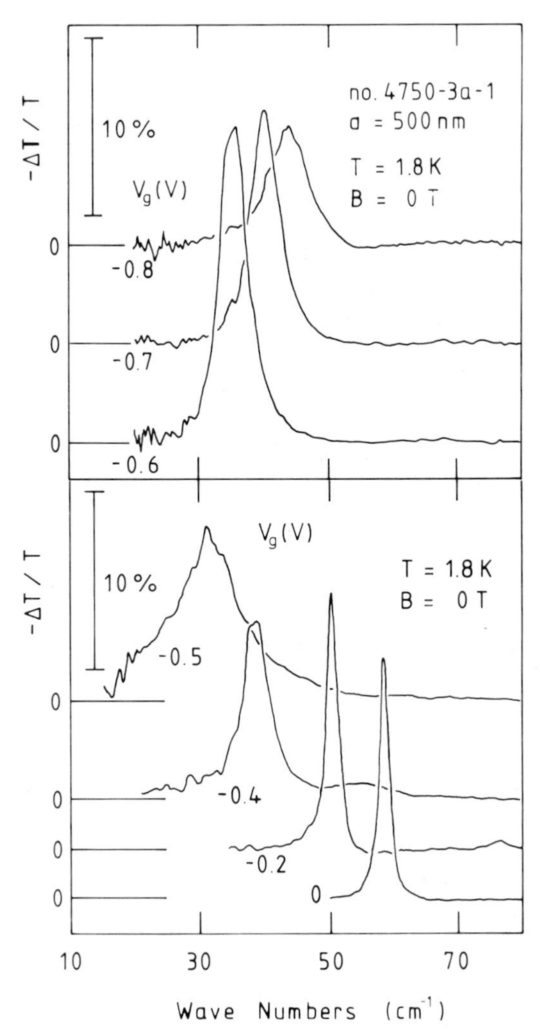

Fig. 3. Infrared transmission spectra of a GaAs-AlGaAs heterojunction with laterally periodic gate of period a=500 nm. The lower part displays spectra in the 2D regime ($V_g \geq V_d$=-0.5V), the upper part in the 1D regime. From Ref. 3.

3. ONE-DIMENSIONAL QUANTUM WIRES ON INDIUMANTIMONIDE

Infrared excitations in the 1D regime have also been studied intensively on laterally periodic MOS-structures on InSb as shown in Fig. 1(b). Normalized spectra[16] for light polarized parallel and perpendicular to the inversion channels are depicted in Fig. 4(a,b) at various voltages $\Delta V_g = V_g - V_t$ above threshold in the absence of a magnetic field. In parallel polarization we observe Drude type spectra corresponding to the free motion of electrons along the channels. In perpendicular polarization we detect resonances between distinct subbands. The dashed line in Fig. 4(b) represents a classical fit to the spectrum for gate voltage $\Delta V_g = 50$ V from which we obtain a mobility $\mu = 16\ 000\ cm^2/Vs$ and density $\overline{N}_s = 5.3 \times 10^{11}\ cm^{-2}$.

Spectra in a magnetic field B=5.4 T are shown in Fig. 4(c,d). For both directions of the light polarization we now observe the same spectral shapes. The resonances are similar to cyclotron resonances of a homogeneous 2D electron system. This is explained by the high magnetic field strength and the related Landau radius $\ell = 11$ nm which is much less than the channel width $W \simeq 100$ nm. In fact, we observe subband-shifted cyclotron resonances.[16] In strong magnetic fields the cyclotron frequency ω_c clearly exceeds the resonance frequency Ω in the absence of magnetic fields. The dashed line in Fig. 4(d) is a cyclotron resonance fit, again for voltage $\Delta V_g = 50$ V, using the 2D Drude magnetoconductivity[17] ($\omega_c \gg \Omega$) and taking into account the occupation of three 2D subbands i=0 to 2. This yields values $\mu = 20\ 000\ cm^2/Vs$ and $\overline{N}_s = 5.4 \times 10^{11} cm^{-2}$ for the ground 2D subband i=0 which roughly agree with the ones obtained in zero magnetic fields. Though the spectra are normalized with the transmittance at threshold voltage, the signal measured in parallel polarization in Fig. 4(c) is about a factor of two smaller than the one in perpendicular polarization in Fig. 4(d). This is explained by the polarizing effect of the metal stripes which act as a parallel wire grid.

Resonance positions versus magnetic field strength[16] are summarized in Fig. 5(a) for three gate voltages ΔV_g. The inset once more presents the data in low magnetic fields on an expanded scale. Transmittance spectra could not be taken close to the reststrahlen band $\bar{\nu} = 183$–$194\ cm^{-1}$ of InSb indicated by the dashed horizontal lines. In the absence of magnetic fields we observe resonance transitions between 1D subbands with resonance frequencies that increase with gate voltage, i.e., number of free electrons induced into the inversion channels. This increase of resonance

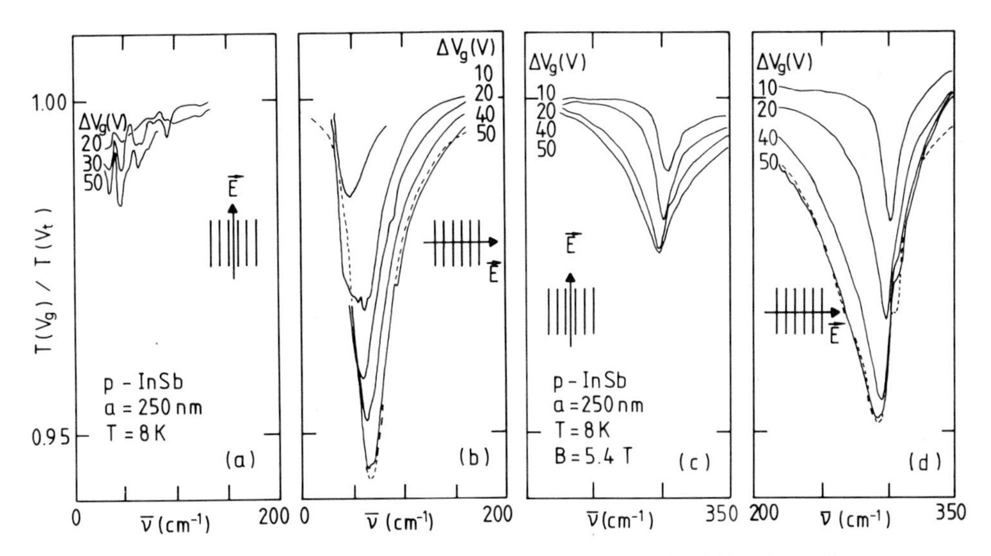

Fig. 4. Intersubband resonances (a,b) and subband-shifted cyclotron resonances (c,d) for 1D inversion channels at various gate voltages ΔV_g above threshold. Light polarization is parallel and perpendicular to the wires. The dashed lines represent Drude fits to the ΔV_g=50 V spectra. From Ref. 16.

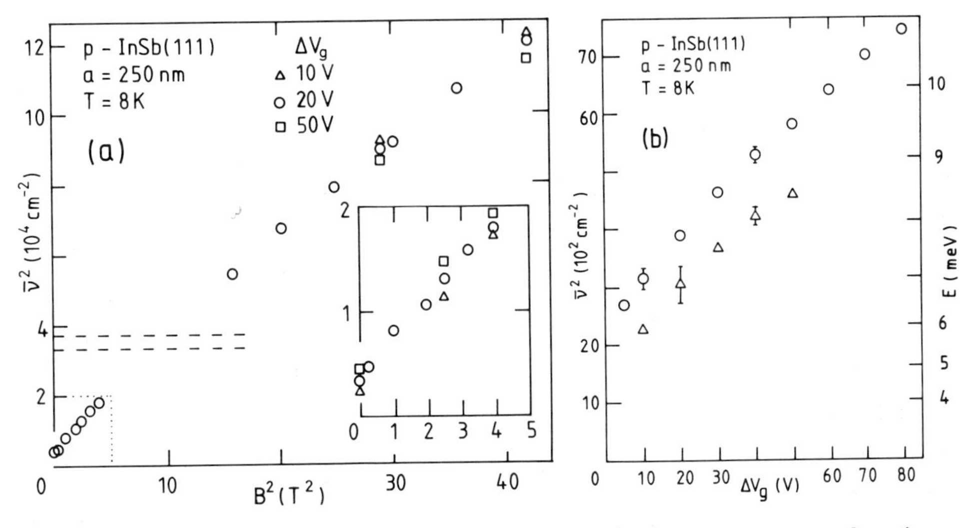

Fig. 5. (a) Positions of 1D subband-shifted cyclotron resonances for three gate voltages ΔV_g. The inset shows the data in low magnetic fields on an enlarged scale in the same units. (b) Intersubband resonance positions (B=0) vs gate voltage. From Ref. 16.

energies is depicted in Fig. 5(b) for two samples of different resonance energies.[16] In strong magnetic fields above the reststrahlen band, we have subband-shifted cyclotron resonance with frequencies that slightly decrease with increasing gate voltages. This decrease of the cyclotron frequency compares well with the corresponding one observed in a homogeneous 2D electron gas and is a consequence of band nonparabolicity.[17] The data in Fig. 5(b) taken in the absence of magnetic fields allow us to extrapolate the measured resonance energies to zero gate voltage $\Delta V_g=0$. Thus we determine the subband spacing due to the lateral confinement for vanishing electron density $N_\ell \to 0$. For the two particular samples, we obtain subband spacings $\hbar\omega_0=5.3$ meV and 6.0 meV, respectively.

To demonstrate that depolarization shifts are of much less importance on the InSb structure than for the GaAs device, at least at low densities, and to determine the single electron subband spacings at finite densities, the quasi-static resistance R and its derivative dR/dV_g have been measured[4] in the direction along the stripes. Experimental traces of derivatives dR/dV_g taken at various voltages ΔV_g and a constant current of 1 μA are shown in Fig. 6. In this figure we also assign subband indices $n^\pm$ to the maxima. There is some weaker structure in the traces that is not due to 1D subband quantization. The shoulders at magnetic fields $B \simeq 4$-5 T and voltages $\Delta V_g=10$ to 16 V are due to SdH oscillations of the 2D regions between the wire grid and the contacts. An additional oscillation occurs for voltages $\Delta V_g \gtrsim 40$ V at magnetic fields $B \approx 5$ T and is attributed to an oscillation of the i=1 subband. For various gate voltages ΔV_g, values 1/B of oscillation maxima are given in Fig. 7 versus quantum number $n^\pm$. The highest populated hybrid subband can directly be read from Fig. 7. To give an example, subband n=4 is the highest populated one at voltages $\Delta V_g=44$ V and 52 V.

In Fig. 8 intersubband resonance energies (circles) are compared with subband spacings (squares) deduced from the magnetoresistance oscillations.[4] The FIR energies exceed the subband spacings and increase more pronounced with gate voltage, i.e., increasing electron density as a consequence of depolarization. The oscillations have been evaluated in EMA as well as in k·p-approximation. In the limit $N_\ell \to 0$ both approaches yield the same subband spacing $\hbar\omega_0 \simeq 9$ meV which agrees with the extrapolated FIR resonance energy: Both nonparabolic and depolarization effects vanish in the limit $N_\ell \to 0$. For the gate voltage $\Delta V_g=40$ V we deduce a shift $\hbar\omega_{dep}=11.6$ meV from Fig. 8 if we compare the FIR data with

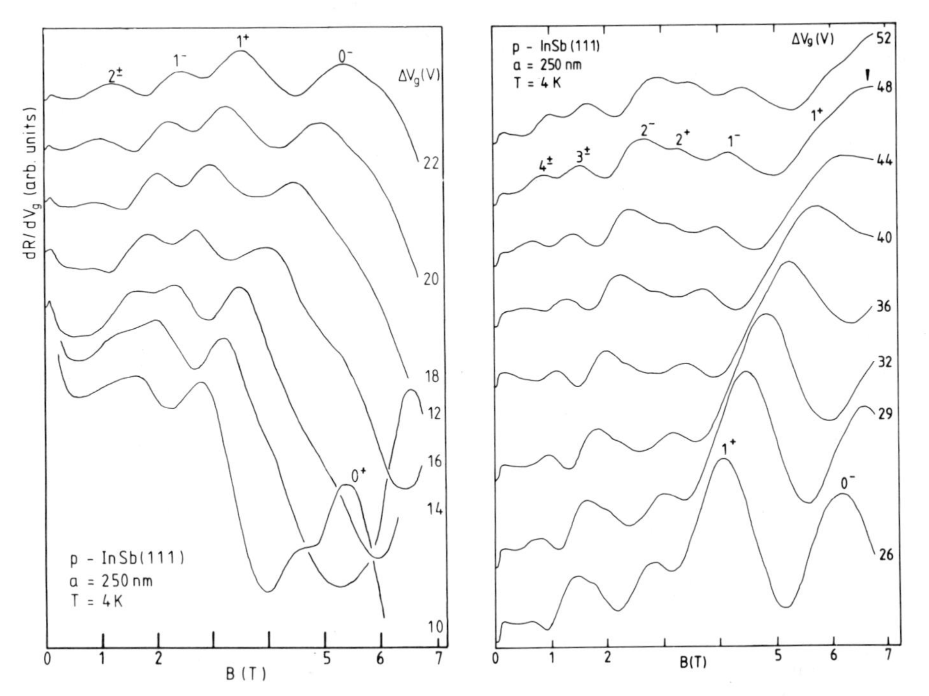

Fig. 6. Derivatives of the magnetoresistance along the inversion channels at various gate voltages ΔV_g above threshold. Oscillation maxima are marked by their subband indices. From Ref. 4.

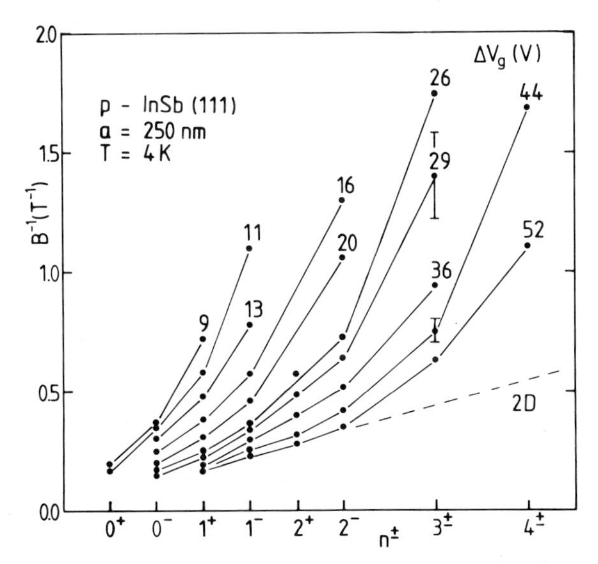

Fig. 7. Fan-chart 1/B vs subband indices of magnetoresistance maxima for various gate voltages ΔV_g. The dashed line for ΔV_g=52 V represents the limiting 2D straight line approached for high magnetic fields in which the electrons behave 2D. From its slope the areal electron density N_c in a channel can be deduced.

the k·p values. The classical depolarization frequency gives a value $\hbar\omega_{dep}$=17 meV with the nonparabolic mass m^*=0.020 m_e, the linear density N_ℓ=3x10^6 cm^{-1}, and the width W=85 nm all obtained from the theoretical description of the magnetoresistance oscillations, as well as from the dielectric constant $\bar{\varepsilon}$=10. The classical value clearly exceeds the experimental one and a similar result is obtained from the Hartree result, however, the theoretical value reduces if one takes into account coupling of wires and screening by the gate.[15]

The electronic width W of the inversion channels is depicted in the inset of Fig. 8 and is obtained from the simple relation $W = N_\ell/N_c = \frac{2}{3}W_F$

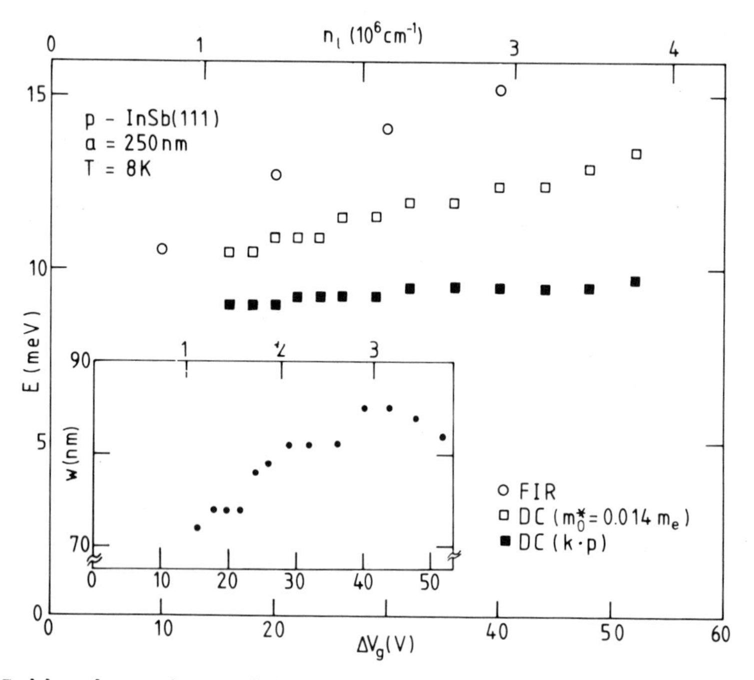

Fig. 8. Subband spacings of 1D inversion channels vs gate voltage ΔV_g or linear density N_ℓ deduced from DC magnetoresistance oscillations using the EMA (open squares) and the k·p-approximation (closed squares). For both approximations the linear densities N_ℓ are almost identical. The inset shows the channel width W. Intersubband (FIR) energies (open circles) exceed the subband spacings due to the depolarization shift.

which gives the width of subband wave functions averaged over all populated 1D subbands. This relation can be derived within the parabolic well approximation for the case when many subbands are populated. The electron densities N_ℓ and N_c are taken from the fitting procedure of the magnetoresistance oscillations and the limiting straight line in the fan-charts of Fig. 7, respectively.

4. QUANTUM DOTS ON INDIUMANTIMONIDE

Spectroscopy of dots[8] is carried out with an optically pumped FIR laser with linearly polarized light. The relative change of transmittance is recorded versus the strength of the magnetic field. Spectra for various laser energies $\hbar\omega$ and gate voltages ΔV_g are shown in Fig. 9. The spectra are almost independent of the polarization direction in the plane as is expected by virtue of sample preparation. Spectra for energy $\hbar\omega$=10.4 meV resemble cyclotron resonances of a homogeneous 2D gas but the resonance magnetic fields are already shifted considerably ($\Delta B \simeq 0.4$ T) to lower field strengths. This directly reflects the spatial quantization in the confining lateral potential. For energy $\hbar\omega$=7.6 meV, we no longer observe a distinct resonance maximum at finite fields but a monotonic decrease of the relative transmittance when the magnetic field is increased. This is indeed expected from the classical conductivity[8] when the quantization energy of the lateral potential approximately coincides with the laser energy ($\omega \simeq \Omega$). For the energy $\hbar\omega$=3.2 meV, we again observe distinct but weak resonances at $B \simeq 1.5$ T. These ω_- resonances are characteristic of a

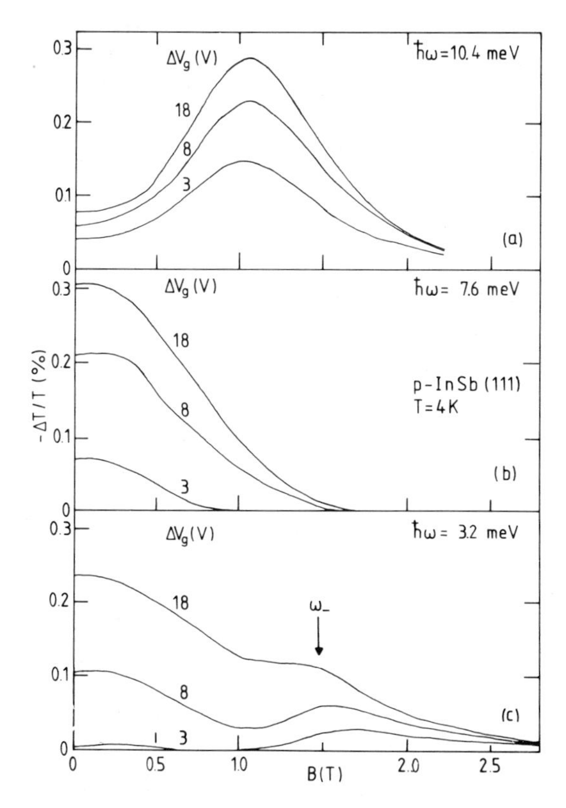

Fig. 9. Far-infrared spectra of quantum dots for three laser frequencies ω and three gate voltages ΔV_g. (a) ω_+ resonance at $B \simeq 1$ T for a laser frequency above the quantization frequency ω_0, (b) traces for $\omega \simeq \omega_0$, (c) ω_- resonance at $B \simeq 1.5$ T for $\omega < \omega_0$. From Ref. 8.

system confined in both lateral directions. From cyclotron-resonance fits of spectra taken in strong magnetic fields ($\omega_c \gg \Omega$) we obtain for gate voltage ΔV_g=8 V above threshold an electron number n_0=9±1 and a dot radius r_F=66 nm at the Fermi energy. For reasons we do not yet understand in detail, the number of electrons saturates at voltages above ΔV_g=20 V for the present sample. Experimental resonance positions for voltage ΔV_g=8 V are given in Fig. 10 together with theoretical curves. At

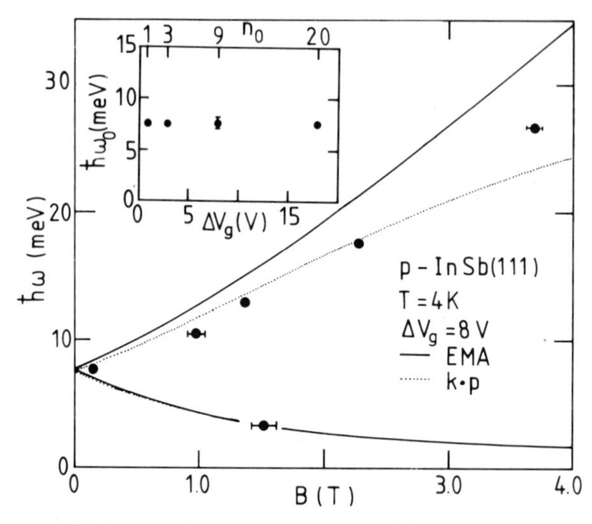

Fig. 10. Zeeman splitting of resonance positions in quantum dots. The inset gives the dependence of the quantization energy $\hbar\Omega$ on voltage ΔV_g and electron number n_0, respectively. The solid lines are calculated from the effective mass approximation (EMA). The dotted lines are obtained from a k·p-approximation ignoring spin splitting.

the highest energy ($\hbar\omega$=26.6 meV) there is a shift ΔB=0.8 T between the experimental and the result of the effective mass approximation. This shift is only qualitatively explained by the the k·p-approximation since we ignore spin splitting. The influence of band nonparabolicity is less important for the lower branch. For lower magnetic fields, the EMA provides an almost quantitative description and we can estimate a quantization energy $\hbar\omega_0$=(7.5±1) meV. This value agrees with the one which we already deduced from the shape of the $\hbar\omega$=7.6 meV spectra in Fig. 9.

As evident from the inset of Fig. 10, the quantization energy does not much depend on electron number. This provides strong evidence that collective depolarization modes[18] which might be expected to become important at higher electron numbers are strongly suppressed in our devices. In fact, macroscopic electric fields are effectively screened by the NiCr Schottky gate since it is evaporated in very close vicinity to the electron systems. For the same reason, we do not expect strong electromagnetic coupling of dots. The independence of the excitation energy on electron number is clearly distinct from the situation in real atoms if one compares, e.g., hydrogen and helium. In our artificial atoms, however, the electrons are not bound in a Coulomb potential but in a parabolic well. This fact and the large InSb dielectric constant $\varepsilon=17$ strongly modify the electron-electron interaction.[19,20] In particular, no influence is detectable in "helium" since ground and excited singulet states S=0 are shifted by the same energy Δ.[19] There is no exact cancellation for higher electron numbers and we expect shifts of fractions of the energy $\Delta=(\pi Ry^{*}\cdot\hbar\Omega)^{1/2}$ with effective Rydberg Ry^{*}. The smallness of the parameter Δ together with the rather broad lines may explain our experimental result.

5. MAGNETORESISTANCE IN A LATERAL SUPERLATTICE

In high mobility GaAs-AlGaAs heterojunctions with a moderate periodic modulation of the 2D electron density Weiss et al. recently observed a new type of magnetoresistance oscillations that directly reflect the effect of the periodic density modulation.[10] In their experiment they induce the density modulation via the persistent photoeffect by illuminating the sample with a light grating. With devices as in Fig. 1(a) tunable electrostatic modulation of the carrier density is achieved by just applying gate voltages $V_g>V_d$. In suitably shaped Hall bar samples, that allow four point probing of the magnetoresistance component ρ_{xx} perpendicular to the density grating (see coordinates in Fig. 1) these novel magnetoresistance oscillations are directly observable as illustrated in Fig. 11. These oscillations have the following characteristics. They usually are observed in the resistance ρ_{xx} at magnetic fields below the SdH-oscillations and are much weaker as well as phase shifted in ρ_{yy}. They are periodic in 1/B and yield maxima in ρ_{xx}, whenever the classical cyclotron diameter $2R_c=2m^{*}v_F/(eB)=(m+0.25)a$ with m=1,2,..., where a is the period of the potential modulation. They occur at magnetic fields where $\mu B>1$ and where one has approximately the relation $\rho_{xx}\simeq\sigma_{yy}B^2/(N_s^2e^2)$, i.e., resistance perpendicular to the density grating reflects conductance parallel to the grating.

The origin of these novel oscillations was independently associated to the effect of the density superlattice on the magnetic band structure by Winkler et al. (Ref. 7) and Gerhardts et al. (Ref. 11). In a simple approach where the effect of a weak periodic potential modulation on the magnetic band structure is considered by perturbation theory[21] one can see that Landau levels transform into Landau bands since the periodic potential lifts the degeneracy of Landau states with different center

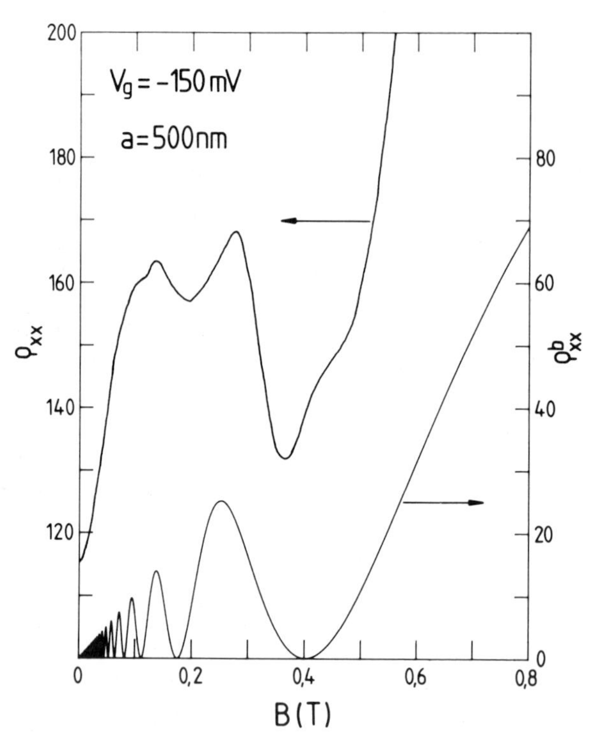

Fig. 11. Measured magnetoresistance ρ_{xx} in a GaAs-AlGaAs-heterojunction with periodically microstructured gate (left scale) in comparison to the calculated oscillatory band contribution ρ^b_{xx} (right scale). From Ref. 7.

coordinate x_0. As illustrated in Fig. 12 the bandwidth varies with energy and has a maximum at the Fermi energy when $2R_c=(m+0.25)a$. In addition to the scattering induced conductivity that usually controls ρ_{xx} for $\mu B \gg 1$ one obtains a band conductivity σ^b_{yy} reflecting the bandwidth of the Landau bands. With reasonable approximations one can calculate σ^b_{yy} analytically for the regime $\mu B>1$. In Fig. 11 the associated band resistivity ρ^b_{xx} thus calculated with experimentally determined parameters is compared to the experimental observation and is found to yield the correct period, phase, and amplitude.[7] Since the dispersion of the Landau bands

$\partial E/\partial k_y = \hbar v_y$ gives rise to an $\vec{E}\times\vec{B}$ drift velocity v_y it can be understood that a semiclassical calculation[22] yields the identical expression for the oscillatory conductivity σ_{xx}^{b} that has been calculated by Winkler et al.[7] Essential for the appearance of the novel resistance oscillations is the condition $\mu B>1$ at the ratio $2R_c/a=1$, meaning that the orbiting electron must test at least one superlattice period a without being scattered. Insofar these magnetoresistance oscillations directly reflect the effect of a lateral superlattice potential on the magnetoresistance.

6. CONCLUSION

The above discussion has served to demonstrate that laterally periodic nanostructures in which the confining potential is modulated via

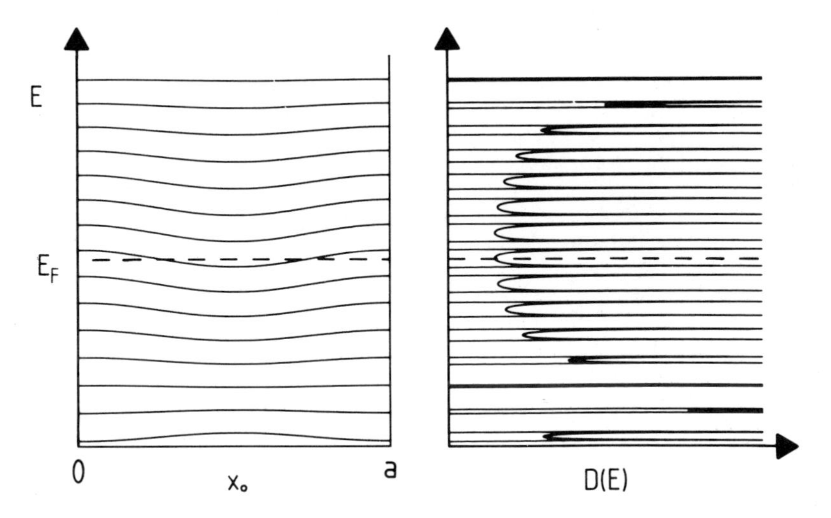

Fig. 12. Sketch of the dispersion of Landau bands $E(x_0)=E(\hbar k_y/eB)$ (left) and the corresponding density of states D(E) (right) in the absence of broadening. From Ref. 7.

field effect are excellent devices to investigate effects of lateral quantum confinement on both electronic transport properties as well as on infrared excitations. With field effect lateral confining dimensions can be well controlled in the 100 nm regime and give rise to lateral quantization energies of several meV. Since electrostatic confinement preserves good device mobilities these quantization energies can be much larger than the level broadening. The novel electronic properties of quantum wires, quantum dots, as well as lateral superlattices thus become accessible for detailed experimental studies.

ACKNOWLEDGEMENTS

The work summarized here was carried out in close collaboration with J. Alsmeier, F. Brinkop, W. Hansen, K. Ploog, Ch. Sikorski, and R. Winkler. We also wish to thank C. W. J. Beenakker, A. V. Chaplik, R. R. Gerhardts, D. Heitmann, and D. Weiss for stimulating discussions. Financial support of the Volkswagenstiftung and the Deutsche Forschungsgemeinschaft is gratefully acknowledged.

REFERENCES

1. For a recent review see, e.g. "Physics and Technology of Submicron Structures", H. Heinrich, G. Bauer, and F. Kuchar, ed., Springer Series in Solid-State Sciences, Vol. 83, Springer, Berlin (1988).

2. W. Hansen, M. Horst, J. P. Kotthaus, U. Merkt, Ch. Sikorski, and K. Ploog, Phys. Rev. Lett. 58:2586 (1987).

3. J. P. Kotthaus, W. Hansen, H. Pohlmann, M. Wassermeier, and K. Ploog, Surf. Sci. 196:600 (1988).

4. J. Alsmeier, Ch. Sikorski, and U. Merkt, Phys. Rev. B 37:4314 (1988).

5. F. Brinkop, W. Hansen, J. P. Kotthaus, and K. Ploog, Phys. Rev. B 37: 6547 (1988).

6. J. P. Kotthaus, in "Proceedings of th 19th International Conference on the Physics of Semiconductors", W. Zawadzki, ed., Vol. 1, p. 47, Polish Academy of Sciences, Warsaw (1988).

7. R. W. Winkler, J. P. Kotthaus, and K. Ploog, Phys. Rev. Lett. 62:1177 (1989).

8. Ch. Sikorski and U. Merkt, Phys. Rev. Lett. May (1989).

9. J. P. Kotthaus, Physica Scripta T 19:120 (1987).

10. D. Weiss, K. v. Klitzing, K. Ploog, and G. Weimann, Europhys. Lett. 8:179 (1989).

11. R. R. Gerhardts, D. Weiss, K. v. Klitzing, Phys. Rev. Lett. 62:1173 (1989).

12. K. F. Berggren, T. J. Thornton, D. J. Newson, and M. Pepper, Phys. Rev. Lett. 57:1769 (1986).

13. S. E. Laux, D. J. Frank, and F. Stern, Surf. Sci. 196:101 (1988).

14. W. Que and G. Kirczenow, Phys. Rev. B 37:7153 (1988); 39:5998 (1989).

15. A. V. Chaplik, Superlattices and Microstructures, to be published.

16. U. Merkt, Ch. Sikorski, and J. P. Kotthaus, Superlattices and Microstructures 3:679 (1987).

17. U. Merkt, M. Horst, T. Evelbauer, and J. P. Kotthaus, Phys. Rev. B 34:7234 (1986).

18. W. Que and G. Kirczenov, Phys. Rev. B 38:3614 (1988).

19. A. V. Chaplik, private communication.

20. G. W. Bryant, Phys. Rev. Lett. 59:1140 (1987).

21. A. V. Chaplik, Solid State Commun. 53:539 (1985).

22. C. W. J. Beenakker, Phys. Rev. Lett. 62:2020 (1989).

EMISSION PROCESS IN QUANTUM–STRUCTURES

Erich Gornik

Walter Schottky Institut der
TU München
D–8046 Garching, Am Coulombwall

ABSTRACT

Different emission processes based on fundamental excitations in low dimensional systems are discussed. Subband transitions in GaAs/GaAlAs heterostructures and quantum wells provide a level system which can be tuned by technological parameters. For subband energies below the optical phonon energy weak subband emission has been observed by excitation through carrier heating and tunnel injection. The tunnel injection mechanisms and the requirements for a subband laser are discussed.

Ballistic electron transport in combination with a lateral periodic potential bears the possibility to obtain stimulated emission in a semiconductor analogous to the free electron laser. A first step into this direction is demonstrated by the observation of spontaneous emission from electrons moving ballistically over a sinusoidal grating in high mobility GaAs/GaAlAs heterostructures. Two–dimensional plasma oscillations are also a candidate to generate coherent far infrared radiation.

INTRODUCTION

Radiative transitions between subband levels of confined carriers in semiconductors have attracted the attention of many investigators. Since the pioneering work of Esaki and Tsu[1] on negative differential conductivity in superlattices the realization of coherent oscillators became feasible. Direct absorption experiments between subbands in GaAs–quantum wells have demonstrated large values of dipole matrix elements making this system attractive as an active medium for infrared and far infrared generation. Especially in the far infrared range there are only a few solid state sources. The most successful of these the p–Ge laser[4-6] requires the use of liquid helium temperatures and a magnetic field, which limits its wide application.

Science and Engineering of One- and Zero-Dimensional Semiconductors
Edited by S.P. Beaumont and C.M. Sotomajor Torres
Plenum Press, New York, 1990

The inter–subband optical transitions have many unique features as compared to the usual valence – to conduction band transitions. The absorption and the gain coefficient are quite sharp, the subband to subband energy gaps as well as their bandwidth can be tuned. The subband system in superlattices can be treated as a man made narrowband gap material with almost arbitrarily controllable parameters.

In order to access the merits of the inter–subband transitions for lasers, two problems remain to be solved: one is the method of pumping for population inversion, the other is the reduction of nonradiative recombination rates. Kazarinov and Suris[7] were the first who examined the possibility of the amplification of electromagnetic waves in a semiconductor superlattice. Since then several concepts for superlattice infrared sources have been published[8-10] predicting extremely high gain[10] and quantum efficiencies of up to 10 % [8,9].

Subband emission has been observed for energies below the optical phonon energy in GaAs–superlattices induced by carrier heating[11,12]. The tunnel injection mechanisms in superlattices for the first time demonstrated by Capasso et al.[13] provides the possibility to obtain inversion between subbands. Recently spontaneous subband emission has been observed excited by tunnel injection in a superlattice by Helm et al.[14].

A somewhat different approach treats the problem more from a classic point of view, investigating the interaction of narrow energy electron beams in periodic structures. Ballistic electron transport in semiconducting heterostructures with "narrow" energy distribution was observed by Levi et al.[15] and Heiblum et al.[16]. The success of creating "electron beams" in semiconductors has stimulated several authors to predict a new type of solid state source[17,19]: Since the bandgap of the superlattice varies periodically in space, the electrons "sense" during their motion an effective periodic electrostatic potential, which plays the role of an electrostatic "wiggler". The existence of a stream like distribution of carriers bears the possibility to excite coherent plasma oscillations[17] or induce stimulated emission between the superbands of a superlattice[19]. A first step into this direction is demonstrated by the observation of spontaneous Smith–Purcell emission from a system where electrons move ballistically over a sinusoidal lateral potential induced by plasma etching[20]. The emitted frequency is found to be correlated with the drift velocity. The possibility to obtain emission from plasma oscillations[21] will be discussed briefly.

SUBBAND EMISSION

The first observation of electric subband emission was demonstrated in Si–MOSFETs[22]. The emission was generated by electrically heating the carrier gas in the 2–dimensional inversion channel. The detected intensity was extremely weak ($\approx 10^{-11}$ W) and coupled out by the sample edges due to a stripline type of geometry. A broadband Ge–detector in combination with different narrowband filters was used to analyze the radiation. The intensity on electric field behaviour was consistent with an electron temperature model[23].

Subband emission from a multiquantum well structure of GaAs/GaAlAs was reported in 1981[11]. The sample consisted of 20 double layers (400 Å GaAs, 390 Å GaAlAs). The radiation resulting from subband transitions is expected to propagate along the interface with polarization perpendicular to it. Therefore the two surfaces of the sample were coated with Al to channel the radiation along the interface and thus to emit from the ends of the sample. Since the subband transition energy is not tunable for a given sample, a tunable n–InSb detector was used to analyze the radiation.

Fig. 1 shows the detector response as a function of the tuning magnetic field; the GaAs/GaAlAs multilayer sample was pulsed with different electric fields. While the first peak in the spectra is due the detector characteristic (which shifts with bias) the peak at 2,3 T corresponding to 18 meV is identified as narrowband emission in connection with subband transitions. In Ref. 11 it is argued that the observed radiation is due a plasma–like intersubband emission of the $1 \rightarrow 0$ transition. This argument was based on the fact that the quantum wells were doped to $n_S = 8 \times 10^{11}$ cm^{-2} per layer. The emission along the layers senses the plasmon mode of the system.

The observation of far infrared emission due to transitions between subbands in GaAs/GaAlAs superlattices coupled out by a metallic grating was recently reported by Helm et al.[12]. A superlattice consisting of 400 Å wide GaAs wells and 50 Å GaAlAs ($x = 0.3$) barriers with a total thickness of 6 μm was used, nominally undoped, with a background doping of 1×10^{15} cm^{-3}. The relatively wide wells were chosen in order to keep several subbands below the optical phonon energy ($\hbar\omega_{lo} = 36$ meV). Significant radiative emission is expected only for transition energies well below $\hbar\omega_{lo}$. The nonradiative relaxation rate is at least two orders of magnitude smaller for electrons not capable of emitting optical phonons as demonstrated recently[24,25]. In the experiment the electron distribution is heated by an electric field in the plane of the layers. The frozen out carrier in the wells are first excited to the lowest subband where an electron temperature distribution is established resulting in a small population of the higher subbands. The outcoupling of the radiation is enhanced as compared to previous reports[11,22] by using an optimized metallic grating structure of 30 μm period. The emission spectrum was analyzed by a novel technique: a tunable filter consisting of n–InSb placed in a magnetic field in

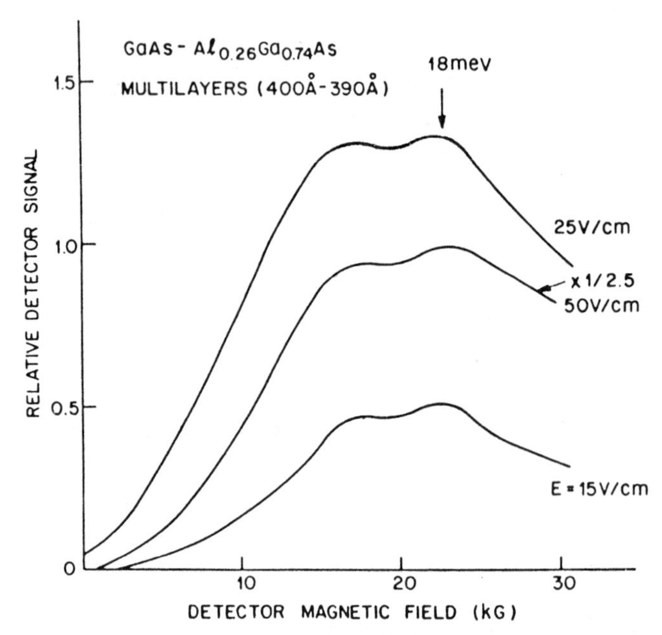

Fig. 1. Detector signal versus magnetic field for a n-InSb-detector. The GaAs/GaAlAs multilayer emitter is pulsed with different electric fields along the channel. The energy of the subband peak is indicated by an arrow (after Ref. 11).

combination with a high sensitivity Si–bolometer allowed the detection of the weak emission.

Fig. 2 shows the observed spectra. In the upper part the emission without the grating structure is shown. Only radiation due to transitions between donor states (2p = 1s) is observed. The sample with a 30 μm grating shows an additional line of comparable size as the donor line. The assignment of the transitions is depicted in the insert of Fig. 2. It should be noted that the coupling efficiency of the grating is quite high since the subband peak is comparable to the impurity peak. The second subband lies about 10 meV above the 2p donor state. Thus it is much less populated at a given electron temperature indicating a higher external emission efficiency for the subband emission since the transition probabilities are about the same.

This experiment demonstrates the feasibility of exciting subband emission in quantum wells. The interesting excitation mechanism which bears the possibility for population inversion is tunnel injection. Several proposals have suggested schemes to obtain inversion through tunnel injection[8-10]. Most of them assume unrealistic nonradiative relaxation rates. The relaxation times for subband transitions above $\hbar\omega_{lo}$ are in the order of 7 – 10 ps[24] while for transitions below $\hbar\omega_{lo}$ times in the order of 500 ps[25] are found. This times have to be compared with the radiative transition times between subband given by

$$\tau_{sp} = \frac{\epsilon h c^3}{2n^3\mu^2\omega^2}, \qquad \mu^2 = \frac{\hbar^2 e^2}{2m_0\omega} f_{ij}$$

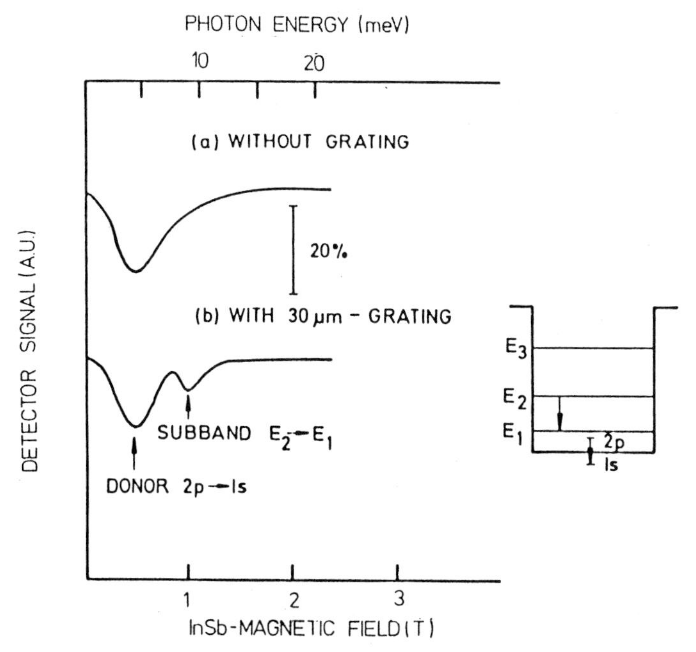

Fig. 2. Detector signal versus magnetic field of InSb-filter and corresponding photon energy. The observed 2p - 1s donor transition and the $E_2 - E_1$ subband transition are indicated in the insert (after Ref. 12).

where n is the refractive index and f_{ij} the oscillator strength. For energies below $\hbar\omega_{lo}$ τ_{sp} is in the order of 10^{-7}sec (f_{ij} is assumed to be unity) thus two order of magnitude larger than the nonradiative transition times. For subband splittings above $\hbar\omega_{lo}$ the internal quantum efficiency will be even two orders of magnitude lower thus making the generation of significant population inversion very unlikely.

In the following we want to consider a tunnel structure as proposed by Liu[10] shown in Fig. 3 and discuss the possibility for population inversion. A common feature to all proposed devices is current injection by resonant tunneling from one side of the active layer to the upper subband and resonant tunneling from the lower subband to the other side of the active layer. For a population inversion to take place it is required that the two resonant energies will not match and that there will be tunneling only in a narrow energy band centered at each resonant energy. However, in a realistic situation there exist also leaky nonresonant tunneling electrons from the upper level to the adjacent layer. This situation is depicted in Fig. 4 and described by the following set of rate equations:

$$\frac{dn_2}{dt} = \frac{J}{e} - \frac{n_2}{\tau_{21}} - \frac{n_2}{\tau_{nr}}, \qquad \frac{dn_1}{dt} = \frac{n_2}{\tau_{nr}} - \frac{n_1}{\tau_r}$$

where J is the current density, τ_r, τ_{nr} are the resonant and nonresonant tunneling times through the adjacent layer. τ_{21} is the dominant relaxation time between subbands. In a steady state situation the population inversion is

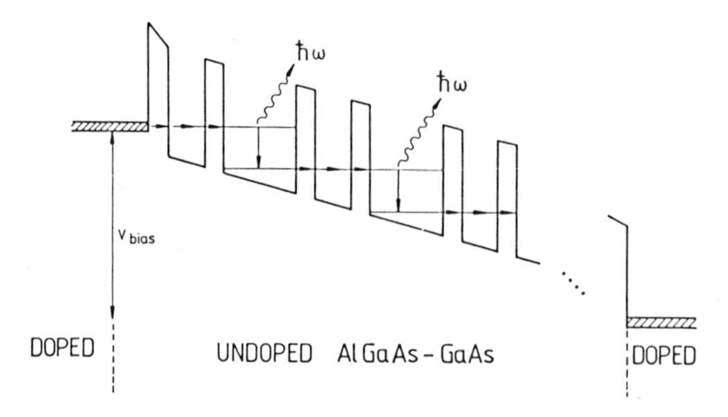

Fig. 3. Tunnel-structure suggested by Liu[10] under bias conditions. The heavily doped contact areas are hatched, the emission process occur in the wide wells.

$$n_2 - n_1 = \frac{J}{e}\left(\frac{\tau_{nr}}{\tau_{nr}+\tau_{21}}\right)(\tau_{21} - \tau_r).$$

The times τ_{nr} and τ_r can be influenced by the thickness of the second barrier. Assuming $\tau_{21} \gg \tau_{nr}$ and $\tau_{21} \gg \tau_r$ the population inversion becomes limited by τ_{nr}: $n_2 - n_1 \approx (J/e)\tau_{nr}$. However, in a situation which is achievable by choosing a proper thickness of the second barrier τ_{21} will be shorter than τ_{nr}, while $\tau_r \ll \tau_{21}$. In this case

$$n_2 - n_1 = \frac{J}{e}\tau_{21}$$

and the inversion is only controlled by the inter–level relaxation. Since the ratio τ_{21}/τ_{sp} is very small a high current density will be required to obtain gain. For a current density of 100 A/cm^2 and τ_{21} in the order of 500 ps a population inversion of $3 \times 10^{11} cm^{-2}$ can be achieved, sufficient for lasing[8 10]. Higher current densities do not increase the population inversion since Auger processes become dominant reducing the nonradiative lifetime further[9,26]. With the above current density emission intensities

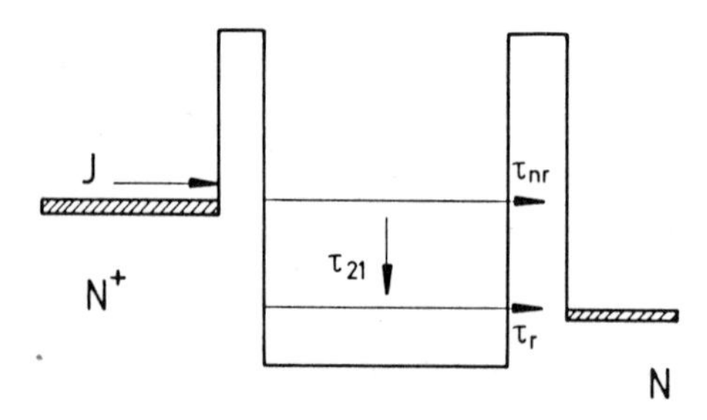

Fig. 4. Schematic structure for creating population inversion in quantum wells by tunnel injection. The basic relaxation times are indicated.

in the order of mW/cm^2 are expected. For subband transition energies above the optical phonon energy current densities in the order of $10^4 A/cm^2$ are necessary to obtain a significant inversion. However, lasing action may still not be possible due to the influence of Auger processes. The low internal quantum efficiencies of $\approx 10^{-2}$ to 10^{-3} are the reason that weak subband emission by tunnel injection has been observed only very recently[14]. Multilayer systems as shown in Fig. 3 with high tunnel current densities have not been realized so far.

All the possibilities of band structure engineering will be necessary to reduce the nonradiative lifetimes. One way to reduce the optical phonon relaxation rate is to reduce the dimensionality of the system. For a zero dimensional tunnel structure the optical phonon lifetime is expected to be increased by several orders of magnitude, thus making lasing possible in the whole range of infrared and far infrared frequencies.

EMISSION FROM BALLISTIC ELECTRONS

The combination of ballistic transport with a one–dimensional periodic potential opens the possibility for a new class of solid state devices. Since the pioneering work of Esaki and Tsu[1] none of the proposals to obtain stimulated emission[7,19] in this system have been realized yet. The most recent work by Botton and Ron, following the line of Gover and Yariv[27] predict stimulated emission from ballistic electrons in a superlattice in the energy band around 100 meV. For current densities of $10^4 A/cm^2$ and an average electron speed of $\approx 5 \times 10^7 cm/s$ a gain in the order of $\approx 0.4\ cm^{-1}$ is predicted. This is a quite low gain, considering that the length of the device could not exceed several mm.

A first step into this direction is demonstrated by the observation of spontaneous emission from a system where electrons move ballistically over a sinusoidal lateral

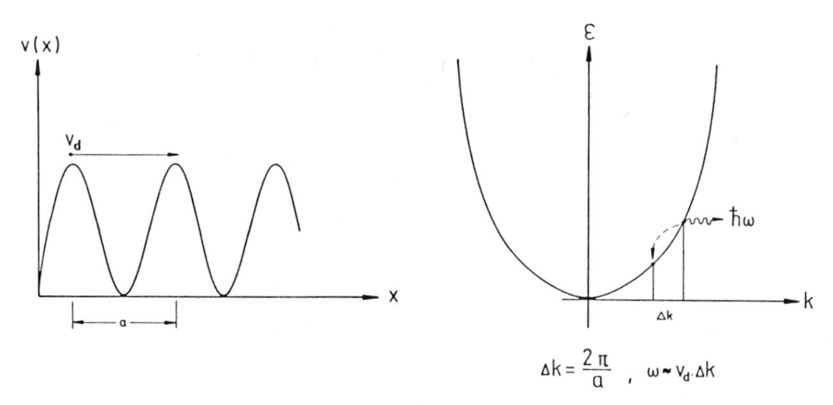

Fig. 5. Schematic of the Smith-Purcell emission. A carrier moves in a periodic potential (left). The energy transition induced by the grating (Δk) is shown in the right part.

potential[20]. The basic principle of the effect is demonstrated in Fig. 5: A periodic surface potential with the period a induces a periodic potential acting on the 2D electrons. This periodic potential induces an energy loss in a drifting carrier gas via photon emission, with the emitted frequency given by $\omega = v_d \cdot \Delta k$ ($\Delta k = 2\pi/a$).

For a degenerate carrier distribution as present in a GaAs/GaAlAs heterostructure with densities above $10^{11}cm^{-2}$ the situation is somewhat more complicated. In this situation the emission frequency is proportional to a velocity which is the sum of the drift and the Fermi–velocity: $\omega \approx (v_d + v_F) \cdot \Delta k$.

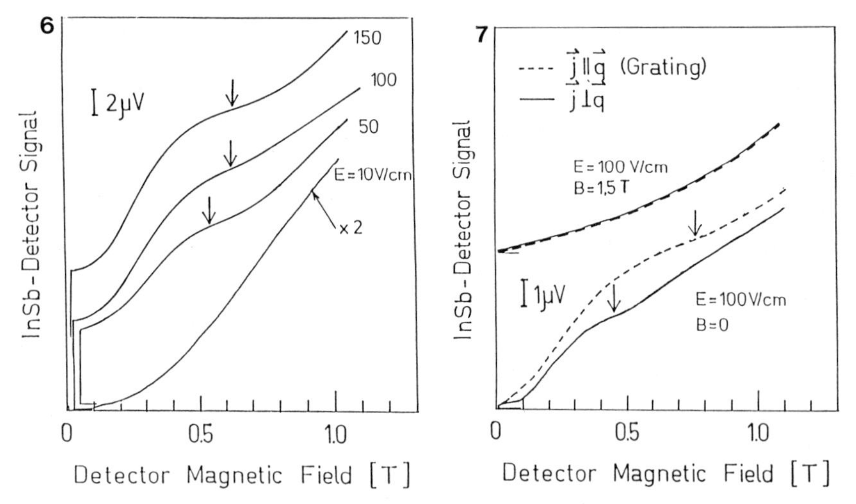

Fig. 6. Emission spectra from a GaAs/GaAlAs-sample with the grating orientation $\vec{q} \| \vec{j}$ measured with a tunable InSb-detector for different electric fields. The arrows show the broad cut off point of the radiation which is shifting to higher energies with increasing field (after Ref. 20).

Fig. 7. Emission spectra as in Fig. 6. A Comparison for the $\vec{j} \| \vec{q}$ and $\vec{j} \perp \vec{q}$ geometries. A magnetic field normal to the interface destroys the selective emission for both configurations (upper curves).

A periodic grating with a period (a) was plasma etched into the top layer of a GaAs/GaAlAs heterostructure. From Shubnikov–de Haas oscillations the height of the periodic potential is estimated to be of the order of 1 meV. Two different sample geometries were realized: One sample had the current direction parallel to the grating wavevector $\vec{q} = \Delta\vec{k}$ ($\vec{j} \| \vec{q}$). In this configuration the carriers "feel" the grating. The second sample had the current perpendicular to $\vec{q}$ ($\vec{j} \perp \vec{q}$).

The emission spectra were analyzed with a magnetically tunable n–InSb detector[31] (tunability 60 cm^{-1} per T, resolution 8 cm^{-1}). Fig. 6 shows spectra for several electric fields applied in the plane of the 2D layer. For electric fields above 30 V/cm a broad well defined structure appears which shifts to higher frequencies with increasing field indicated by arrows. To demonstrate that the observed emission is due to the grating induced potential a comparison of the emission for $\vec{q}\|\vec{j}$ and $\vec{q}_{\perp}\vec{j}$ is shown in Fig. 7. In both orientations radiation due to the grating induced loss (Smith–Purcell effect[32]) is observed. However, it is evident that the spectrum for $\vec{q}\|\vec{j}$ is shifted to higher energies. For $\vec{q}_{\perp}\vec{j}$ the emission comes mainly from carriers moving with the Fermi–velocity perpendicular to the current direction. From this experiment it is directly evident that emission from drifting carriers is observed. The difference between the $\vec{q}\|\vec{j}$ and $\vec{q}_{\perp}\vec{j}$ configuration is just the drift velocity. In the present experiment a drift velocity of 1.7 x 10^7cm/s is derived from the frequency shift between the two configurations.

As an additional proof that the emission is due to the high carrier velocity the velocity is reduced with a magnetic field perpendicular to the 2D plane. The selective emission disappears for both configurations (upper curves in Fig. 7) for a magnetic field of 1.5 T. From an analysis of the width of the spectra a rather "low" carrier temperature is estimated in the order of 50 K. This indicates that the distribution is rather narrow and streaming like. However, this emission process has to be extended to samples with considerable lower electron densities. In this case the spectra should be dominated by the drift velocity and the influence of the Fermi–velocity should be small. Experiments along these lines are in process.

Another possibility to obtain stimulated emission is the excitation of surface plasmon modes[28]. If the drift velocity of the carrier gas is equal to the phase velocity of a certain plasmon oscillation a direct conversion of drift energy into a plasmon oscillation with a fixed wave vector is possible. This system does not require a grating structure for generating emission: the plasmon wave in the drifting system represents an oscillating change density that is able to radiate. For the excitation of stimulated radiation a grating would be necessary in order to amplify the spontaneous density modulations coherently, but also other methods of feedback (e.g. resonant cavities) are possible. The phase velocities of plasmon waves in inversion channels of GaAs/GaAlAs structures range from ≈ 10^7cm/s (for very low densities) to 10^8cm/s. Velocities in the order of 5 x 10^7cm/s have been observed for ballistic electrons [15,16] while average drift velocities do not exceed 2 x 10^7cm/s at low temperatures. The generation of stimulated plasmon emission seems feasible with carrier densities below $10^{11}$$cm^{-2}$ ($\omega_p^2 \propto n_s$) and grating periods in the order of 500 Å. In this situation the carriers have to move ballistically over several periods of the grating.

However, until now only spontaneous far infrared emission from two–dimensional plasmons in GaAs/GaAlAs heterostructures has been observed with gratings of μm–periods[29,30]. Intensities in the order of 10^{-8} W/cm^2 have been achieved.

Acknowledgement

The work is supported by Projekt I/61.840, Stiftung Volkswagenwerk, Darmstadt, and by Siemens Corporation (Sonderforschungsbereich TU–München), München.

REFERENCES

1. L. Esaki, R. Tsu, IBM J. Res. Develop. 61 (1970).
2. L.C. West, S.J. Eglash, Appl. Phys. Lett. 46:1153 (1985).
3. B.F. Levine, A.Y. Cho, J. Walker, R.J. Mallik, D.A. Kleinmann, D.L. Sivco, Appl. Phys. Lett. 52:1481 (1988).
4. A.A. Andronov, I.V. Zverev, V.A. Kozlov, Yu.N. Nozdrin, S.A. Pavlov, V.N. Shastin, Sov. Phys.–JETP Lett. 40:804 (1984).
5. S.Komiyama, N. Iizuka, Y. Akasaka, Appl. Phys. Lett. 47:958 (1985).
6. K. Unterrainer, M. Helm, E. Gornik, E.E. Haller, J. Leotin, Appl. Phys. Lett. 52:564 (1988).
7. R.F. Kazarinov, R.A. Suris, Sov. Phys. Semicond. 5:707 (1971).
8. Perng–fei Yuh, K.L. Wang, Appl. Phys. Lett 51:1404 (1987).
9. Perng–fei Yuh, K.L. Wang, Phys. Rev. B37:1328 (1988).
10. H.C. Liu, J. Appl. Phys. 63:2856 (1988).
11. E. Gornik, R. Schawarz, D.C. Tsui, A.C. Gossard, W. Wiegmann, Solid State Commun. 38:541 (1981).
12. M. Helm, E. Colas, P. England, F. DeRosa, S.J. Allen Jr., Appl. Phys. Lett. 53:1714 (1988).
13. F. Capasso, K. Mohammed, A.Y. Cho, Appl. Phys. Lett. 48:478 (1986).
14. M. Helm et al., preprint
15. A.F. Levi, J.R. Hayes, P.M. Platzmann, W. Wiegmann, Phys. Rev. Lett. 55:2071 (1985).
16. M. Heiblum, M.I. Nathan, D.C. Thomas, C.M. Knoedler, Phys. Rev. Lett. 55:2200 (1985).
17. V. Gruzinskis et al., Solid State Electron. 31:345 (1988).
18. A.M. Belyantsev et al., Solid State Electron. 31:379 (1988).
19. M. Botton, A. Ron, Appl. Phys. Lett. 54:418 (1989).
20. E. Gornik, R. Christanell, R. Lassnig, W. Beinstingl, K. Berthold, G. Weimann, Solid State Electron. 31:751 (1988).
21. R.A. Höpfel, E. Gornik, Surface Science 142:375 (1984).
22. E. Gornik, D.C. Tsui, Phys. Rev. Lett. 37:1425 (1976).
23. E. Gornik, D.C. Tsui, Solid State Electron. 21:139 (1978).
24. A. Seilmeier, H.J. Hubner, G. Abstreiter, G. Weimann, W. Schlapp, Phys. Rev. Lett. 59:1345 (1987).
25. D.Y. Oberli, D.R. Wake, M.V. Klein, J. Klem, T. Henderson, H. Morkoc, Phys. Rev. Lett. 59:696 (1987).
26. S. Borenstain, J. Katz, Phys. Rev. B, preprint.
27. A. Grover, A. Yariv, J. Appl. Phys. 16:121 (1978).
28. E. Gornik, R.A. Höpfel, AEÜ 37:213 (1983).
29. R.A. Höpfel, E. Gornik, A.C. Gossard, W. Wiegmann, Physica B 177 & 178:646 (1983).
30. N. Okisu, Y. Sambe, T. Kobayshi, Appl. Phys. Lett. 48:776 (1986).
31. E. Gornik, W. Müller, F. Kohl, IEEE Trans. on Microwave Theory and Techniques MTT 22:991 (1974).
32. S.J. Smith, E.M. Purcell, Phys. Rev. 92:1069 (1953).

MAGNETORESISTANCE AND MAGNETOCAPACITANCE IN A TWO-DIMENSIONAL ELECTRON GAS IN THE PRESENCE OF A ONE-DIMENSIONAL SUPERLATTICE POTENTIAL

Dieter Weiss

Max-Planck-Institut für Festkörperforschung, Heisenbergstraße 1
D-7000 Stuttgart 80, Federal Republic of Germany

A novel type of magnetoresistance oscillation is observed in a two-dimensional electron gas in a high mobility GaAs-AlGaAs heterostructure with a holographically induced lateral periodic modulation in one direction. The modulation arises due to the persistent photoconductivity of the samples at low temperatures. The experiments show that the 1/B periodicity of the additional oscillations is determined by the carrier density N_s and the period a of the grating, reflecting the commensurability of cyclotron diameter and modulation period. The key to the explanation of the novel magnetotransport oscillations is an oscillatory linewidth of the modulation broadened Landau bands. To demonstrate this we have performed magnetocapacitace measurements in order to obtain direct information about the density of states of the modulated two-dimensional electron gas

INTRODUCTION

At low temperatures the magnetoresistance of a degenerate two-dimensional electron gas (2-DEG) exhibits the well known Shubnikov-de Haas (SdH) oscillations reflecting the discrete nature of the degenerate Landau energy spectrum [1]. A superimposed one-dimensional periodic potential lifts the degeneracy of the Landau levels and leads to a novel type of magnetoresistance oscillation periodic in 1/B as long as the period of the modulation is small compared to the mean free path of the electrons [2]. The periodicity of these oscillations is governed by an interesting commensurability problem owing to the presence of two length scales , the period a of the potential and the cyclotron radius R_C at the Fermi energy [2,3]. In selectively doped AlGaAs-GaAs heterostructures a persistent increase in the two-dimensional electron density is observed at temperatures below T=150K if the device is illuminated with infrared or visible light. This phenomenon is usually explained on the basis of the properties of DX-centers which seem to be related to a deep Si donor. The increase in the electron density depends on the photon flux absorbed in the semiconductor so that a spatially modulated photon flux generates a modulation in the carrier density. In our measurements a holographic illumination of the heterostructure at liquid helium temperatures is used to produce a periodic potential

Science and Engineering of One- and Zero-Dimensional Semiconductors
Edited by S.P. Beaumont and C.M. Sotomajor Torres
Plenum Press, New York, 1990

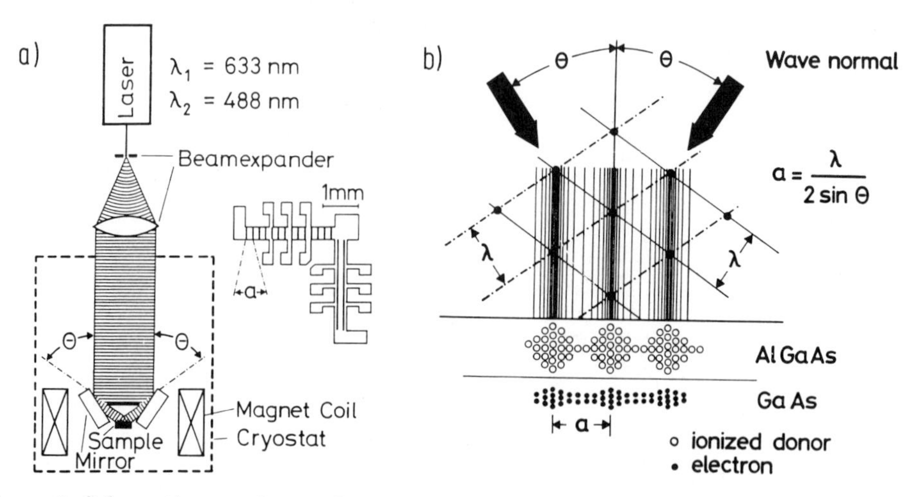

Figure 1. Schematic experimental set up and top view of the L shaped sample geometry with sketched interference pattern (a). Principle of holographic illumination by spatial modulation of the concentration of ionized donors in the AlGaAs layer and of electrons in the 2-DEG using two interfering laser beams (b). The interference pattern is shown schematically – the period a is determined by the wavelength λ and the angle Θ

with a period on the order of the wavelength of the interfering beams, a method first used by Tsubaki et al. [4]. The potential modulation obtained by this technique is on the order of 1 meV where the Fermi energy E_F in our samples is typically 10 meV.

In the first part of this contribution the experiment displaying the novel magnetoresistance oscillations is briefly reviewed followed by a brief discussion of the Landau energy spectrum in the presence of a superimposed one-dimensional potential and a sketch of the theory explaining the observed oscillatory magnetoresistance. In the last section the modification of the energy spectrum is experimentally demonstrated by magnetocapacitance measurements reflecting directly the thermodynamic density of states (DOS) at the Fermi energy.

MAGNETORESISTANCE OSCILLATIONS

The experiments were carried out using conventional AlGaAs-GaAs heterostructures grown by molecular beam epitaxy with carrier densities between $1.5 \cdot 10^{11}\mathrm{cm}^{-2}$ and $4.3 \cdot 10^{11}\ \mathrm{cm}^{-2}$ and low temperature mobilities ranging from $0.23 \cdot 10^{6}\mathrm{cm}^2/\mathrm{Vs}$ to $1 \cdot 10^{6}\mathrm{cm}^2/\mathrm{Vs}$. Illumination of the samples increases both the carrier density and the mobility at low temperatures. We have chosen an L-shaped geometry (sketched on the right hand side of Fig.1a) to investigate the magnetotransport properties parallel and perpendicular to the interference fringes. Some of the samples investigated have an evaporated semi-transparent NiCr front gate (thickness $\approx$ 8nm) in order to vary the carrier density as well as to carry out magnetocapacitance measurements after holographic illumination. A sketch of the experiment exploiting the persistent photoconductivity to periodically modulate the positive background charge in the AlGaAs-layer is shown in Fig.1a and 1b. We used either a 5mW HeNe laser ($\lambda = 633$nm) or a 3mW Argon-Ion laser ($\lambda = 488$nm) mounted on top of the sampleholder. The expanded laser beam entered the sampleholder through a quartz window and a shutter ensuring typical illumination times of about 100 ms. Two mirrors mounted close to the sample were used to create two interfering plane waves. The advantage of this kind of 'microstructure engineering' is its simplicity and the achieved high mobility of the microstructured sample due to the absence of defects introduced by the usual pattern transfer techniques [5].

The result of standard magnetoresistance measurements carried out perpendicular ($\rho_\perp$) and parallel ($\rho_{||}$) to the periodic modulation is shown in Fig.2. In addition to the usual Shubnikov-de Haas oscillations appearing at about 0.5T additional oscillations

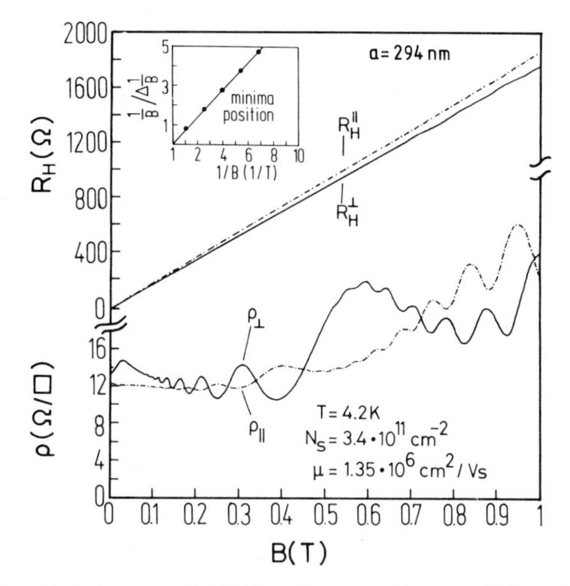

Figure 2. Magnetoresistivity ρ and Hall resistance R_H parallel and perpendicular to the interference fringes. The positions of the minima of $\rho_\perp$ are plotted in the inset demonstrating the 1/B periodicity of the novel oscillations

become visible at even lower magnetic fields. While pronounced oscillations of this new type dominate $\rho_\perp$ at low magnetic fields, weaker oscillations with a phase shift of 180^0 relative to the $\rho_\perp$ data are visible in the $\rho_{||}$ measurements. No additional structure appears in the Hall resistance. The novel oscillations are, analogous to SdH oscillations, periodic in 1/B as is displayed in the inset of Fig.2. As the temperature is increased from 2.2K to 4.2K the SdH oscillations are strongly damped whereas the additional oscillations are apparently unaffected. The periodicity is obtained from the minima of $\rho_\perp$, which can be characterized by the commensurability condition

$$2R_c = (\lambda - \frac{1}{4})a, \qquad \lambda = 1, 2, 3, ..., \tag{1}$$

between the cyclotron diameter at the Fermi level, $2R_c = 2v_F/\omega_c = 2l^2k_F$, and the period a of the modulation. Here $k_F = \sqrt{2\pi N_s}$ is the Fermi wavenumber, $l = \sqrt{\hbar/eB}$ the magnetic length, and $\omega_c = \hbar/ml^2$ the cyclotron frequency with the effective mass $m = 0.067m_0$ of GaAs. For magnetic field values satisfying Eq.(1) minima are observed in $\rho_\perp$. The periodicity $\Delta(1/B)$ can easily be deduced from Eq.(1)

$$\Delta\frac{1}{B} = e\frac{a}{2\hbar k_F} \tag{2}$$

The validity of Eq.(1) has been confirmed by performing these experiments on different samples, by changing the carrier density with an applied gate voltage, and by using two laser wavelengths in order to vary the period a [2]. This is demonstrated in Fig.3 where the periodicity $\Delta(1/B)$ (displayed as carrier density $n = \frac{e}{\pi\hbar}(\Delta\frac{1}{B})^{-1}$) is plotted as a function of the carrier density N_S and the period a. The solid lines correspond to Eq.(2). Recently similar magnetoresistance oscillations have also been observed in conventionally microstructured samples by Winkler et al. [6].

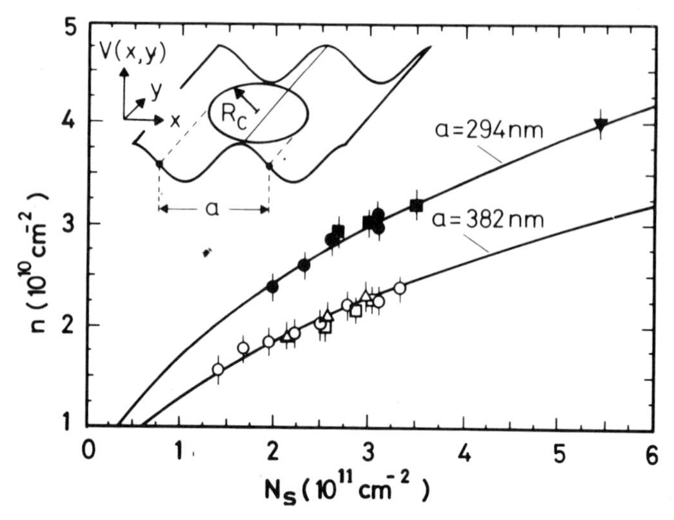

Figure 3. $n = \frac{e}{\pi\hbar}(\Delta\frac{1}{B})^{-1}$ versus N_s. Full symbols correspond to a laser wavelength λ=488nm, open symbols to λ=633nm and different symbols represent different samples. The solid lines are calculated using the condition that the cyclotron orbit diameter $2R_c$ is equal to an integer multiple of the interference period a, as is sketched in the inset

SKETCH OF THE THEORY

Since the theory of these novel type of oscillation is discussed in detail by Gerhardts in this volume I restrict myself to a rough sketch of it. A periodic potential, e.g., in x direction $V(x) = V_0 cos(Kx)$ with $K = 2\pi/a$ lifts the degeneracy of the Landau levels, and yields eigenstates $|x_o n\rangle$ which carry current in the y-direction,

$$\langle x_o n|v_y|x_o n\rangle = -\frac{1}{m\omega_c}\frac{d\varepsilon_n}{dx_o} = \frac{1}{\hbar}\frac{d\varepsilon_n}{dk_y} , \tag{3}$$

where x_0 is the center coordinate $x_0 = -l^2 k_y$ and n is the Landau level (LL) index. On the other hand $\langle x_o n|v_x|x_o n\rangle = 0$, which is the origin of the anisotropic transport coefficients observed. The width of the Landau bands oscillates, which is most easily understood within first order perturbation theory with respect to V_o. This yields for the energy spectrum plotted in Fig.4

$$\varepsilon_n^{(1)}(x_o) = \hbar\omega_c(n+\frac{1}{2}) + V_o \, cos(Kx_o) \, e^{-\frac{1}{2}X} \, L_n(X) , \tag{4}$$

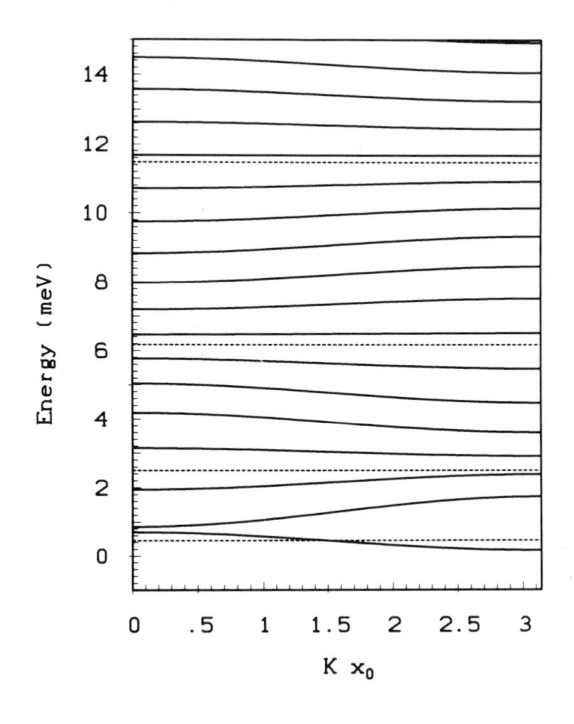

Figure 4. Calculated energy bands $\epsilon_n^1(x_0)$ for B=0.5T, $V_0 = 1$meV, $a = 100$nm and material parameters of GaAs. The dotted lines correspond to energies of zero band width

where $X = \frac{1}{2}K^2l^2$, and exhibits qualitatively the same features as the exact energy spectrum [3]. For fixed magnetic field (fixed X), the Laguerre polynominal $L_n(X)$ oscillates as a function of its index n [7]. The flat band condition, $L_n(X) = 0$, can be expressed in terms of the cyclotron radius $R_C = l\sqrt{2n+1}$, and is identical with Eq.1 [3]. The flat band energies are indicated for $\lambda = 1, ..., 4$ as dotted lines in Fig.4. In a real physical system the flat bands are of course collision broadened and described by a linewidth Γ. The origin of the oscillations is that the wavefunctions, having a spatial extent of approximately $2R_C$, sense effectively the average value of the periodic potential over an interval of length $2R_C$. The nonvanishing matrix elements $\langle x_o n|v_y|x_o n\rangle$ lead to an additional contribution to the conductivity σ_{yy} which becomes important for high-mobility systems [3]. It is this contribution of current carrying states at the Fermi level, which has no counterpart in σ_{xx}, which accounts for the anisotropy of the transport coefficients and leads to oscillations with minima if flat bands occur at the Fermi energy. Since σ_{yx} shows no noticeable oscillations and since $\sigma_{yx}^2 \gg \sigma_{xx}\sigma_{yy}$, one has $\rho_{xx} \approx \sigma_{yy}/\sigma_{yx}^2$, $\rho_{yy} \approx \sigma_{xx}/\sigma_{yx}^2$, and the minina of σ_{yy} coincide with those of ρ_{xx}. The result of a calculation carried out in the Kubo formalism is shown in Fig.5 where the calculated $\rho_{xx} = \rho_\perp$ and $\rho_{yy} = \rho_\parallel$ curves are compared to experimental ones. In this experiment a higher modulation amplitude has been achieved than in previously published data taken from the same sample material [3]. The resulting higher amplitudes of the novel oscillations are therefore well described using a modulation amplitude $V_0 = 0.6meV$ in the calculations which is twice as large as in Ref. [3]. Details of the calculation are discussed by Gerhardts in this volume. The novel oscillations of $\rho_\perp$ are nicely reproduced by the calculation, which for $\rho_\parallel$ essentially yields the Drude result (independent of B). The temperature dependence of the novel oscillations is much weaker than that of the SdH oscillations, since the relevant energy is the distance between flat bands, which is much larger than the mean distance between adjacent bands. A similar model, based on the Boltzmann transport theory, has been used by Winkler et al. [6] to explain the additional oscillations. A semiclassical picture explaining the effect as a resonance between the periodic cyclotron orbit motion and the oscillating $\vec{E} \times \vec{B}$ drift of the orbit center induced by the potential grating has been suggested by Beenakker [8]. Such a semiclassical approximation, however, does not describe the whole physics of the observed phenomena as will be shown in the next section.

MAGNETOCAPACITANCE EXPERIMENTS

The oscillations of the LL linewidth, mentioned above, should in turn lead to oscillations of the density of states. Since for fixed magnetic field, the number of states per Landau band is fixed, oscillations of the band width should lead to oscillations of the

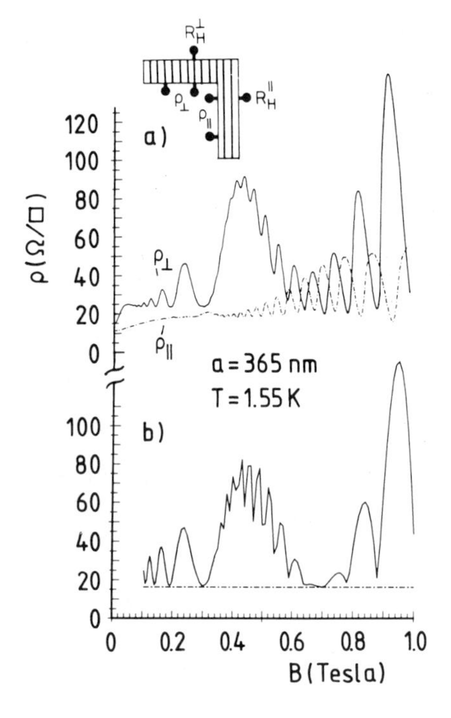

Figure 5. Magnetoresistivities for current perpendicular ($\rho_\perp$, solid lines) and parallel ($\rho_\parallel$, dash dotted lines) to the interference fringes, as indicated in the inset, for a sample with $N_S = 3.39 \cdot 10^{11} cm^{-2}$, $\mu = 1.1 \cdot 10^6 cm^2/Vs$, and $a = 365nm$, (a) measured and (b) calculated for temperature T=1.55K, using V_0=0.6meV. Calculation after Gerhardts Ref.[3]

peak density of states with maxima for flat bands satisfying Eq.(1). This behaviour is demonstrated experimentally in magnetocapacitance measurements. The capacitance between the semi-transparent gate and the 2-DEG is measured by applying an ac-voltage between the gate and one channel contact and measuring the out-of-phase ac-current with lock-in techniques. The oscillations of the capacitance as a function of the magnetic field are directly connected to the DOS at the Fermi energy [9,10,11,13]. In a homogeneous 2-DEG it has been shown experimentally that the LL linewidth (due to collision broadening) Γ has a magnetic field dependence of the form $\Gamma \propto B^\alpha$ with

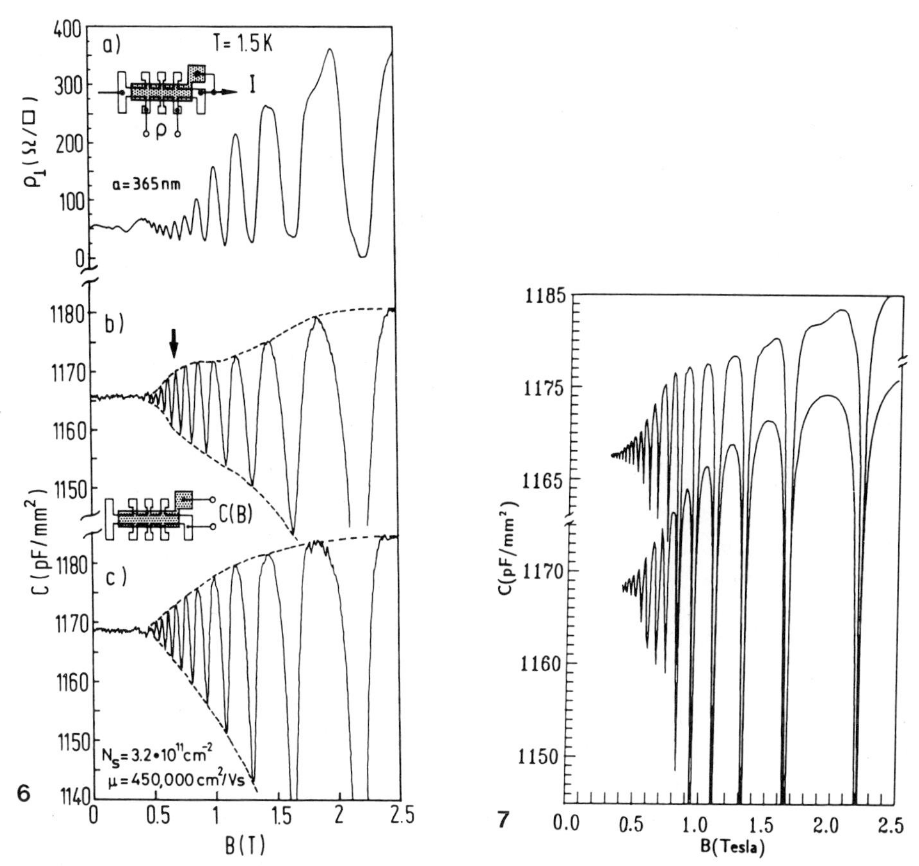

Figure 6. Measured magnetoresistivity $\rho_\perp$ (a) and magnetocapacitance (b) of a modulated sample compared to the capacitance of an essentially unmodulated sample (c). The arrow corresponds to the magnetic field value fulfilling Eq.1 for $\lambda = 1$. The insets sketch the measurements

Figure 7. Calculated magnetocapacitance versus magnetic field for $N_S = 3.2 \cdot 10^{11} cm^{-2}$ and $a = 365nm$. A B-independent linewidth is chosen to be $\Gamma = 0.3meV$. The upper curve is for $V_0 = 0.7meV$, and the lower one for the weak modulation $V_0 = 0.1meV$. After Zhang Ref.[16]

$0 \leq \alpha \leq 0.5$ [12,13], whereas theoretically $\Gamma \propto \sqrt{B}$ is expected for short range scatterers and a B-independent Γ for long range scatterers [14,15]. Since the Landau degeneracy is proportional to B, the peak values of the DOS in the individual LL's and, as a consequence, the peak values of the capacitance, are expected to increase monotonically, with a structureless envelope, with increasing magnetic field. On the other hand the envelope of the magnetocapacitance minima decreases monotonically with B due to the increasing LL separation $\hbar\omega_c$. In Fig.6 the magnetocapacitance data after an initial holographic illumination (90ms long) (b) is compared with the capacitance measured after an additional illumination which essentially smears out the periodic modulation (c). In Fig.6a the magnetoresistivity $\rho_\perp$ measured under the same experimental conditions as the magnetocapacitance data in Fig.6b is shown. The carrier density in Fig.6c has been adjusted to the same value as before the additional illumination using a negative gate voltage. In contrast to Fig.6c, where the magnetocapacitance behaves as one usually observes in a 2-DEG, the capacitance oscillations in Fig.6b display a pronounced modulation of both the minima and maxima which is easily explained from the energy spectrum plotted in Fig.4. At about 0.69T (marked by an arrow) where the cyclotron diameter at the Fermi level equals three quaters of the period a, $\rho_\perp$ in Fig.6a displays the last minimum ($\lambda = 1$) corresponding to the last flat band condition. Therefore the magnetocapacitance values near 0.69T are approximately equal in Fig.6b and Fig.6c. If now the magnetic field is increased, broader Landau bands are swept through the Fermi level, and cause the nonmonotonic behaviour visible in Fig.6b. At higher magnetic fields the level broadening saturates and the usual LL degeneracy again raises the DOS in a LL with increasing field. It should be mentioned that the modulation effect is observed for different angles between the one-dimensional modulation and the long axis of the Hall bar as is expected for a thermodynamic quantity in contrast, of course, to the magnetoresistivity. In order to check the magnetocapacitance data theoretically microscopic calculations of the density of states based on a generalisation of the well known selfconsistent Born approximation have been performed by Zhang [16]. Parts of this theory are also discussed by Gerhardts (this volume) and therefore I only present the results obtained from such calculations. In Fig.7 calculated magnetocapacitance data for a modulated (upper curve) and an essentially unmodulated 2-DEG are shown which are in good agreement with the experimental data. The collision broadening used in the calculations was $\Gamma = 0.3$meV in agreement with previous magnetocapacitance measurements carried out on homogeneous samples with similar mobility [13].

I am indepted to K. v.Klitzing and R.R. Gerhardts for many discussions and comments, D. Heitmann for helpful suggestions and C. Zhang for performing numerical calculations. I also want to thank K. Ploog and G. Weimann for providing me with high quality samples, and S. Bending for a critical reading of the manuscript. The work was supported in part by the Bundesministerium für Forschung und Technologie, West Germany.

REFERENCES

[1] L. Shubnikov, W. J. de Haas, *Leiden Commun.* **207a**, **207c**, **207d**, **210a** (1930)

[2] D. Weiss, K. v. Klitzing, K. Ploog, G. Weimann *Europhys. Lett.* **8**, 179 (1989; also in *The Application of High Magnetic Fields in Semiconductor Physics*, ed. G. Landwehr, Springer Series in Solid-State Sciences (Berlin), to be published

[3] R. R. Gerhardts, D. Weiss, K. v. Klitzing; *Phys. Rev. Lett.* **62**, 1173 (1989)

[4] K. Tsubaki, H. Sakaki, J. Yoshino, Y. Sekiguchi, *Appl. Phys. Lett.* **45**, 663 (1984)

[5] For recent work in this field see *Physics and Technology of Submicron Structures*, Vol. 83 of Springer Series Solid-State Sciences, ed. by H. Heinrich, G. Bauer, and F. Kuchar (Springer, Berlin, 1988)

[6] R. W. Winkler, J. P. Kotthaus, K. Ploog; *Phys. Rev. Lett.* **62**, 1177 (1989)

[7] M. Abramowitz and I. A. Stegun, ed., *Handbook of Mathematical Functions,* (Dover Publications, New York (1972))

[8] C.W.J. Beenakker, preprint

[9] T. P. Smith, B. B. Goldberg, P. J. Stiles, M. Heiblum, *Phys. Rev.* **B32**, 2696 (1985)

[10] V. Mosser, D. Weiss, K. v. Klitzing, K. Ploog, G. Weimann, *Solid State Commun.* **58**, 5 (1986)

[11] V. Gudmundsson, R. R. Gerhardts, *Phys. Rev.* **B35**, 8005 (1987)

[12] see e.g.: E. Gornik, R. Lassnig, G. Strasser, H. L. Störmer, A. C. Gossard, W. Wiegmann, *Phys. Rev. Lett.* **54**, 1820 (1985); J. P. Eisenstein, H. L. Störmer, V. Narayanamurti, A. Y. Cho, A. C. Gossard, *Phys. Rev. Lett.* **55**, 1820 (1985) 539 (1985)

[13] D. Weiss, K. v. Klitzing in *High Magnetic Fields in Semiconductor Physics*, ed. G. Landwehr, Springer Series in Solid-State Sciences **71**, (Berlin), 1987

[14] T. Ando, Y. Uemura, *J. Phys. Soc. Jpn.* **36**, 959 (1974)

[15] R. R. Gerhardts, *Z. Physik* **B21**, 285 (1975)

[16] D. Weiss, C. Zhang, R.R. Gerhardts, K. v. Klitzing, G. Weimann, submitted to *Phys. Rev. B*

THEORY OF GROUNDSTATE AND TRANSPORT PROPERTIES OF A TWO-DIMENSIONAL ELECTRON GAS IN THE PRESENCE OF A UNIDIRECTIONAL PERIODIC MODULATION AND A MAGNETIC FIELD

Rolf R. Gerhardts

Max-Planck-Institut für Festkörperforschung, Heisenbergstr. 1
D-7000 Stuttgart 80, Federal Republic of Germany

1. INTRODUCTION

In this paper we report recent theoretical work on groundstate and transport properties of a two-dimensional electron gas (2DEG) subjected to both a periodic modulation in one direction, say the x-direction, and a perpendicular homogeneous magnetic field B in z-direction, while translation invariance is maintained in y-direction.

Strong modulation can be obtained in gated structures where the gate electrode covers a linear grating produced by some etching procedure [1], so that the distance between the plane of the 2DEG and the gate is a periodic function of x. Applying a gate voltage to deplete the 2D channel becomes most effective where the distance is smallest. Since with increasing gate voltage the modulation increases whereas the average density N_S of the 2DEG decreases, the amplitude V_0 of the effective modulation potential can become as large as or larger than the Fermi energy E_F ($E_F = \hbar^2 k_F^2/2m$, $k_F = \sqrt{2\pi N_S}$), and eventually the 2DEG may split into quasi one-dimensional stripes [1]. In these strongly modulated systems interesting non-linear sreening effects are expected for zero magnetic field [2,3] and even more in strong magnetic fields [3,4]. These non-linear sreening effects have important implications for the density of states (DOS) and possibly for the relevant localization mechanism in a 2DEG under the conditions of the integer quantum Hall effect [4,5,6].

In the present paper we will consider, however, only the case of weak modulations, $V_0 \ll E_F$, in which the linear sreening approximation is an excellent one for $B = 0$. Furthermore, we will consider only weak magnetic fields with cyclotron energy $\hbar\omega_c \ll E_F$, so that many Landau levels (LL's) are occupied, and we will assume the effect of the magnetic field on the screening to be negligible in this situation.

Weakly modulated systems, where the modulation was produced by a holographic technique, have recently been investigated by Weiss et al. [7,8]. Exploiting the persistent photoconductivity effect in $Al_xGa_{1-x}As - GaAs$ heterojuntions at Helium temperatures, the sample was illuminated with the fringe pattern of two interfering laser beams and a periodically modulated density of charged donors was created in the $Al_xGa_{1-x}As$ layer. With this technique, both the average density N_S and the ($B = 0$) mobility μ_0 of the 2DEG increase with the modulation. From the increase ΔN_S of the density and the sample geometry one estimates $V_0/E_F \leq 0.1$, i.e. a weak modulation, for the experiments of Weiss et al. (see the Appendix). Nevertheless, interesting new effects are observed in both the magnetorésistance [7,8] and the magnetocapacitance [9,10] of these systems, which reflect the commensurability of the cyclotron radius of electrons at the Fermi level with the modulation period, and which are discussed in Sect.4 and Sect.3, respectively. In the following Sect.2 we define the theoretical model, and in Sect.5 we conclude with a short summary and the discussion of some open questions.

Science and Engineering of One- and Zero-Dimensional Semiconductors
Edited by S.P. Beaumont and C.M. Sotomajor Torres
Plenum Press, New York, 1990

2. THE MODEL

In order to keep the theory simple, we use the following approximation procedure [4,3]. We start with a three-dimensional model of an unmodulated heterostructure (without magnetic field) in the effective mass approximation,

$$[-\frac{\hbar^2}{2m}\nabla^2 + V_c(z) - E_\alpha]\Psi_\alpha = 0 \ , \tag{1}$$

and assume strong confinement, so that only the lowest subband with energies $E_\alpha = E_0 + \varepsilon_\alpha$ and eigenfunctions $\Psi_\alpha(\vec{r}) = \psi_\alpha(x,y)\chi_0(z)$ is partially occupied. In the following we assume

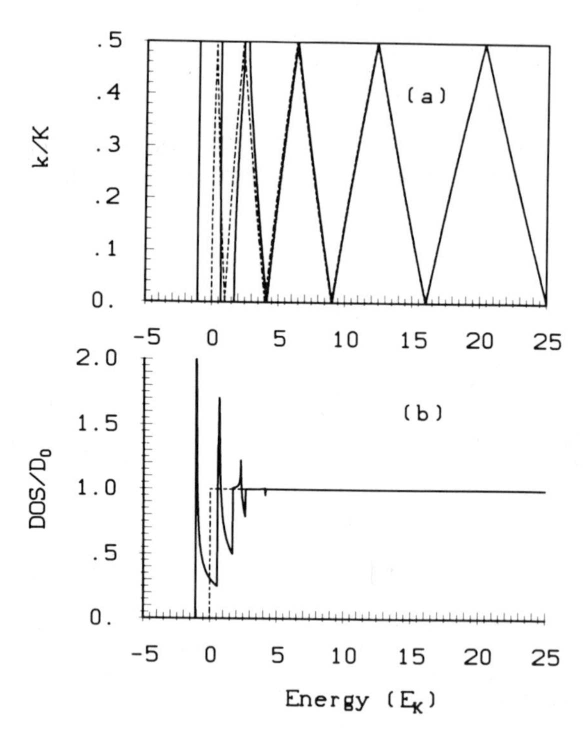

Fig. 1. (a) Bloch energy spectrum in x-direction, (b) two-dimensional density of states, for a 2DEG with modulation potential of eq.(7), where $V_o = 2E_K$ is the amplitude and $a = 2\pi/K$ is the period of the modulation, $E_K = \hbar^2K^2/2m$, $B = 0$.

that neither the magnetic field (in z-direction) nor the modulation in x-direction changes the self-consistently determined confinement potential V_c and, consequently, E_0 and $\chi_0(z)$. This is not strictly true. For a theory of magnetocapacitance, for instance, one needs to consider the effect of the Landau density of states on V_c [11]. The coupling of x- and z-direction by the modulation, on the other hand, may not be negligible for a 2DEG in an accumulation layer, where the higher electrical subbands are nearby in energy. For a 2DEG in an inversion layer the present approximation should, however, be sufficient.

We assume that a periodical arrangement of positive charges at some distance from the 2DEG gives rise to an 'external' modulation potential $\delta\bar{V}^{ext}(x,z)$ with period a in x-direction. An explicit example is given in the Appendix. Solving Poisson's equation with the induced electron density fluctuation

$$\delta n(x,z) = \chi_0^2(z)\delta N(x), \quad \delta N(x) = \sum_K \delta N_K e^{iKx}, \tag{2}$$

and taking average values with respect to $\chi_0(z)$ according to $\delta V^{ext}(x) = \int dz \chi_0^2(z) \delta \bar{V}^{ext}(x,z)$, one obtains for the total modulation potential

$$\delta V_K = \delta V_K^{ext} + \frac{2\pi e^2}{\kappa K} F(K) \delta N_K, \tag{3}$$

where κ (= 12.4 for GaAs) is a background dielectric constant assumed constant throughout the sample, and

$$F(K) = \int dz \int dz' \chi_0^2(z) \chi_0^2(z') \, exp(-K|z-z'|) \tag{4}$$

is a form factor [12], with $F(K) \approx 1$ if $2\pi/K$ is much larger than the spatial extent of $\chi_0(z)$. In actual calculations we will take this limit. If we include a vector potential $\vec{A} = (0, Bx, 0)$ in the Hamiltonian and exploit the translational invariance in y-direction, $\psi_\alpha(x,y) = L_y^{-\frac{1}{2}} exp(ik_y y) \phi_\alpha(x)$ with L_y a normalization length, Schrödinger's eq. (1) reduces to the one-dimensional form

$$[-\frac{\hbar^2}{2m} \frac{d^2}{dx^2} + \frac{m}{2} \omega_c^2 (x - x_o)^2 + V(x) - \varepsilon_\alpha] \, \phi_\alpha(x) = 0 \ , \tag{5}$$

with $V(x) = \sum_{K \neq 0} \delta V_K exp(iKx)$ the total screened modulation potential, $\omega_c = eB/(mc)$ the cyclotron frequency, $x_o = -l^2 k_y$ a center coordinate, and $l = (m\omega_c/\hbar)^{-\frac{1}{2}}$ the magnetic length. For $B = 0$, the parabolic potential $\frac{m}{2}\omega_c^2(x - x_o)^2$ reduces to the free electron energy in y-direction, $\hbar^2 k_y^2/2m$. For $B \neq 0$, the periodic modulation potential lifts the degeneracy of the Landau energies $\varepsilon_n = \hbar\omega_c(n + \frac{1}{2})$ of the unmodulated system and leads to Landau bands with the symmetry

$$\varepsilon_\alpha = \varepsilon_n(x_o) = \varepsilon_n(x_o + a) = \varepsilon_n(-x_o). \tag{6}$$

Thus, the energy spectrum needs to be calculated only for $0 \leq Kx_o < \pi$. The corresponding wavefunctions satisfy $\phi_{n,x_o+a}(x) = \phi_{n,x_o}(x - a)$.

3. GROUNDSTATE PROPERTIES

In the following we assume a weak modulation and weak magnetic fields, and we assume that the linear screening approximation is sufficient and magnetic field effects on the screening are neglected. To be specific, we take a fixed (screened) modulation potential

$$V(x) = V_0 \, cos Kx \ , \ K = 2\pi/a \ , \tag{7}$$

with $V_0 \ll E_F$ and $\hbar\omega_c \ll E_F$. Typical numbers will be $E_F \approx 11meV$, $V_0 \approx 0.5meV$, $a \approx 300nm$, $B < 1T$, i.e., $\hbar\omega_c < 1.73meV$, $l > 26nm$.

3.1 *Zero magnetic field*

For $B = 0$, Schrödinger's equation (5) with (7) reduces to the well known Mathieu problem [13]. The Bloch energy spectrum $\varepsilon_n(k_x)$ is easily calculated numerically and in y-direction one has the free energy spectrum $\hbar^2 k_y^2/2m$. The DOS can be written as

$$D(E) = D_o \sum_n \frac{2}{\pi} \int_0^{\pi/a} \frac{dk_x}{K} \frac{\Theta(E - \varepsilon_n(k_x))}{\sqrt{[E - \varepsilon_n(k_x)]/E_K}} \ , \tag{8}$$

where $\Theta(E)$ is the unit step function, and $D_o = m/(\pi\hbar^2)$ the constant free electron DOS of the 2DEG. Figure 1 shows numerical results for the Bloch energies in x-direction and for the DOS of the modulated 2DEG.

The parameters are chosen according to the experiment [7,8,14] as $a = 382nm$ and $V_0 = 0.3meV$, the energy scale is $E_K = \hbar^2 K^2/2m = 0.154meV$, so that the Fermi energy would appear in Fig.1b at 74. The dotted lines in Fig.1 indicate the results for the unmodulated system, i.e. a backfolded parabola in Fig.1a. We see the well-known feature of the Mathieu problem [13] that the band gaps decrease very rapidly with increasing band index. For energies $E > 4V_0$ no noticeable effect of the modulation is left.

3.2 *Finite magnetic field*

For $B \neq 0$ we start with the oscillator functions $\phi_n^{(0)}(x-x_o) = \langle x|x_o n\rangle^{(0)}$ of the unmodulated Landau system and calculate the matrix elements of $V(x)$ analytically (for $n' \leq n$ and with $X = \frac{1}{2}K^2 l^2$):

$$\langle x_o' n'|V|x_o n\rangle^{(o)} = \delta_{x_o,x_o'} (\frac{n'!}{n!})^{\frac{1}{2}} X^{\frac{n-n'}{2}} e^{-\frac{1}{2}X} L_{n'}^{(n-n')}(X)\, V_0\, Re(e^{iKx_o} i^{n-n'})\,, \tag{9}$$

where $L_n^{(\alpha)}(X)$ is a Laguerre polynominal [13]. The diagonal matrix elements yield the energy spectrum to first order in V_0 [15],

$$\varepsilon_n^{(1)}(x_o) = \varepsilon_n + U_n\, cosKx_o, \quad \varepsilon_n = \hbar\omega_c(n+\frac{1}{2}), \quad U_n = V_0\, e^{-\frac{1}{2}X} L_n(X)\,, \tag{10}$$

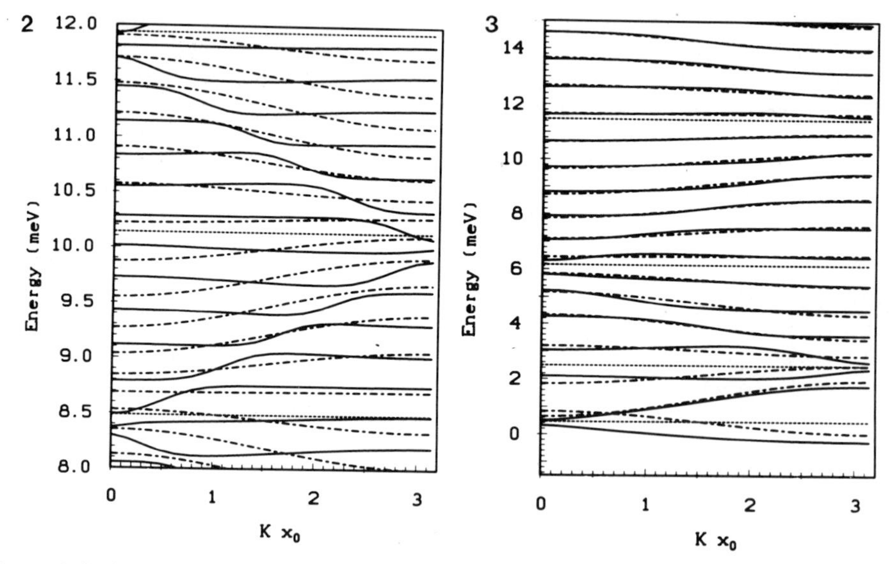

Fig. 2. Calculated energy bands $\varepsilon_n(x_o)$ for $B = 0.5T$, $V_0 = 1.5meV$, $a = 100nm$, and material parameters of GaAs ($m = 0.067m_o$); solid lines: numerical diagonalization, thick broken lines: first order approximation of eq.(10), thin dotted lines: energies of zero band width

Fig. 3. Same as Fig.2, but for $B = 0.15T$ and the restricted energy interval $8meV < E < 12meV$.

which is shown by the thick broken lines in Figs.2 and 3. The solid lines are obtained from an exact numerial diagonalization of the Hamiltonian matrix using a sufficiently large basis set. As seen from Fig.2, the first order approximation becomes very good for large band index n, i.e., for high energies. The energy values where this becomes true increase, however, with decreasing values of B. For instance, near $E = 11meV$ eq.(10) is a good approximation for $B = 0.5T$ (Fig.2), but a poor one for $B = 0.15T$ (Fig.3); and in the limit $B \to 0$ the first order approximation is not reliable. For small magnetic fields, the exactly calculated bands exhibit a zig-zag structure which is located along the backfolded free energy parabola $\varepsilon(k_y) = \hbar^2 k_y^2/2m$ in the Brillouin zone of width $\Delta k_y = a/l^2$ (note $x_o = -l^2 k_y$), and reflects the coupling of the motion in x- and in y-direction by the magnetic field. Such a structure has been obtained and explained previously [3] for a weak magnetic field and strong modulation.

Qualitatively, the most important feature of the energy spectrum is the oscillation of the width of the Landau bands $\varepsilon_n(x_o)$ as a function of n. Formally, this arises from the oscillatory dependence of the Laguerre polynominals on their index n. The energy values $\varepsilon_\lambda = \hbar\omega_c(n_\lambda + \frac{1}{2})$ at which the bands become flat are obtained approximately by solving the asymptotic formula for the zeroes of Laguerre polynominals [13],

$$L_n(X) = 0 \;\; if \;\; X = X_\lambda^{(n)} \approx \frac{1}{4}[\pi(\lambda - \frac{1}{4})]^2/(n + \frac{1}{2}) \; , \;\; \lambda = 1, 2, \dots \; . \tag{11}$$

for n, and are indicated in Figs.2 and 3 by the horizontal dotted lines. In terms of the cyclotron radius $R_n = l\sqrt{2n+1}$, eq.(11) reads

$$2R_n = a\,(\lambda - \frac{1}{4}) \tag{12}$$

and expresses the fact that the expectation value of the potential $V(x)$ becomes zero if the extent of the wavefunction equals (apart from the phase factor one quarter) an integer multiple of its period.

Since the number of states per Landau level depends only on the magnetic field, an oscillation of the level width has an effect on the DOS, which is easily calculated for the first order approximation (10),

$$D(E) = \frac{2}{2\pi l^2}\int_0^a \frac{dx_o}{a}\sum_n \delta(E - \varepsilon_n(x_o)) \approx \frac{1}{\pi^2 l^2}\sum_n \frac{\Theta(U_n^2 - (E - \varepsilon_n)^2)}{\sqrt{U_n^2 - (E - \varepsilon_n)^2}} \; , \tag{13}$$

and exhibits the van Hove-type singularities typical for a one-dimensional stucture. The center height (and the average height) of an individual LL is inversely proportional to its width. In Figs.4 and 5 (solid curves) we show the thermodynamic DOS

$$D_T(\mu) = \frac{\partial N_S}{\partial \mu} = \int dE\; D(E)\,[-\frac{d}{dE}\, f(E)] \; , \tag{14}$$

with $f(E) = [exp(\frac{E-\mu}{k_B T}) + 1]^{-1}$ the Fermi function, calculated from (13) for different values of magnetic field and temperature. In contrast to the $B = 0$ case, we now obtain drastic effects of the modulation up to high energies. Whereas the Landau quantization for unmodulated systems leads to oscillations with constant amplitude, indicated in Fig.4a by the dash-dotted lines, the modulated systems show a modulation of this amplitude, with maxima of the amplitude when (12) is satisfied. Figure 5 shows very clearly that this quantum beat effect can exist only if the fundamental Shubnikov-de Haas type oscillations can be resolved ($\hbar\omega_c \geq 4k_BT$).

Direct information about the thermodynamic DOS can be obtained from magnetocapacitance measurements [16], and quite recent results of Weiss et al. [9,10] show indeed a corresponding beat effect in the magnetocapacitance, in full agreement with theory.

4. MAGNETOTRANSPORT

We describe magnetotransport in the framework of Kubo's formula for the conductivity tensor, which can be expressed as [17]

$$\sigma_{\mu\nu}(\omega) = -i[\chi_{\mu\nu}(\omega) - \chi_{\mu\nu}(0)]/\omega \; , \;\; \chi_{\mu\nu}(0) = \delta_{\mu\nu}e^2 N_S/m, \tag{15}$$

in terms of the susceptibility

$$\chi_{\mu\nu}(\omega) = -\frac{2e^2}{L_x L_y}\sum_{\alpha,\alpha'}\frac{f(E_\alpha) - f(E_{\alpha'})}{E_\alpha - E_{\alpha'} + \hbar\omega}\langle\alpha|v_\mu|\alpha'\rangle\langle\alpha'|v_\nu|\alpha\rangle, \tag{16}$$

where $|\alpha\rangle$ and E_α are the exact orbital eigenstates and eigenvalues of the 2D Hamiltonian including the modulation potential, and $\omega = \omega + i0^+$ is understood. For a microscopic evaluation of scattering rates and line broadening effects, one should include the potentials of randomly distributed impurities in the Hamiltonian and finally take the average of $\sigma_{\mu\nu}$ over the impurity configurations. We adress these questions below. Here we concentrate on the most striking new effects in modulated systems related to the bandwidth oscillations discussed in Sect.3.

4.1 *Simple damping theory*

For a discussion of the consequences of the linewidth oscillations on the transport properties we evaluate the conductivity within a simple damping approximation, replacing the frequency ω by i/τ, where $\tau^{-1} = \gamma/\hbar$ has the meaning of a scattering rate. Then, using the eigenstates $|x_o n\rangle$ of (5), we obtain

$$\sigma_{\mu\mu} = \frac{2e^2\hbar}{2\pi l^2}\int_0^a \frac{dx_o}{a}\sum_n \left\{ -\frac{1}{\gamma}\frac{df}{dE}(\varepsilon_n(x_o))\,|\langle x_o n|v_\mu|x_o n\rangle|^2 \right.$$
$$\left. - \sum_{n'(\neq n)} \frac{f(\varepsilon_n(x_o)) - f(\varepsilon_{n'}(x_o))}{\varepsilon_n(x_o) - \varepsilon_{n'}(x_o)} \frac{\gamma|\langle x_o n|v_\mu|x_o n'\rangle|^2}{[\varepsilon_n(x_o) - \varepsilon_{n'}(x_o)]^2 + \gamma^2} \right\}, \quad (17)$$

$$\sigma_{yx} = \frac{2e^2\hbar}{i2\pi l^2}\int_0^a \frac{dx_o}{a}\sum_{n,n'} \frac{f(\varepsilon_n(x_o)) - f(\varepsilon_{n'}(x_o))}{[\varepsilon_n(x_o) - \varepsilon_{n'}(x_o)]^2 + \gamma^2}\langle x_o n|v_y|x_o n'\rangle\langle x_o n'|v_x|x_o n\rangle. \quad (18)$$

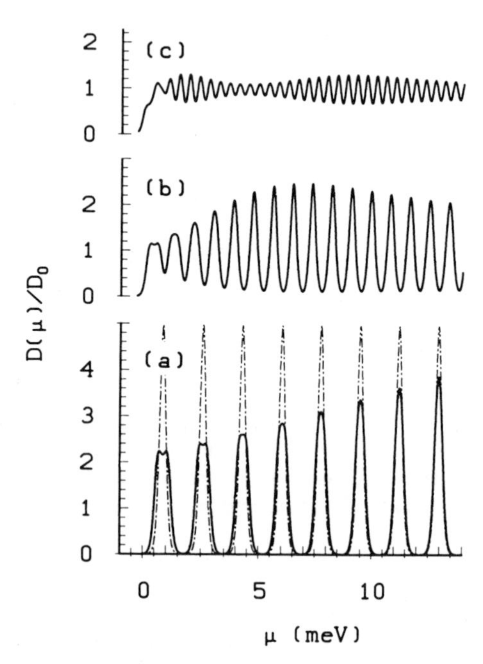

Fig. 4: Thermodynamic DOS vs. chemical potential for temperature $T = 1K$ and magnetic field $B = 1T$ (a), $0.5T$ (b), and $0.25T$ (c). The modulation is given by $V_o = 0.3meV$, $a = 382nm$. The dash-dotted curves in (a) are for $V_o = 0$.

For the homogeneous Landau system the velocity matrix elements are off-diagonal, the first term in (17) gives no contribution, and the other terms are easily evaluated to yield the classical Drude result,

$$\sigma^D_{\mu\mu} = \sigma_o/[1 + (\omega_c\tau)^2]\,, \quad \sigma^D_{yx} = \omega_c\tau\;\sigma^D_{xx}\,, \quad \sigma_o = e^2 N_S \tau/m. \quad (19)$$

Without any scattering, the current would be perpendicular to the driving electric field, and for high-mobility systems, $\omega_c\tau \gg 1$, $\sigma^D_{\mu\mu} \propto \frac{1}{\tau}$ becomes small.

For the modulated system, however, the situation is very different and the first term of (17) does contribute to σ_{yy}, since now the eigenstates carry current in y-direction (a local Hall drift), and the velocity operator v_y has finite diagonal matrix elements

$$\langle x_o n|v_y|x_o n\rangle = -\frac{1}{m\omega_c}\frac{d\varepsilon_n}{dx_o} = \frac{1}{\hbar}\frac{d\varepsilon_n}{dk_y} \,. \tag{20}$$

Moreover, for high-mobility samples, i.e. for small scattering rate ($\gamma \ll |\varepsilon_n(x_o) - \varepsilon_{n\pm1}(x_o)| \approx \hbar\omega_c$), the first contribution to σ_{yy}, call it $\Delta\sigma_{yy}$, becomes large ($\propto \tau$, similar to σ_0) and eventually dominates the second one being comparable in magnitude with σ_{xx}, for which the first term in (17) still vanishes since $\langle x_o n|v_x|x_o n\rangle = 0$. Thus, for a high-mobility 2DEG the magnetoconduc-

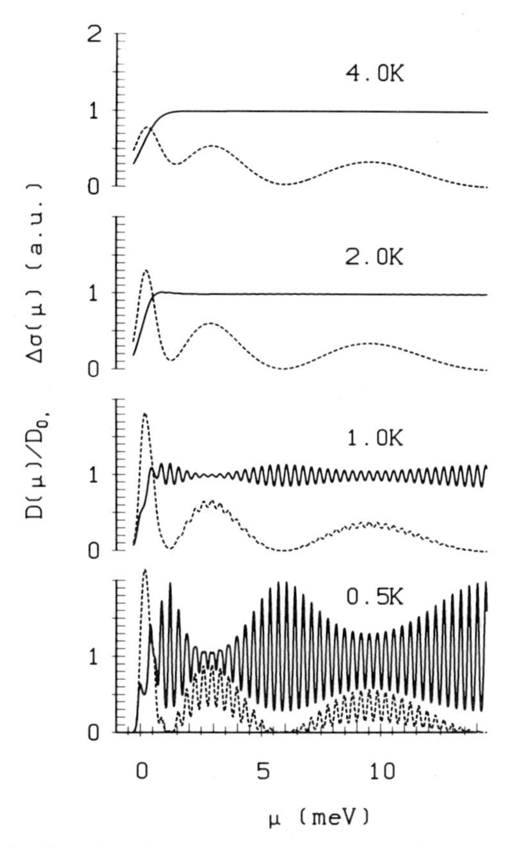

Fig. 5. Solid lines: as in Fig.4 but for $B = 0.2T$ and different temperatures. The dotted lines depict the corresponding novel contributions $\Delta\sigma_{yy}$ to the conductivity (first term in eq.(17)).

tivity tensor becomes very anisotropic. Moreover, the additional term $\Delta\sigma_{yy}$ arising from the diagonal matrix elements (20) exhibits oscillations as, for varying magnetic fields, Landau bands with different band index pass the Fermi level. If the Landau band at the Fermi energy is flat, i.e. satisfies the condition (12), the matrix elements (20) for states $|x_o n\rangle$ near the Fermi energy are small and $\Delta\sigma_{yy}$ exhibits a minimum. For the small magnetic field values of interest, the band index n of the $\varepsilon_n(x_o)$-band closest to the Fermi level can be estimated as $n+\frac{1}{2} = E_F/\hbar\omega_c = \frac{1}{2}l^2k_F^2$, and the cyclotron radius at the Fermi level becomes $R_n = l^2k_F$. Inserting this into (12), we obtain minima of $\Delta\sigma_{yy}$ if the condition

$$\frac{c}{B_\lambda} = \frac{ea}{2\hbar k_F} (\lambda + \phi) , \quad \lambda = 1, 2, ..., \tag{21}$$

is satisfied for $\phi = -\frac{1}{4}$. Maxima are expected at B_λ-values satisfying (21) with ϕ somewhat smaller than $+\frac{1}{4}$, owing to the exponential prefactor in (10).

In Fig.5, $\Delta\sigma_{yy}$ (dashed lines) is plotted for a fixed value of the magnetic field ($B = 0.2T$) and for several temperatures as a function of the Fermi energy (the chemical potential), together with the thermodynamic DOS. In contrast to the modulation-induced beating effect in the latter, the large-period oscillations of $\Delta\sigma_{yy}$ survives at higher temperatures where the short-period SdH-type oscillations are no longer resolved. The reason is that the relevant energy is the distance of flat bands near the Fermi energy, which can be estimated from (11) as $\Delta\varepsilon \sim 2E_F/(\lambda - \frac{1}{4})$, and is considerably larger than the distance between adjacent bands, being of the order $\hbar\omega_c$ (cf. Figs.2 and 3).

This theory nicely explains the novel magnetotransport oscillations first observed by Weiss et al. [7,8,14]. If we adapt our model parameters to these experiments, evaluate (17) and (18), and calculate resistivities,

$$\varrho_{xx} = \sigma_{yy}/[\sigma_{yx}^2 + \sigma_{xx}\sigma_{yy}] , \quad \varrho_{yy} = \sigma_{xx}/[\sigma_{yx}^2 + \sigma_{xx}\sigma_{yy}] , \tag{22}$$

we find that σ_{yx} exhibits no noticeable oscillations and that $\sigma_{yx}^2 \gg \sigma_{xx}\sigma_{yy}$ holds for ($0.1T < B < 1T$). As a consequence, ϱ_{xx} reflects the oscillating behavior of σ_{yy}, whereas ϱ_{yy} shows essentially no structure. The Hall resistance ϱ_{xy} was also calculated and shows the linear B-dependence of the Drude result, without any noticeable oscillations, consistent with experiment [7,8].

Figure 6 shows resistivities calculated from formulae (17) and (18). In Fig.6a the system parameters have been chosen according to the experiment [14]. In order to demonstrate the dependence of the novel oscillations on several parameters, part (b) and (c) of Fig.6 exhibit results for a system with a density N_S and a period a which are smaller than the corresponding values in Fig.6a by factors 2 and $\sqrt{2}$, respectively. In agreement with the formula (21), which first was extracted empirically from the experiments [7,8], the position of the novel oscillations is the same as in (a), whereas the period of the SdH oscillations is larger by a factor 2.

Similar magnetoresistance oscillations in 'conventionally' microstructured samples have recently been reported by Winkler et al. [18] and explained along very similar theoretical ideas. The calculations are restricted to the first order approximation (10) and, in addition, use of the quasi classical large-n limit

$$e^{-\frac{1}{2}X} \; L_n(X) \approx \pi^{-\frac{1}{2}} \; (nX)^{-\frac{1}{4}} \; cos(2\sqrt{nX} - \frac{\pi}{4}) \; . \tag{23}$$

was made. In this limit the first term in (17) may be evaluated as (with $R_F = l^2 k_F$)

$$\Delta\sigma_{yy} \approx \frac{e^2}{2\pi\hbar} \, \frac{V_o^2}{\gamma\hbar\omega_c} \, \frac{4}{ak_F} \, cos^2(2\pi\frac{R_F}{a} - \frac{\pi}{4}) \tag{24}$$

The first order approximation is sufficient in the magnetic field region $B > 0.2T$ but fails in the limit $B \to 0$ (cf. Fig.3) where it yields oscillations of $\Delta\sigma_{yy}$ with diverging amplitudes. Winkler et al. compensate this deficiency by the neglect of the $\sigma_{xx}\sigma_{yy}$-term in the denominator of (22), and put $\varrho_{xx} \approx \sigma_{yy}/\sigma_{yx}^2 \approx \sigma_{yy} \cdot B^2/(N_S e)^2$, which is also incorrect for $B \to 0$ [cf. (19)]. Our numerical calculations are free of these deficiencies.

Figure 5 shows that the magnetotransport oscillations should persist in a range of temperatures and magnetic field values where the quantum beat effects in the DOS are no longer resolved. This is indeed in agreement with the experimental findings [10], which resolve for a given temperature the magnetoresistance oscillations down to much smaller magnetic fields than the magnetocapacitance oscillations. Thus, one might argue that the magnetotransport oscillations are not a genuine quantum effect and may be understood in classical terms. Indeed, recently Beenakker [20] has presented a quasi classical calculation of the mean-square Hall-drift velocity at the Fermi energy which, when inserted into the diffusion constant, explains (via Einstein's relation) the novel oscillations of σ_{yy}, too. From the classical point of view it seems, however, not possible to understand the common origin of the novel magnetoresistance oscillations and the beating effect observed in the magnetocapacitance oscillations.

4.2 *Self-consistent damping theory*

The major shortcoming of the simple damping approximation is the poor treatment of SdH oscillations, which show up only in the additional term $\Delta\sigma_{yy}$, whereas they cancel in the other terms, essentially as for the unmodulated case, where this approximation yields the classical Drude result (19). The reason for this shortcoming is the asssumption of a constant relaxation

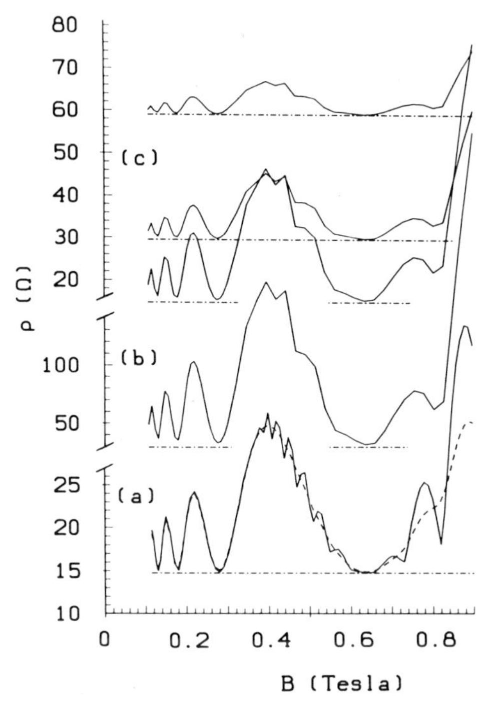

Fig. 6. Calculated resistivities ϱ_{xx} (solid lines) and ϱ_{yy} (dash-dotted lines) for temperature $T = 2.2K$; (a) parameters as in experiment [14]: electron density $N_S = 3.16 \cdot 10^{11} cm^{-2}$, period $a = 382nm$, scattering rate $\gamma = 0.013meV$ (corresponding to mobility $1.3 \cdot 10^6 cm^2/Vs$), and potential amplitude $V_0 = 0.3meV$, the dashed line indicates ϱ_{xx} for $T = 4.2K$; (b) as in (a) but for $N_S = 1.58 \cdot 10^{11} cm^{-2}$ and $a = 270nm$; (c) as in (b) but for $V_0 = 0.1meV$ and $\gamma = 0.0065$, 0.013, and $0.026meV$, respectively (from bottom to top).

rate γ, whereas a microscopic quantum mechanical theory of collision broadening [19,17,12] yields a linewidth of LL's and a transport relaxation rate exhibiting the same SdH-type oscillations as the DOS. We can include such effects in addition to the modulation broadening by a simple ansatz for the impurity-averaged Green's function $G^-_{x_o n}(E) = [E - \varepsilon_n(x_o) - \Sigma^-(E)]^{-1}$ with a quantum-number-independent self-energy determined by the self-consistency equation

$$\Sigma^-(E) = \Gamma_o^2 \sum_n \int_0^a \frac{dx_o}{a} G^-_{x_o n}(E) = \sum_n \frac{\Gamma_o^2}{\{[E - \varepsilon_n - \Sigma^-(E)]^2 - U_n^2\}^{\frac{1}{2}}}, \tag{25}$$

where a high-energy cut off $n \leq 2E_F/\hbar\omega_c$ is used, and both $G^-_{x_o n}(E)$ and $\Sigma^-(E)$ are analytical functions with non-negative imaginary parts in the complex halfplane $\text{Im}E < 0$. For the unmodulated system, $U_n \equiv 0$, and with $\Gamma_o^2 = \frac{2}{\pi}\hbar\omega_c\frac{\hbar}{\tau}$ eq.(25) reduces to the usual self-consistent Born approximation (SCBA) for δ-potential impurities, where τ is the corresponding lifetime for zero magnetic field [17,12]. The solution of (25) yields the DOS (including spindegeneracy)

$$D(E) = \frac{2}{2\pi l^2}\frac{1}{\pi}\mathrm{Im}\left[\frac{\Sigma^-(E)}{\Gamma_o^2}\right] = \frac{1}{\pi^2 l^2}\sum_n \frac{\sqrt{w_n - \eta_n}}{\sqrt{2}\, w_n}, \tag{26}$$

with $\eta_n = [E - \varepsilon_n - \Delta(E)]^2 - U_n^2 - [\frac{1}{2}\Gamma(E)]^2$, $w_n = \{\eta_n^2 + \Gamma(E)^2[E - \varepsilon_n - \Delta(E)]^2\}^{\frac{1}{2}}$, and $\Sigma^-(E) = \Delta(E) + \frac{i}{2}\Gamma(E)$. In the limit $\Gamma_o \to 0$, eq.(26) reduces to (13). The thermodynamic DOS and the magnetocapacitance have been calculated in this approximation, and a good understanding of the corresponding experimental results was obtained [9,10].

Since the self-energy is assumed independent of the quantum numbers, in the corresponding approximation for the current-current correlation function vertex corrections must be neglected [17], and eq.(16) should be replaced by

$$\begin{aligned}\chi_{\mu\nu}(\omega) \;=\; & -\frac{2e^2}{2\pi l^2}\int_{-\infty}^{+\infty} dE\; f(E) \int_0^a \frac{dx_o}{a} \sum_{nn'} \frac{1}{2\pi i}[G_{x_on}^-(E) - G_{x_on}^+(E)] \\ & \cdot[\langle x_on|v_\mu|x_on'\rangle G_{x_on'}^+(E+\hbar\omega)\langle x_on'|v_\nu|x_on\rangle \\ & \qquad -\langle x_on|v_\nu|x_on'\rangle G_{x_on'}^-(E-\hbar\omega)\langle x_on'|v_\mu|x_on\rangle].\end{aligned} \tag{27}$$

The evaluation of the resistivities within this approach takes into account both the well known SdH oscillations and the novel oscillations, and is now in progress.

5. SUMMARY

The periodic modulation of a 2DEG in one direction produces interesting new effects in the presence of a perpendicular magnetic field, even if both, modulation and field, are small. When many Landau levels are occupied, the classical cyclotron diameter of electrons at the Fermi level is larger than the period of the modulation, and commensurability effects leading to an oscillatory dependence of the width of the modulation-broadened LL's on the band index n become observable. This is the common origin of the novel magnetoresistance oscillations and interference patterns in the magnetocapacitance oscillations observed recently. The theory discussed in this paper successfully explains the main features of the experiments, but leaves some questions open.

The strong positive magnetoresistance observed experimentally in ϱ_{xx} at very small magnetic fields ($B < 0.02T$) is out of the scope of the present theory.

Preliminary results of the self-consistent damping approach discussed in Sect.4.2 indicate that the anti-phase oscillations observed experimentally in ϱ_{yy} can be reproduced by the microscopic theory of the scattering rate. Comparison with the experiment indicates, however, that the transport scattering rate is very different from the line width of the LL's, just as for unmodulated systems [16]. This emphasizes the importance of long-range scatterers, which lead to a more complicated self-energy than the simple ansatz (25).

Finally we want to mention that the role of screening effects, which is very crucial and non-trivial for strong magnetic fields, has not yet been investigated carefully in the case of small modulations and small magnetic fields considered in this paper.

I am grateful to D. Weiss and K. v.Klitzing for many discussions on the subject and for providing me with information on many additional experimental results, and I appreciate stimulating discussions with J. Hajdu in the early stages of this work. I am also grateful to C. Beenakker for sending a preprint [20] on his attempt to explain the novel magnetoresistance oscillations classically. Last but not least, I gratefully acknowledge many clarifying discussions with C. Zhang, who works on the numerical evaluation of the self-cosistent damping approximation. The work was supported in part by the Bundesministerium für Forschung und Technologie, West Germany (under grant No. NT-2718-C).

APPENDIX

We assume that the modulation potential, with period $a = 2\pi/K$, seen by the 2DEG centered at the x-y-plane ($z = 0$) is created by a periodically modulated density of positive charges centered around a plane $z = -D$. To be specific, we assume ionized donors in a layer of thickness $2d$ with a harmonic density modulation

$$\delta n_K(\vec{r}) = \frac{1}{2d}\,\delta N_D\,\Theta(d - |z + D|)\,cosKx\ ,$$

where δN_D is the modulation amplitude of the area density of positive charges. For $z > -D + d$, this produces a contribution $\delta V(\vec{r}) = \delta V^{ext}\,e^{-Kz}\,cosKx$ to the potential energy of the electrons, with

$$\delta V^{ext} = -\frac{2\pi e^2}{\kappa K}\,e^{-KD}\,\frac{sinhKd}{Kd}\,\delta N_D\ ,$$

where we assumed the same static dielectric constant κ throughout the sample. If we take typical parameters from the experiment [8,14], $a = 382nm$, $2d = 33nm$, and $D \approx 60nm$ (assuming the 2DEG centered about $10nm$ above the $Al_xGa_{1-x}As - GaAs$ interface), we may safely neglect the finite thickness d of the donor layer, whereas $exp(-KD) \approx 0.4$. If we completely neglect the extent of the 2DEG in z-direction, and assume linear screening with the Thomas-Fermi dielectric function of the 2DEG [4,3], $\epsilon(K) = 1 + 2/(a_B K)$, where $a_B = \kappa\hbar^2/(me^2) = 9.78nm$ is the effective Bohr radius of $GaAs$, we obtain for the induced modulation of the electron density

$$\delta N_S = [1 - 1/\epsilon(K)]\,exp(-KD)\,(Kd)^{-1}\,sinh(Kd)\,\delta N_D\ ,$$

and for the amplitude of the effective, screened potential $V_o = \pi\frac{e^2}{\kappa}a_B\delta N_S$. With the above numbers and the effective Rydberg energy $e^2/(2\kappa a_B) = 5.93meV$, we obtain the estimates $\epsilon(K) \approx 13.5$, $\delta N_S \approx 0.4\delta N_D$, and $V_o \approx 3.6meV \cdot [\delta N_S \cdot 10^{-11}cm^2]$. From the total increase $\Delta N_S = 8 \cdot 10^{10}cm^{-2}$ of the electron density during illumination [8] we get an upper limit $\delta N_D \leq \Delta N_S$, and from the sample structure we estimate $\delta N_S \approx 0.4 \cdot \delta N_D$, which yields as an upper limit $V_o^{>} \leq 1.2meV$ and $\delta N_S \approx 3 \cdot 10^{10}cm^{-2}$. To reproduce the amplitude of the novel ϱ_{xx} oscillations with the simple damping calculation, we need an amplitude $V_o \approx \frac{1}{4}V_o^{>}$ for the experiment of Refs. [8,14], and about $V_o = \frac{1}{2}V_o^{>}$ for experiments performed at a lower temperature [10]. This is reasonable if during the holographic illumination a temperature-dependent diffusion of electrons between the donors takes place.

REFERENCES

[1] W. Hansen, M. Horst, J. P. Kotthaus, U. Merkt, Ch. Sikorski, and K. Ploog, *Phys. Rev. Lett.* **58**, 2587 (1987)

[2] U. Wulf, *Phys. Rev. B* **35**, 9754 (1987); *B* **38**, 2187 (1988)

[3] U. Wulf and R. R. Gerhardts, in *Physics and Technology of Submicron Structures*, eds.: H. Heinrich, G. Bauer, and F. Kuchar, Springer Series in Solid-State Sciences, Vol. **83** (Spriger-Verlag,Berlin, 1988), p. 162

[4] U. Wulf, V. Gudmundsson, and R.R. Gerhardts, *Phys. Rev. B* **38**, 4218 (1988)

[5] V. Gudmundsson and R. R. Gerhardts, in *High Magnetic Fields in Semiconductor Physics II*, edited by G. Landwehr, Springer Series in Solid-State Sciences (Springer-Verlag, Berlin, to be published)

[6] A. L. Efros, *Solid State Commun.*, in press; ibid. **65**, 1281 (1988); **67**, 1019 (1988)

[7] D. Weiss, K. v. Klitzing, K. Ploog, and G. Weimann, in *High Magnetic Fields in Semiconductor Physics II*, ed. G. Landwehr, Springer Series in Solid-State Sciences (Springer-Verlag, Berlin, to be published)

[8] D. Weiss, K. v. Klitzing, K. Ploog, and G. Weimann, *Europhys. Lett.* **8**, 179 (1989)

[9] D. Weiss, C. Zhang, R. R. Gerhardts, K. v.Klitzing, and G. Weimann, *Phys. Rev. B*, submitted

[10] D. Weiss, *this volume*

[11] V. Gudmundsson and R. R. Gerhardts, *Phys. Rev. B* **35**, 8005 (1987)

[12] T. Ando, A. B. Fowler, and F. Stern, *Rev. Mod. Phys.* **54**, 437 (1982)

[13] M. Abramowitz and I. A. Stegun, (eds.), Handbook of Mathematical Functions, Dover Publications, New York (1972)

[14] R. R. Gerhardts, D. Weiss, and K. v.Klitzing, *Phys. Rev. Lett.* **62**, 1173 (1989)

[15] A. V. Chaplik, *Solid State Commun.* **53** , 539 (1985)

[16] see, e.g., D. Weiss and K. v.Klitzing, in *High Magnetic Fields in Semiconductor Physics*, edited by G. Landwehr, Springer Series in Solid-State Sciences Vol. **71** (Springer-Verlag, Berlin, 1987), p. 57

[17] see, e.g., R.R. Gerhardts, *Z. Physik B* **22**, 327 (1975); and further references therein

[18] R. W. Winkler, J. P. Kotthaus, and K. Ploog, *Phys. Rev. Lett.* **62**, 1177 (1989)

[19] T. Ando and Y. Uemura, *J. Phys. Soc. Jpn.* **36**, 959 (1974)

[20] C.W. J. Beenakker, *preprint*

UNDERSTANDING QUANTUM CONFINEMENT IN ZERO-DIMENSIONAL NANOSTRUCTURES: OPTICAL AND TRANSPORT PROPERTIES

Garnett W. Bryant

McDonnell Douglas Research Laboratories
P. O. Box 516
St. Louis, Missouri, USA 63166

INTRODUCTION

In zero-dimensional semiconductor nanostructures with motion confined in all directions, electronic states are discrete. In contrast, the spectrum of single-particle states in a quantum well or quantum-well wire is a set of subbands of two- or one-dimensional states, respectively. Each subband is a continuum of states. Because the single-particle spectrum for a zero-dimensional quantum box is discrete rather than a continuum, understanding confinement effects in these systems presents unique challenges not addressed for wells and wires.

The near band-edge optical properties of a semiconductor system are determined by the exciton states of the system. In a zero-dimensional structure, the exciton states are determined by a competition between the confinement effects and the correlation of the electron-hole pair induced by the Coulomb interaction. The single-particle levels, with splittings which scale as $1/L^2$ where L is the size of the quantum box, are determined by the confinement. The Coulomb energies scale as $1/L$. As the confinement in a zero-dimensional structure increases, the Coulomb-induced mixing of states becomes more difficult, even though the Coulomb energies increase, and electrons and holes become frozen in the lowest energy single-particle states. Freeze-out of motion in the confined dimensions also occurs in quantum wells and quantum-well wires when the confinement is increased. However, Coulomb-induced correlation of the unconfined motion in a well or wire is enhanced by the confinement. One unique challenge to understanding excitons confined in boxes is to correctly describe the transition from the correlated exciton in the bulk limit to the uncorrelated pair in the limit of complete confinement. Determining the length scale for this transition is necessary to accurately model the optical response of boxes because excitons and uncorrelated pairs have different linear and nonlinear responses.

The effects of three-dimensional confinement in microcrystallites[1-3] have been observed by use of exciton luminescence. However, exciton luminescence of quantum box nanostructures made by laterally confining motion in two-dimensional semiconductor quantum wells[4-9] does not yet provide conclusive evidence for three-dimensional confinement in quantum-box structures. Recently Reed et al.[10] investigated transport through a three-dimensionally

Science and Engineering of One- and Zero-Dimensional Semiconductors
Edited by S.P. Beaumont and C.M. Sotomajor Torres
Plenum Press, New York, 1990

confined quantum well to provide another characterization of the quantum box that might reveal the discrete density of confined states. Reed et al. investigated a structure (see Fig. 1(a)) which had an n = 1 quantum-well resonance deep in the confined well and an excited state (n = 2) resonance near the top of the confining barrier. Resonant tunneling through the n=2 quantum-well resonance was observed. At high temperature, the current/voltage characteristics are similar to the characteristics for normal resonant tunneling through two-dimensional wells and barriers. At low temperatures (T ~ 1.0 K) additional fine structure was superimposed on the normal RT characteristics (see Fig. 1(b)). The lowest two fine-structure peaks were split by bias of 80 meV. Other peaks were split by 50 meV, which is consistent with estimates for the splitting of the quantum-well subbands into discrete levels by lateral confinement.

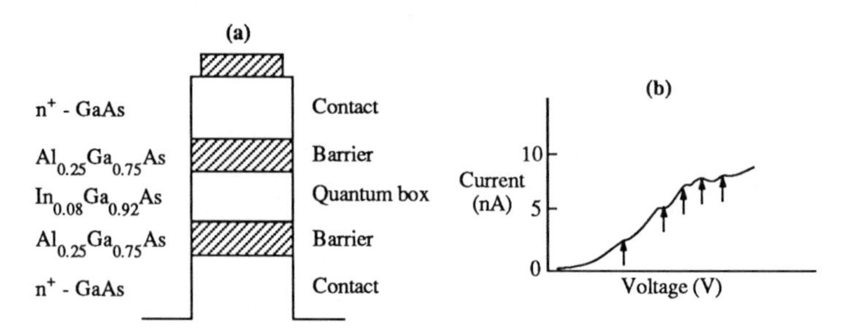

Fig. 1. (a) Schematic of the nanostructure used to observe quantum box resonant tunneling. (b) Low-temperature (T~1.0 K) current/voltage characteristics of an isolated quantum box nanostructure. Tunneling occurs through the n = 2 quantum well resonance.

Quantum box resonant tunneling (QBRT) differs from normal RT in two important respects. First, the density of lateral states is discrete in QBRT. Second, there is no analog to the conservation of momentum rule. If the lateral confinement potential is different for different regions (contacts, barriers and box), then the laterally quantized states will have different energies and wave functions in the different regions. A lateral state in one region can couple to a lateral state in the adjacent region if the overlap between the states is nonzero. An electron initially in a specific source contact-lateral-level can tunnel through any box-lateral-level which couples to the contact-lateral-level due to wave-function-mixing at the interfaces. The tunneling is a multichannel process rather than a collection of single channel processes for a continuum of subband states as in quantum well tunneling. A unique challenge to understanding QBRT is to determine whether these two differences can explain the fine structure observed in QBRT.

In this chapter these challenges will be explored and the optical and transport properties of quantum box systems will be discussed. Two other related topics will be presented. The modeling of quantum boxes will be described to point out issues which must be addressed to incorporate the relevant physical effects. The exploitable properties of quantum boxes will be those properties which persist when the boxes are fabricated in arrays for use in devices. The modification of the properties of isolated quantum boxes that occurs when the boxes are in an array where interbox tunneling is possible will be discussed.

MODELING QUANTUM BOXES

Several issues must be addressed to accurately model quantum boxes. Unlike quantum wells and wires which have many charge carriers, quantum boxes will only contain a few carriers. For example, a box with 10 nm sides will contain less than one particle when the charge density is less than 10^{18} cm^{-3}. The consequences of having only a few particles in a box must be included in any model for quantum boxes. For example, exchange and screening are ineffective in quantum boxes.[11] While the nonlinear optical response of wells has contributions from exchange, screening and band filling, only band filling is important for the nonlinear optical response of small boxes.[11]

One simplifying approximation used for quantum wells which can be extended to quantum boxes is the effective mass approximation. The effective mass approach is adequate for quantum wells which are wider than 1 nm. Typical quantum box dimensions are greater than 1 nm.

If the quantum box and the confining barrier are made from materials with significantly different dielectric constants, then image charges will make an important contribution to the electrostatic potential acting on charges inside the box.[3] For microcrystallites in glasses and for clusters in air, the dielectric mismatch can be large. For semiconductor nanostructures made from GaAs boxes (ϵ = 13) and AlGaAs barriers ($10 < \epsilon < 13$) the mismatch is much smaller and usually ignored.

In principle, several boundary effects should be modeled carefully. In practice simplifying approximations can often be made. Quantum boxes which are fabricated from two-dimensional quantum wells (with width w) by confining lateral motion typically have sides of length L with L $\gg$ w. These boxes are thin plates or disks and are typically circular. However, the choice of a shape is not important because shape effects are minimal for structures with the same cross-sectional area.[12] Barriers for isolated quantum boxes are often assumed to be infinite. Leakage of the carriers out of the box and into the surrounding barrier is only important in very small ($L \lesssim 5$ nm) structures.[13] The effective width L of the confining structure, as determined by the depth of sidewall depletion or other patterning to define the confining potential, can be much less than the physical dimension of the structure. The magnitude of confinement effects is a sensitive function of L, so L must be modeled accurately.

EXCITONS CONFINED IN QUANTUM BOXES

Exciton states are determined by solving the electron-hole effective-mass Schrodinger equation by use of a variational approach, as has been done for excitons in bulk and in quantum wells. The electron-hole interaction is the Coulomb interaction screened by the background dielectric constant. Details of these calculations for excitons confined in boxes and references to related work are given in Ref. 14-16.

The ground-state energy E_{GS} of the confined heavy-hole exciton in a square, two-dimensional (w=0) box is shown in Fig. 2. Since w = 0, no confinement energy due to perpendicular motion is included. Results are qualitatively the same when the box has a finite thickness as long as $w \lesssim L$ (i.e., $w \lesssim 5$ nm which is typical of quantum boxes). Results are also qualitatively the same for light-hole excitons. Effective masses and dielectric constant for GaAs have been used. The energies are scaled by the exciton effective Rydberg R_μ = 3.03 meV.

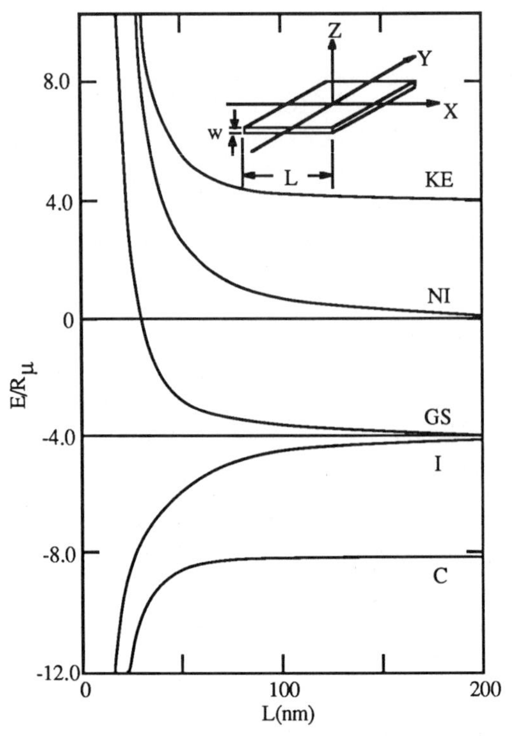

Fig. 2. The ground-state (GS) energy of a heavy-hole exciton confined in a square box with sides of length L and width w = 0. The exciton kinetic energy (KE) and Coulomb energy (C), the energy of a noninteracting electron-hole pair (NI), and the interaction energy (I) are shown. The insert shows the configuration of the quantum box.

The length scales for the transition from a correlation-dominated exciton to a quantum-confined exciton are revealed by Fig. 2. At large L ($L \gtrsim 100$ nm), the exciton ground-state energy approaches the energy for the unconfined two-dimensional exciton, $- 4R_\mu$. The shift from the limiting value is less than five per cent for $L \gtrsim 100$ nm. At small L ($L \lesssim 10$ nm), the confinement energy dominates the Coulomb energy and the ground-state energy varies as $1/L^2$. The finite barriers of a real quantum box will not confine the exciton when $L \lesssim 5$ nm. Thus for $L \lesssim 5$ nm, the $1/L^2$ scaling will break down in a real quantum box and the exciton ground-state energy will approach the energy for an exciton in the barrier material.

The confinement energy E_{NI} of an uncorrelated, noninteracting, electron-hole pair in the lowest energy box states and the interaction energy E_I, which is the shift in pair energy that occurs when the electron-hole interaction is included and the pair becomes a correlated exciton ($E_{GS} = E_{NI} + E_I$) are shown in Fig. 2. The exciton kinetic energy E_{KE} and Coulomb energy E_C ($E_{GS} = E_{KE} + E_C$) are also shown. For large boxes ($L \gtrsim 100$ nm) the exciton is much smaller than the box and the exciton ground-state energy E_{GS} is affected little by confinement. The exciton kinetic and Coulomb energies show a weak dependence on L for larger boxes similar to the weak dependence of E_{GS}. In contrast, E_{NI} displays the most rapid variation as L decreases. For large L, E_{KE} is much larger than E_{NI}. Most of the kinetic energy of an exciton in a large box is the kinetic energy due to correlation as an exciton and only a small part is due to quantum confinement. As L decreases, E_{KE} increases more rapidly than E_C decreases, indicating that the kinetic energy effects are more sensitive to confinement than are Coulomb interactions. Moreover, as L decreases, E_{KE} approaches E_{NI}, indicating that the correlation effects are becoming less important and that confinement makes the more important contribution. The direct Coulomb energy (E_C for an uncorrelated pair) makes the dominant contribution to E_I for small boxes. Less than 10 per cent of E_I is due to correlation for $L \lesssim 10$ nm, but more than half of E_I is due to correlation for $L \gtrsim 50$ nm.

The electron-hole separation R is 10.9 nm for a free, two-dimensional heavy-hole exciton. The free-exciton radius is comparable to the size of box ($L \sim 10$ nm) where the confinement is complete and is an order-of-magnitude smaller than the L (~100 nm) where confinement effects begin. For $L \gtrsim 100$ nm, R is unaffected by the confinement. For $L \lesssim 10$ nm the confined exciton is much smaller than the free exciton and R approaches the radius for an uncorrelated electron-hole pair. The value of R, as determined with a wave function which contains the correlation effects but no confinement effects, actually increases as L decreases. Box confinement rather than correlation effects shrinks the exciton as L decreases.

OPTICAL RESPONSE OF QUANTUM BOXES

The dependence of exciton energies on box size reveals three box size regimes with qualitatively different exciton properties. For large boxes ($L \gtrsim 100$ nm) excitons are bulk like. For small boxes ($L \lesssim 10$ nm) excitons are weakly correlated electron-hole pairs localized in the same region by confinement rather than Coulomb attraction. In the intermediate regime the exciton is correlated but localized by confinement and sensitive to box size. The quantum box optical response near the exciton resonance is also qualitatively different in each size regime.

The magnitude of the optical response is determined by the oscillator strength[14]

$$f = \frac{C}{(E_{ex} - E_o)} \left| \int \Psi_{ex} (\vec{r}_e, \vec{r}_e) \, d^3 r_e \right|^2 \quad ,$$

where C includes the intracell momentum matrix element, E_o and E_{ex} are the ground state and excited state (exciton or uncorrelated electron-hole pair) energies and Ψ_{ex} is the excited state wave function with the electron and hole evaluated at the same position. The integral is over the box.

Oscillator strengths for heavy-hole excitons are shown in Fig. 3. As L decreases, f vanishes because the energy denominator increases as $1/L^2$ due to the confinement energy. Similarly f decreases as w decreases because the well confinement energy varies as $1/w^2$ (no well confinement energy is included in the calculation for $f(w = 0)$, so $f(w = 0)$, as calculated, is larger than $f(w \neq 0)$).

For large boxes the energy denominator is weakly dependent on L, so the wave function integral determines the behavior of f. The oscillator strength for a fully correlated exciton f_{ex} is very different from the oscillator strength f_p for an uncorrelated electron-hole pair in the lowest energy pair-state (see Fig. 3). For an exciton confined in a narrow quantum disk (as in Fig. 1), $\Psi(\vec{r}_e, \vec{r}_e) \sim 1/\sqrt{AA_{ex}}$ where $A = L^2$ is the two-dimensional area of the disk and A_{ex} is the area of the exciton. Consequently,

$|\int \Psi(\vec{r}_e,\vec{r}_e)\, d^2r_e|^2 \sim L^2/A_{ex}$. For a fully correlated exciton, the oscillator strength is calculated from the coherent superposition of the amplitudes for absorbing a photon in each unit cell. Thus f_{ex} increases as L^2. For the uncorrelated electron-hole pair,

$$\Psi(\vec{r}_e,\vec{r}_e) \sim \frac{4}{L^2} \cos^2(kx_e) \cos^2(ky_e) \quad ,$$

and $\int \Psi(\vec{r}_e,\vec{r}_e)\, d^2r_e = 1$. Consequently, f_p is independent of L for large L. Excitons confined in boxes belonging to the intermediate size regime (with $L^2 \gtrsim A_{ex}$) are correlated and f_{ex} is much larger than f_p due to the coherence of the exciton state. Excitons in small boxes (with $L^2 \lesssim A_{ex}$) would have oscillator strengths much less than f_p if the excitons were fully correlated. However, the exciton becomes uncorrelated in small boxes so the oscillator strength converges to f_p as L decreases.

The oscillator strength per box, f/L^2 is shown in Fig. 4. For large L, f/L^2 is independent of L. For small L, there is a strong enhancement of the normalized f. The enhancement in f is due to the increased electron-hole overlap that arises from the box confinement and is not due to coherent superposition effects which result from correlation. This large enhancement in f/L^2 in the limit of complete quantum confinement has motivated others to suggest that quantum boxes should have enhanced optoelectronic properties.[11,17-19] There is substantial enhancement due to confinement in f/L^2 even for $L \gtrsim 10$ nm.

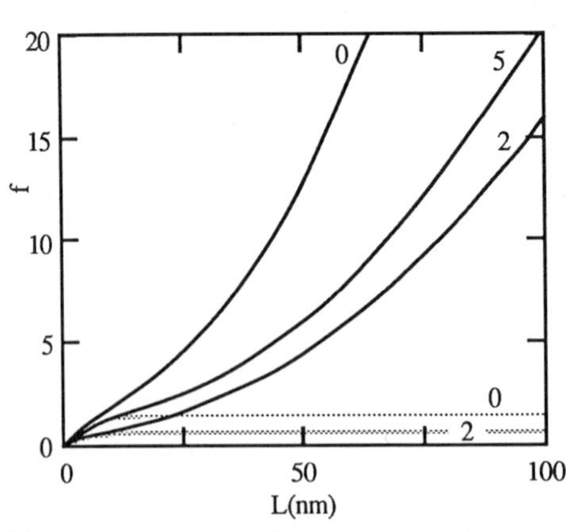

Fig. 3. Oscillator strength for the heavy-hole exciton confined in a square box (w = 0, 2, and 5 nm as indicated). The solid curves are for correlated excitons. The dashed curves are for uncorrelated electron-hole pairs in the lowest energy pair-states.

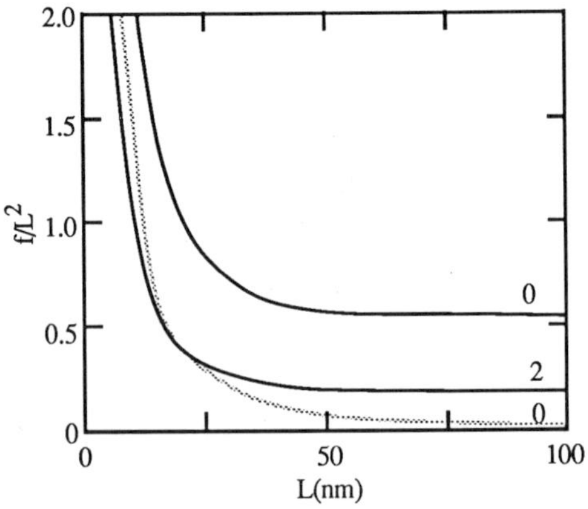

Fig. 4. Exciton oscillator strength normalized by area, L^2, of the box. The curves are the same as in Figure 3.

The enhancement in optical properties for small boxes is due to the quantum confinement. The optical response, especially the nonlinear response[20-23] should be enhanced for intermediate box sizes because of the coherence of the exciton. A simple argument shows why enhanced nonlinearities are possible.[22] For the third order susceptibility, $\chi^{(3)} \sim f^2$. Since $f \sim L^2$ for $L^2 \gtrsim A_{ex}$, $\chi^{(3)} \sim L^4$. The third order susceptibility per box is $\chi^{(3)}/L^2 \sim L^2$. Thus in the intermediate size regime, enhancements of $\chi^{(3)}$ should occur by increasing L. This enhancement of $\chi^{(3)}$ in the intermediate regime due to coherence effects must break down for very large boxes with bulk-like excitons which are perturbed weakly by confinement.[23] In the intermediate regime, the confined exciton is a localized excitation which couples strongly with light. In large boxes, bulk-like excitons become extended excitations with well-defined wave vectors which couple to photons with the same wave vector. The coupling produces polaritons rather than the radiative decay of the exciton states.[23]

NANOSTRUCTURE ARRAYS

Useful devices will be made not from isolated boxes but from arrays of boxes. Boxes in arrays should also have enhanced optical response if boxes in an array can localize states. Augmented plane wave calculations[24] have been performed to determine the conduction band structure of quantum box arrays. The arrays are two-dimensional, periodic square arrays (with lattice constants A_x) of two-dimensional circular nanostructures (radius R). The third dimension is assumed to be a narrow well with one occupied sub-band. The effective-mass approximation, with an isotropic mass, is used to describe motion in the absence of the imposed array of potentials. Sharp heterojunctions between well and barrier materials define the box boundaries. The conduction-band offset between well and barrier determines the confining potential.

The band structures of quantum box arrays have been calculated[24] for arrays with lattice constant A_x = 30 nm and box radius R = 8 nm. Different structures will have the same band structure (except for the scale factor $1/\lambda^2$) if other energies and potentials scale as $1/\lambda^2$ and distances scale as λ. The band structure for a quantum-box array defined by a small band offset V ($\lesssim$ 40 meV) is similar to the zone-folded, free-electron band structure. The corresponding bands are easy to identify. However, band mixing, avoided crossings, and distortion are evident for small offsets. Band gaps are present at the zone edges. One quasi-bound state exists for V = 37 meV. As V increases, substantial band-distortion occurs and the correspondence with zone-folded free-electron bands is difficult to identify.

The ordering of the bands changes as V increases. This reordering is necessary to provide the correct degeneracies for quasi-bound states. The ground state in an isolated circular box is non-degenerate. Other bound states are doubly degenerate. Three quasi-bound states (with the correct degeneracies) appear in the array band-structure when V = 74 meV. Six states are bound when V = 148 meV (see Fig. 5). The highest bound state still has dispersion and is not yet doubly degenerate. The dispersion of higher energy states also decreases as V increases. These states merge into quasi-doubly-degenerate flat bands as V increases. For an array with the same lattice constants but made from larger boxes, more states are quasi-bound; however, the states have larger energy dispersion because tunneling between boxes is more likely. For example, for boxes with R = 12 nm and V = 148 meV, ten states are quasi-bound.

The charge densities of the two lowest-energy zone-center states in the box array (V=148 meV) are shown in Fig. 6. The lowest energy quasi-bound state has an s-like charge density in a box. The higher energy quasi-bound states also have definite local atomic character (a p-like state is shown in Fig. 6).

The choice of nanostructure determines the type of states in the band structure. Box arrays possess localized states. Arrays made from other nanostructures have also been tested.[24] Quantum bumps, which are barriers with finite spatial extent, effectively exclude carriers from the barrier regions. When quantum bumps are placed in a two-dimensional array, channels are formed. Low-energy electrons localize at the connections between channels and form bonds between the connections. The dispersion of these states is determined by the width of the channels. The connections become quantum boxes if adjacent bumps touch and block off the channels.

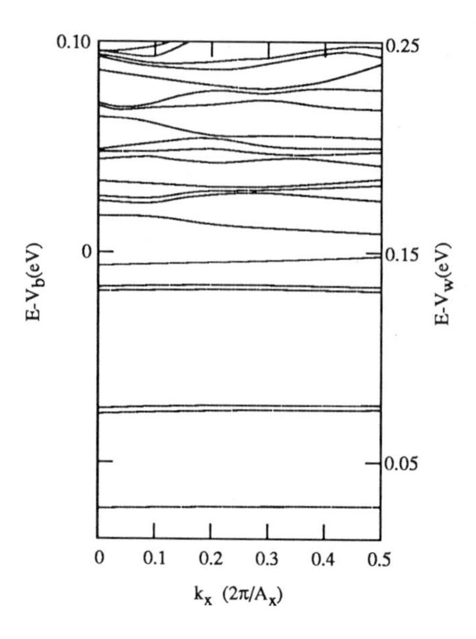

Fig. 5. Band structure in the X-direction for a square quantum box array: $A_x = A_y = 30$ nm, R = 8 nm, V_b - V_w = 148 meV (V_b and V_w are barrier and well band edges).

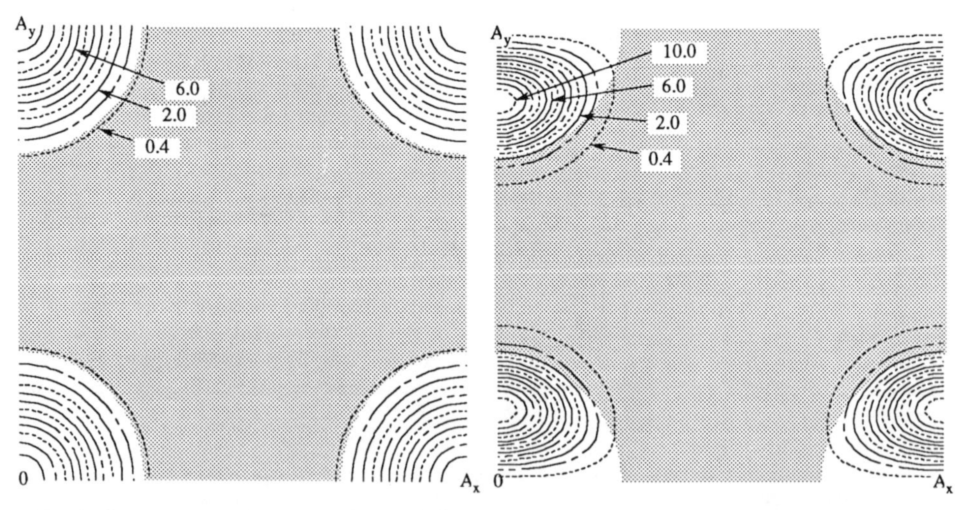

Fig. 6. Charge density of the lowest-energy (left) and first-excited (right) zone-center (k = 0) conduction band state in a quantum box array with V = 148 meV. One unit cell is shown. The barrier region is shaded. The state is normalized such that a plane wave has unit charge density everywhere.

Quantum doughnuts are finite well regions surrounded by barriers with finite thickness which are surrounded by connected well regions. States which exist in the channels between doughnuts, and resonance states sensitive to the size of the doughnut hole, both exist. The energies and dispersion of these states can be controlled by the choice of the inner and outer doughnut radius and the channel width. Structures can be designed so that channeling and resonance states overlap in energy and strongly hybridize.

QUANTUM BOX RESONANT TUNNELING

Quantum box resonant tunneling is a challenge to understand because it is a three-dimensional multichannel process rather than a one-dimensional process as in quantum well tunneling. Consequently, confinement effects must be modeled carefully. The confinement induced splitting of the lateral levels is determined by the sidewall depletion potential. The sidewall depletion potential in each region is approximately parabolic, so the same level splitting applies for all levels in each region. Reed et al. estimated the splitting of levels in the box to be $\Delta E \sim 25$ meV. The barriers and well are undoped and should have similar sidewall depletion depths and confinement energies. Two extreme choices are possible for splittings of levels in the contacts. If the contacts and boxes have similar depletion depths, then they will have similar level splittings. If the depletion depth is much smaller in the contacts, then the level splittings in the contacts will be much smaller. Both choices have been tested to determine which choices produce RT fine structure.

The theory of resonant tunneling through a single channel used to model normal RT has been generalized[25] to study multichannel tunneling in a confined structure. The strength of the coupling of different channels is determined by the overlap between lateral states of different regions. The lateral states of different regions of a cylindrical structure mix only if the levels have the same transverse parity. To obtain a qualitative understanding of level-mixing effects a simple model with four levels for each type of channel (as determined by the transverse parity) has been used. The form of the overlap matrix between levels that mix at an interface is determined by the unitarity requirement. In the simplest model, a single parameter β, the overlap between a level on one side of the interface and the corresponding level on the other side of the interface, determines the matrix. Overlap is complete and no level-mixing is possible when $\beta = 1$. In that case, tunneling is a single channel process. In the extreme of weak confinement in the contact and strong confinement in the box $\beta \sim 0.6$.

The effect of the discrete density of lateral states on QBRT is determined by calculating the RT in the diagonal tunneling approximation (referred to as diagonal because the overlap matrices are diagonal, β=1) so no effects of level-mixing are included. When lateral confinement is much weaker in the contact than in the box, the QBRT fine structure is clearly resolved. One fine structure peak appears in the RT for each occupied contact lateral-level. The fine structure is resolved in this case since a different bias voltage is needed to line up each contact lateral-level with the corresponding box-lateral-level. Adjacent fine structure peaks are separated by a bias of about twice the box level splitting, $2\Delta E$. When the QBRT current/voltage characteristics are calculated in the diagonal tunneling approximation for similar lateral confinement in the box and contacts, individual peaks are broad and cannot be resolved when more than one contact lateral-level occupied.

The effects of lateral-level mixing at the contact-barrier interface on the QBRT are shown in Figs 7-8. Figure 7 shows the RT when the lateral confinement in the contact is much weaker the box lateral-confinement. When only the lowest energy contact lateral-level is occupied, RT fine-structure peaks appear for tunneling from the ground-state contact lateral-level through each box-lateral-level with the same transverse parity, not just for tunneling through the lowest box-lateral-level as occurs in diagonal tunneling. The peaks are separated by $4\Delta E$ because levels with the same parity are split by $2\Delta E$. When the second contact-lateral-level is occupied an additional set of RT peaks appears for tunneling involving states with odd parity. These peaks are also separated by $4\Delta E$, but the peaks are shifted by $2\Delta E$ from the other set of peaks. No additional peaks occur when more than two lateral-level are occupied. When level-mixing occurs, only two contact-levels need be occupied to produce all possible fine structure peaks. When level-mixing is absent one contact level must be occupied for each fine structure peak produced. When the box and contact lateral confinement are similar, the fine structure peaks can not be resolved if more than one contact-level is occupied even with level-mixing included.

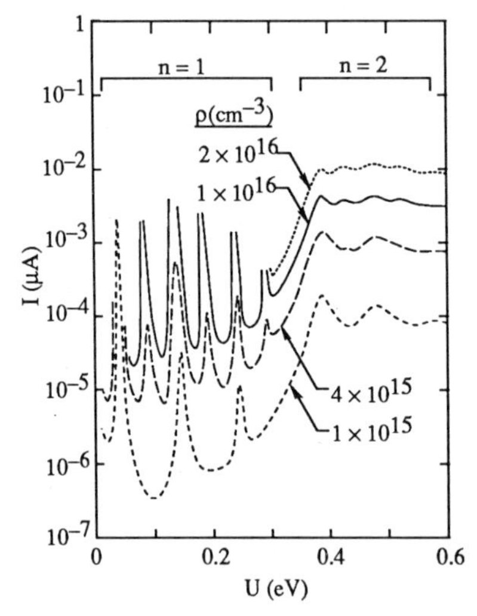

Fig. 7. QBRT current/ voltage characteristic calculated with lateral-level mixing ($\beta = 0.8$) and weak confinement in the contact. Tunneling through the ground state ($n = 1$) and first excited state($n = 2$) in the well are indicated. One, two, three and four contact lateral-levels are occupied, respectively, for increasing contact charge density ρ.

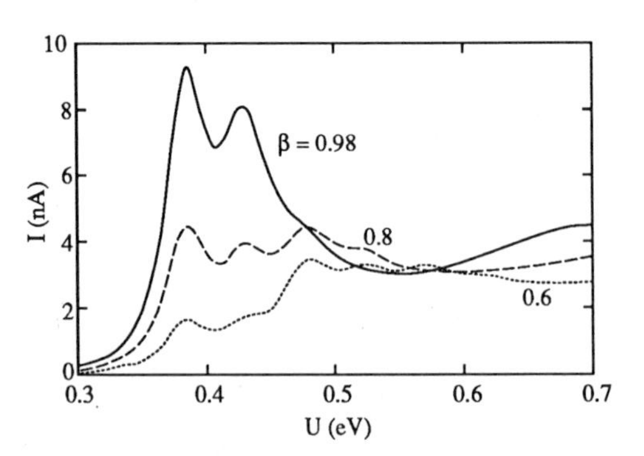

Fig. 8. QBRT current/voltage characterisic calculated with lateral-level mixing, weak contact confinement, and three contact lateral-levels occupied ($\rho = 10^{16} cm^{-3}$).

The QBRT fine structure calculated for weak contact confinement is shown in Fig. 8 as a function of the leveling mixing. The magnitude of the current is consistent with the currents observed by Reed et al. The currents calculated for similar box and contact confinement are too large. Two features of the RT fine-structure depend on the strength of level coupling and have direct relevance to the results of Reed et al. For weak coupling, only the RT fine-structure peaks at low bias are important. As the coupling increases, peaks for RT through higher energy levels become just as important. In the experiment of Reed et al. the fourth fine-structure peak (see Fig. 1) has the largest magnitude. Because the fine-structure peaks at high bias are observable, level mixing should be significant. Reed et al. also observed a difference in bias voltage between the first two fine-structure observed peaks which was roughly twice the difference between other peaks. The large separation between the first two peaks can be explained by large level-mixing. As shown in Fig. 8, for β = 0.6 the second fine-structure peak is small compared to the first two large peaks which are due to tunneling in the channels coupled to the ground-state level.

SUMMARY

Three size regimes characterize the near-band-edge optical response involving excitons. In very large boxes, confined excitons are bulk-like, Coulomb interactions and correlation effects dominant the confinement effects. Photons hybridize with excitons to form polaritons. In very small boxes, confined excitons are uncorrelated electron-hole pairs frozen in the lowest-energy single particle states. Confinement effects are dominant. Oscillator strengths are enhanced from the spatial localization of the electron and hole by the confinement. In the intermediate regime both confinement and correlation effects are important. The excitons are coherent, localized interactions. The localized excitons couple strongly with photons and the oscillation strengths increase with box size due to the coherence.

Quantum box resonant tunneling exhibits fine structure peaks which have been attributed to the discrete density of box states. Calculations of QBRT performed to determine what conditions are necessary to observe lateral confinement effects indicate that lateral confinement in the box should be different from lateral confinement in the contact. When contact and box lateral levels have similar level spacings, the peaks are difficult to resolve. The discrete density of lateral states and the level mixing both produce fine structure in the QBRT. Level mixing provides a more consistent explanation of the observed QBRT fine structure.

ACKNOWLEDGMENTS

This review was performed under the McDonnell Douglas Independent Research and Development program.

REFERENCES

1. C. J. Sandroff, D. M. Hwang, and W. M. Chung, Carrier confinement and special crystallite dimensions in layered semiconductor colloids, Phys. Rev. B, 33:5953 (1986).
2. J. Warnock and D. D. Awschalom, Picosecond studies of electron confinement in simple colored glasses, Appl. Phys. Lett. 48:425 (1986).
3. L. Brus, Zero-dimensional "exciton" in semiconductor clusters, IEEE J. Quantum Electron. QE-22:1909 (1986) and the references therein.
4. M. A. Reed, R. T. Bate, K. Bradshaw, W. M. Duncan, W. R. Frensley, J. W. Lee, and H. D. Shih, Spatial quantization in GaAs-AlGaAs multiple quantum dots, J. Vac. Sci. Technol. B4:358 (1986).
5. K. Kash, A. Scherer, J. M. Worlock, H. G. Craighead, and M. C. Tamargo, Optical spectroscopy of ultrasmall structures etched from quantum wells, Appl. Phys. Lett. 49:1043 (1986).

6. J. Cibert, P. M. Petroff, G. J. Dolan, S. J. Pearton, A. C. Gossard, and J. H. English, Optically detected carrier confinement to one and zero dimension in GaAs quantum well wires and boxes, Appl. Phys. Lett. 49:1275 (1986).
7. H. Temkin, G. J. Dolan, M. B. Panish, and S. N. G. Chu, Low-temperature photoluminescence from InGaAs/InP quantum wires and boxes, Appl. Phys. Lett. 50:413 (1987).
8. Y. Miyamoto, M. Cao, Y. Shingai, K. Furuya, Y. Suematsu, K. G. Ravikumar, and S. Arai, Light emission from quantum-box structure by current injection, Jpn. J. Appl. Phys. 26:L225 (1987).
9. P. M. Petroff, J. Cibert, A. C. Gossard, G. J. Dolan, and C. W. Tu, Interface structure and optical properties of quantum wells and quantum boxes, J. Vac. Sci. Technol. B5:1204 (1987).
10. M. A. Reed, J. N. Randall, R. J. Aggarwal, R. J. Matyi, T. M. Moore, and A. E. Wetsel, Observation of discrete electronic states in a zero-dimensional semiconductor nanostructure, Phys. Rev. Lett. 60:535 (1988). After this chapter was prepared, M. Reed reported a new analysis of the data which indicated that the observed tunneling occurs through the n=1 resonance rather than the n=2 resonance. The qualitative physics of qunatum box resonance tunneling discussed in this chapter is the same for both resonances.
11. S. Schmitt-Rink, D. A. B. Miller, and D. S. Chemla, Theory of the linear and nonlinear optical properties of semiconductor microcrystallites, Phys. Rev. B 35:8113 (1987).
12. G. W. Bryant, Hydrogenic impurity states in quantum-well wires: shape effects, Phys. Rev. B 31:7812 (1985).
13. G. W. Bryant, Hydrogenic impurity states in quantum-well wires, Phys. Rev. B 29:6632 (1984).
14. G. W. Bryant, Excitons in quantum boxes: correlation effects and quantum confinement, Phys. Rev. B, 37:8763 (1988).
15. G. W. Bryant, Electrons and holes in quantum boxes, in: "Interfaces, Quantum Wells, and Superlattices," C. R. Leavens and R. Taylor, eds. Plenum, New York (1988).
16. G. W. Bryant, Excitons in zero-dimensional quantum boxes: correlation and confinement, Comments Condensed Matter Phys. 14:277 (1989).
17. M. Asada, Y. Miyamoto, and Y. Suematsu, Gain and the threshold of three-dimensional quantum-box lasers, IEEE J. Quantum Electron. QE-22:1915 (1986).
18. D. A. B. Miller, D. S. Chemla, and S. Schmitt-Rink, Electroabsorption of highly confined systems: theory of the quantum-confined Franz-Keldysh effect in semiconductor quantum wires and dots, Appl. Phys. Lett. 52:2154 (1988).
19. K. J. Vahala, Quantum box fabrication tolerance and size limits in semiconductors and their effect on optical gain, IEEE J. Quantum Electron. QE-24:523 (1988).
20. T. Takagahara, Excitonic optical nonlinearity and exciton dynamics in semiconductor quantum dots, Phys. Rev. B 36:9293 (1987).
21. L. Banyai, Y. Z. Hu, M. Lindberg, and S. W. Koch, Third-order optical nonlinearities in semiconductor nanostructures, Phys. Rev. B 38:8142 (1988).
22. E. Hanamura, Very large optical nonlinearity of semiconductor microcrystallites, Phys. Rev. B 37:1273 (1988).
23. E. Hanamura, Rapid radiative decay and enhanced optical nonlinearity of excitons in a quantum well, Phys. Rev. B 38:1228 (1988).
24. G. W. Bryant, Electronic band structure of semiconductor nanostructure arrays, Phys. Rev. B (submitted). See references therein for other work.
25. G. W. Bryant, Resonant tunneling in zero-dimensional nanostructures, Phys. Rev. B 39:3145 (1989).

EXCITON-POLARITONS IN QUANTUM WELLS AND QUANTUM WIRES

D. Heitmann, M. Kohl, P. Grambow, and K. Ploog

Max-Planck-Institut für Festkörperforschung
Heisenbergstr. 1, 7000 Stuttgart 80, FRG

$Al_xGa_{1-x}As - GaAs$ multi-quantum-well systems with laterally microstructured cap layers and quantum-well-wire structures with lateral dimensions of 100-200 nm have been prepared. The photoluminescence spectra of these samples exhibit an additional emission on the high energy side of the (n=1)-electron heavy-hole free exciton transition which is strongly polarized. This emission arises from the resonantly enhanced radiative decay of quantum-well exciton polaritons, which is mediated by the microstructure. We will discuss in particular the emission efficiency and dispersion of these excitations.

INTRODUCTION

The recent progress in crystal growth techniques, which made it possible to fabricate layered hetero and superlattice structures with atomically flat resolution, has initiated a broad range of fundamental research and novel applications in many different fields. These systems have unique physical properties which arise from the quasi-two-dimensional (2D) behavior and, in superlattices, from coupling between these 2D layers. Optical investigations, in particular photoluminescence (PL) and reflection measurements, are a powerful tool to study these systems. The PL spectra of quantum-well (QW) systems are governed by efficient intrinsic radiation of free excitons (Ref.1, for a recent review see e.g. Ref.2). One of the challenging topics of current interest are systems of even lower dimensionality, namely 1D quantum wires and 0D quantum dots. These systems have been prepared by starting from a 2D-layered system and employing modern micro and nanometer lithographic techniques[3-12] or by using novel growth techniques.[13] In this paper we will concentrate on the optical investigation of microstructured systems and discuss experiments on the emission and excitation of quantum well exciton polaritons (QWEP). We will show that these excitations give an important contribution to the optical spectra of QW wire systems.[10,11]

SURFACE POLARITONS AND QUANTUM WELL EXCITON POLARITONS

In the following we will discuss 'surface polaritons'.[14] We will use this expression in a general sense for more specified excitations described below. Surface polaritons are the quanta of electrodynamic excitations which exist at the boundary or in thin films of materials when the real part of the dielectric function, $\epsilon_r(\omega)$, is negative. E.g.,

Science and Engineering of One- and Zero-Dimensional Semiconductors
Edited by S.P. Beaumont and C.M. Sotomajor Torres
Plenum Press, New York, 1990

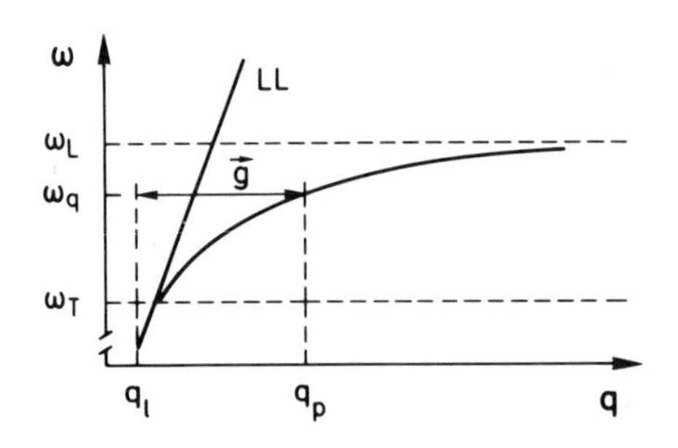

Fig. 1 Schematic diagram of the QWEP dispersion. $|\vec{g}| = 2\pi/a$ is the reciprocal grating vector of the periodic microstructure that couples QWE polaritons of frequency ω_p and wave vector $\vec{q_p}$ with photons of the same frequency and wave vector $|\vec{q_l}| < \omega/c$. LL is the light line $\omega = qc$.

in metal-like systems 'surface plasmon polaritons' (e.g. Ref. 14,15) can be excited below the plasma frequency of the free electron gas. In polar materials 'surface phonon polaritons' exist in the frequency regime between the transversal and longitudinal eigenfrequency, ω_T and ω_L, respectively.[14,16] An equivalent surface excitation in a material with excitonic behavior are surface exciton polaritons.[17,18] The dielectric response of such a material can be described by a Lorentzian resonance

$$\varepsilon(\omega) = \varepsilon_\infty(1 + (\omega_L^2 - \omega_T^2)/(\omega_T^2 + \beta q^2 - \omega^2 - i\omega\Gamma)) \tag{1}$$

where β governs the spatial dispersion, $\vec{q}$ is the wavevector, Γ the damping and ε_∞ the high frequency dielectric function. The dispersion of these excitations is given by[14]

$$q_p = \frac{\omega}{c}\sqrt{\frac{\varepsilon_r(\omega)}{\varepsilon_r(\omega)+1}} \tag{2}$$

where $\vec{q_p}$ is the polariton wavevector in the plane. For a Lorentzian type of response function this dispersion is shown schematically in Fig.1.

There are three important aspects for all types of surface polaritons. (a) In laterally homogeneous systems these excitations are nonradiative due to momentum conservation since the wavevector $|\vec{q_p}| > \omega/c$. The electromagnetic fields decay exponentially from the interface. They do not couple with freely propagating waves ($|\vec{q_x}| < \omega/c$) and thus do not contribute to the response in PL or reflection measurements. However, any spatial modulation of the optical properties in the near field of the interface, in particular a periodic corrugation of the surface with periodicity a, will cause a momentum transfer

$$q_x = q_p \pm n \cdot 2\pi/a, \qquad n = 1,2,3... \; . \tag{3}$$

We consider here electrodynamic excitations propagating in x-direction and grating rules parallel to the y-direction. This is also the experimental arrangement. If $|\vec{q_x}| < \omega/c$, the grating coupler effect of the periodical corrugation gives rise to the radiative decay of the surface polaritons. The radiation is emitted for an angle ϑ with respect to the surface normal and is governed by

$$\frac{\omega}{c}\, sin\, \vartheta = q_x = q_p - n2\pi/a \tag{4}$$

in the case of surface polaritons propagating perpendicular to the grating rules. Grating coupler or the very similar prism arrangements have been used to investigate all different kinds of excitations, e.g. surface plasmon polaritons[14,15], 2D plasmons[19], surface phonon polaritons[16] and surface exciton polaritons[10,11]. In particular, Agranovich

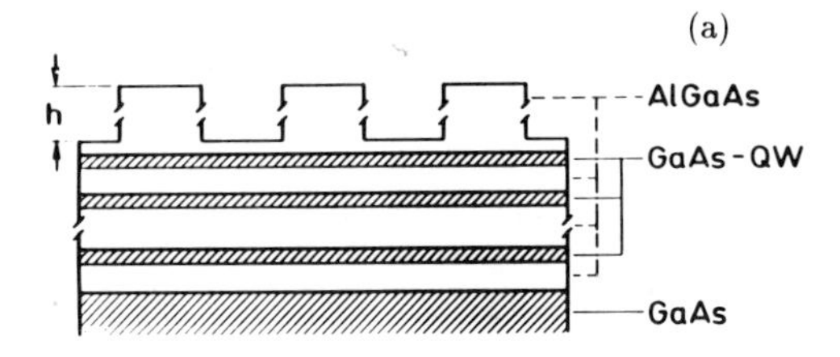

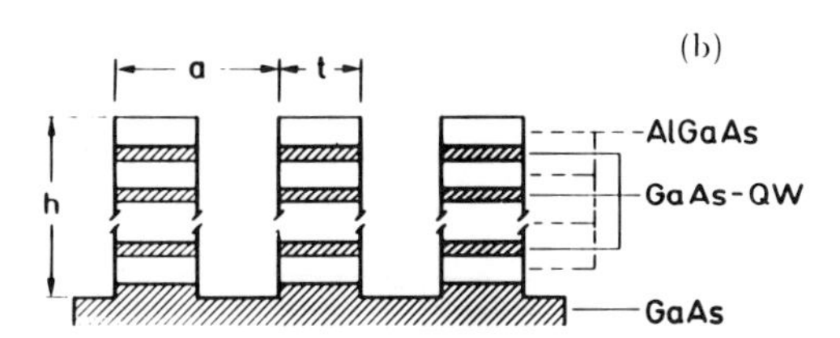

Fig. 2 Schematic configuration of the laterally microstructured samples of type (a) and type (b).

et.al. have pointed out that a corrugation should lead to polariton emission in the photoluminescence.[20] (b) A second important point is that these excitations are associated with a strong, resonantly enhanced field strength at the interface which has been observed in several experiments.[14,15] In particular, the interesting 'Giant Raman' effect has been associated with a surface polariton-type of field strengths enhancement. (c) A third point is that the resonant polariton excitation occurs only for p-polarization, i.e. for excitations where the electric field is polarized in the plane of the polariton wavevector and surface normal. If we observe the radiative decay in the x-y-plane, i. e., perpendicular to the grating rules, the grating does not change the polarization and the emitted radiation is also p-polarized. This polarization selection rule is an important feature to identify surface polariton resonances.

In the special case of QW systems the dielectric response in the frequency regime of excitons can be described by a Lorentzian dependence similarly as in Eq.(1), which, however, is strongly anisotropic. This is due to the different symmetries of the light-hole wave functions on the one side and heavy-hole wave functions on the other and leads to pronounced differences in the optical response for electron light-hole (e-lh) and electron heavy-hole (e-hh) transitions.[21,22] The corresponding surface polariton excitation in QWs are quantum well exciton polaritons (QWEP). The dispersion of QWEP has been calculated by Nakayama et.al..[23,24] It is very similar to the case of the surface exciton polariton dispersion schematically shown in Fig.1. In homogenous systems QWEP do not contribute to PL or reflection spectra, whereas in laterally microstructured QW systems the microstructure mediates the momentum transfer and enables the radiative decay of these excitations. In the following we will discuss experiments where QWEP have been observed in PL and reflection. Very recently QWEP have also been observed by Ogawa et.al.[25] in time-of-flight measurements.

EXPERIMENTS AND DISCUSSION

The samples were grown by molecular-beam epitaxy at 610°C in the following sequence: 200 nm GaAs-buffer layer, a short period GaAs/AlAs superlattice, a 50 nm AlGaAs layer and five GaAs QWs of widths 13 nm, separated by 10 nm AlGaAs barriers. For our investigations we grew two different AlGaAs-cap layers: (a) of 170 nm and (b) of 50 nm thickness. A mask consisting of periodic photoresist lines was prepared by holographic lithography. With reactive ion etching in a $SiCl_4$-plasma nearly rectangular grooves of a depth of 170 nm were etched into the samples. The periodicity a was varied between 250-500 nm, and lateral widths, t, between 100-200 nm were achieved. By this procedure we obtained samples where a grating was etched

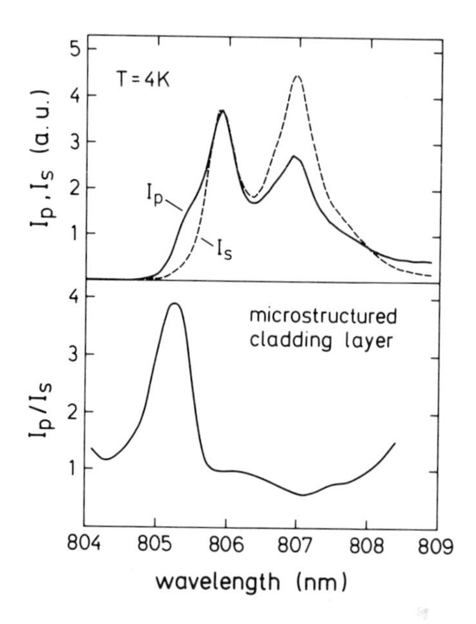

Fig. 3 Experimental luminescence spectra of a microstructured GaAs-MQW sample of type (a) with a = 500 nm and t = 200 nm. I_p and I_s denote the p- and s- polarized luminescence intensity, respectively. The QWEP emission occurs, only in p-polarization, as a shoulder at 805.4 nm on the low wavelength side of the free exciton emission. The excitation intensity was 10 mW/cm^2.(From Ref. 11)

into the thick cap layer (samples of type (a)), leaving the QWs undisturbed. In the other samples (b), rectangular grooves were etched through all QW layers, see Fig.2. The photoluminescence was excited with normal incident light at 647.1 nm from a Kr^+ laser. The luminescent light was collected normally to the sample in a solid angle of 30° opening. The luminescence was analyzed with a 1m Jobin-Yvon monochromator and detected by a photon-counting system. The resolution was set to 0.05 nm.

Figure 3 shows PL-spectra measured on a sample of type (a). The s-polarized spectrum (dashed curve) exhibits two transitions. S-polarization is defined for the electric field vector of the PL radiation being parallel to the grating rules, and thus perpendicular to q_p. The first maximum at 805.9 nm is due to the free e1-hh1 transition. The PL efficiency of this transition is only slightly reduced (by a factor of 1.5) in comparison with the unstructured reference sample and not shifted in position. This indicates that the free QWE is not affected by the etching of the cap layer. The low energy resonance at 807 nm has extrinsic character, as deduced from temperature and intensity dependent measurements. This transition, which is not observable in the unstructured reference sample, arises from etching induced defect states.[10,11]

The most interesting point is that in the p-polarized spectrum an additional PL-contribution at the high energy side of the free QWE transition shows up. This is clearly visible in the pronounced maximum in the lower part of Fig.3, where we show the ratio of p- and s-polarization, I_p/I_s. This additional contribution arises from the grating coupler-induced radiative decay of QWEP. There are three arguments which confirm this interpretation. (1) The polarization dependence of the additional transition corresponds to the selection rules discussed above. (2) No polarization dependence is observed in the unstructured reference samples. (3) The QWEP transition occurs on the high energy side of the free QWE transition. If one assumes that free QWE emission occurs at the transversal frequency ω_T, which is actually known so far only for bulk GaAs,[17] then this energy position would be consistent with a polariton emission at $\omega > \omega_T$ as is expected from the dispersion behavior, see Fig.1. It should be noted that the energy position of the QWEP at frequency $\omega > \omega_T$ in our samples indicates a higher value ($\approx$ 0.3 meV) of the longitudinal-transversal splitting in quantum wells as compared to bulk GaAs (0.1 meV). This is not unexpected because of the enhanced oscillator strengths in 2D-confined systems.

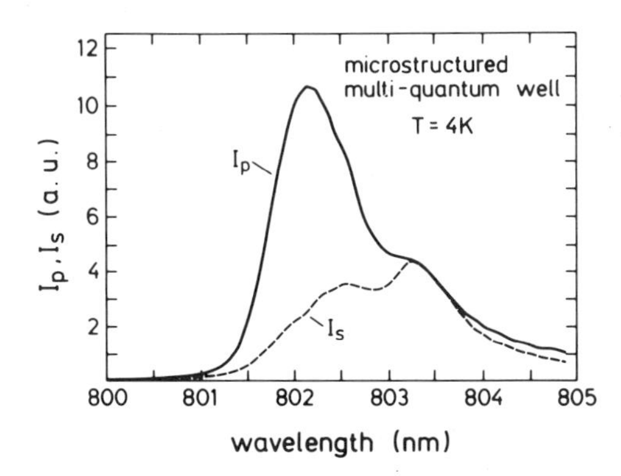

Fig. 4 Experimental luminescence spectra of a microstructured GaAs-MQW sample of type (b) with a = 350 nm and t = 150 nm. I_p and I_s denote the p- and s- polarized luminescence intensity, respectively. The strong transition at 802.1 nm, which only occurs in p- polarization, arises from the radiative decay of QWEP. The excitation intensity was 10 mW/cm^2. (From Ref. 11)

We now like to discuss systems as described in Fig.1b, where we have etched all the way through the five QWs. The width of these QW wires was 150 nm. As we will discuss below, this confinement leads in principle to a 1D quantization of the electronic wavefunctions. However, 150 nm are still too large to resolve 1D quantization phenomena. Here we are only interested in electrodynamic effects. When detecting the PL radiation with the electric field vector parallel to the wires (s-polarization) the spectrum shows two maxima, see Fig.4. The maximum at 802.5 nm is due to the e1-hh1 free QWE transition, as can be deduced from temperature and intensity dependence. The corresponding transition wavelength of about 802.2 nm is shifted by 0.1 nm with respect to the free QWE on the unstructured reference sample. The smallness of this shift indicates that stress-effects[26], which could have been induced by the etching, are negligible in our samples and will not cause any polarization effects. The peak at 803.4 nm arises from etching-induced states, as discussed above. The luminescence efficiency for the free QWE is compared to the corresponding transition in the unstructured reference sample reduced by more than one order of magnitude. In the p-polarized spectrum an additional strong transition shows up on the high energy side of the free QWE transition as in the case of sample (a) and strongly dominates the PL spectrum. We conclude that this p-polarized emission again arises from the radiative decay of QWEP-type of excitations, which thus are also present in QW wire structures. It is important to note that this enhanced p-polarized emission does *not* depend on the polarization of the incident exciting radiation with respect to the wire structure. Temperature-dependent measurements reveal that the QWEP emission can be observed up to more than 60 K. This clearly demonstrates the intrinsic character of the emission.

In Fig. 5 we show optical reflection measurements on a sample of type (b). Light from a halogen lamp was monochromized by a 60 cm double monochromator and directed with 45° angle of incidence onto the sample. The wire structure was perpendicular to the plane of incidence. The spectra in Fig. 5 show some broad features in the regime of the 1e-1lh transition and the le-1hh transition at about 803 nm and 807 nm, respectively. These structures are residual 'interference' effects due to the multi-layer quantum well system with different dielectric functions for the GaAs and the $Al_xGa_{1-x}As$ layers. Because of the non normal incidence these features are different for p- and s-polarizations. Only for the p-polarized reflected light two relatively sharp small dips (see arrows) are observable at 805.7 nm and 807 nm, which are not present in s-polarization or for any polarization on the unstructured reference sample. The position of these dips correspond very well to the positions of the QWEP emission measured in PL on the same sample.[10] We explain the dips in the following way. At wavelengths of about 806 nm p-polarized incident light can couple via the grating coupler effect of the microstructure to QWEP, leading to a decrease of reflectivity.[14,15] The grating coupler-induced QWEP excitation is the reverse process to the emission process described above. If we take the relative depth ($\Delta R/R$ = 3%) of the reflection minimum at 805.7 nm in our spectrum as a measure for the coupling efficiency of QWEP here, then this quantity is comparable with the efficiency that has been achieved for the excitation of surface exciton polaritons on semi-infinite ZnO crystals ($\Delta R/R = 2 - 7\%$).[18]

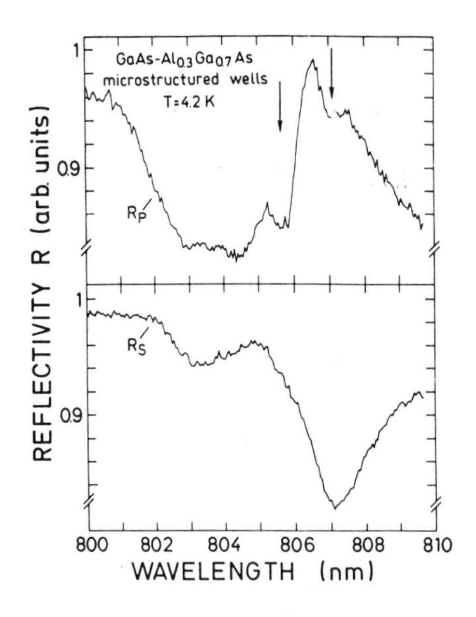

Fig. 5 45° p- and s-polarized reflectivity (R_p and R_s, respectively) on wire structures. In the p-polarized spectrum the grating coupler-induced excitations of QWEP are observed as dips (see arrows) in the reflection. (From Ref. 10)

The PL efficiency for QW-wire structures, as reported by different authors, depends strongly on the materials and the different fabrication processes. For example, in InGaAs-QW systems PL has been observed down to very small lateral dimensions of about 30 nm (e.g. Refs. 4,5,12). For AlGaAs-GaAs QW wires some processes seem to completely quench the free QWE emission even in 500 nm structures.[12] With our process we could observe free QWE emission even in 70 nm small QW wires.[27] The radiative decay of QWEP depends in addition on the sample geometry. This is reflected in our observation that the enhancement of the QWEP luminescence varies for different samples or for different spots on the same sample if the etching is not perfectly homogeneous. These variations arise from a very sensitive balance of 'radiative' and nonradiative damping processes which are responsible for the grating coupler efficiency of surface polaritons. For the radiative decay of surface plasmon polaritons on corrugated metal films, which are, concerning electrodynamics, a very similar system, it was found[15] that for small modulation amplitudes H of the surface the emission from polaritons increases with increasing H. For a certain, 'optimum' value of H a maximum emission rate can be achieved. Then, for still larger modulations, the emission intensity decreases. The reason for the latter is that the emission of photons is an additional, i.e., 'radiative' damping process which diminishes the resonant character and thus the field-strength enhancement of the excitation. The optimized grating coupler efficiency depends in a very sensitive way on the ratio of the 'radiative' damping, which is induced by the grating coupler, and 'internal' damping processes, which in the case of QWEP depend predominantly on the nonradiative recombination of excitons. Thus, we expect that the contribution of QWEP emission in the optical spectra depends strongly on the profile of the microstructures and the quality of the quantum well. The highest polarization ratio I_p/I_s that we have observed is about 6.

From the grating coupler condition (Eq.(4)) we can in principle determine the dispersion $\omega_p(q_p)$ of QWEP by scanning ω and the angle of emission ϑ. However, the dispersion is concentrated between ω_T and ω_L, the LT-splitting, which is only about 0.3 meV in our system. This frequency range is smaller than the linewidth of the observed exciton transitions. Therefore no details of the dispersion can be extracted from the experiment so far. It is nevertheless interesting to consider the dispersion of QWEP in more detail. The dispersion in laterally homogeneous QW systems has been calculated by Nakayama et al.[23,24] In laterally microstructured QW systems, as discussed here, additional effects should arise. If the influence of the lateral confinement is small we expect for instance gaps in the dispersion at $q_p = m \cdot \pi/a$, $m = \pm 1, 2, 3...$ due to the lateral superlattice effect which produces new Brillouin zones. Such lateral superlattice effects have e. g. been observed in the 2D plasmon dispersion of laterally density-modulated 2D electronic systems.[28] If the influence of the lateral structure is stronger, the discretization of the dispersion in laterally confined QW systems can be approximately descibed as 'polaritons in a box'. If the lateral dimensions become very small, i.e. the energy gaps in the polariton model become larger than the exciton-binding energy, we expect a continuous transition to 1D excitons which is in that limit a better description of the excitation. Such a transition from 'polaritons in a box' to confined excitons has been recently very nicely demonstrated for a transition from 3D excitons to 2D excitons in CdTe-CdZnTe superlattices with different well widths.[29]

The confinement energies, energy levels, oscillator strengths and polarization dependences for 1D excitons have been calculated by several authors (e.g. Refs. 30-32 and references therein). To our knowledge, the interaction between neighboring quantum wires, which is present in all experiments that have been reported on 1D excitons so far, and which in a certain sense is a remainder of the polariton model, has not been calculated. In particular, the influence on the polarization dependence would be of great interest, since the experimental polarization dependence might be used as a proof to confirm the 1D character of the exciton (e.g. Ref. 13). Here we would like to note that we have recently succeeded in preparing structures similar to those as in Fig. 2b with a periodictiy of 250 nm and a lateral width of the quantum wires of only 70 nm.[27] Even from these small structures we could observe luminescence of free excitons. Most interestingly, we observed in excitation spectroscopy (PLE) three peaks in the regime of the n=1 hh- and lh-transitions which arise from transitions between quantum confined 1D energy levels in the wire. In particular, a pronounced, and for different 1D-transitions, different polarization dependence was observed.

In conclusion, QWEP emission and excitation was demonstrated in microstructured QW and QW-wire systems. The polariton emission was found to give an important contribution to the PL spectra of these systems showing a characteristic polarization behavior.

We would like to acknowledge financial support from the Bundesministerium für Forschung und Technologie, Bonn.

REFERENCES

[1]R. Dingle, W. Wiegmann, and C.H. Henry, Phys. Rev. Lett. 33: 827 (1974).
[2]C. Weisbuch in "Physics and Application of Quantum Wells and Superlattices", eds. E.E. Mendez and K. von Klitzing, Plenum Press, New York, p. 261 (1987).
[3]K. Kash, A. Scherer, J.M. Worlock, H.G. Craighead, and M.C. Tamargo, Appl. Phys. Lett. 49: 1043 (1986).
[4]J. Cibert, P.M. Petroff, G.J. Dolan, S.J. Pearton, A.C. Gossard, and J.H. English, Appl. Phys. Lett. 49: 1275 (1986).
[5]M.A. Reed, R.T. Bate, K. Bradshaw, W.M. Duncan, W.R. Frensley, J.W. Lee, and M.D. Shih, J. Vac. Sci. Technol. B4: 358 (1986).
[6]H. Temkin, G.J. Dolan, M.B. Panish, and S.N.G. Chu, Appl. Phys. Lett. 50: 413 (1987).
[7]Y. Hirayama, S. Tarucha, Y. Suzuki, and H. Okamoto, Phys. Rev. B. 37: 2774 (1988).
[8]D. Gershoni, H. Temkin, G.J. Dolan, J. Dunsmuir, S.N.G. Chu, and M.B. Panish, Appl. Phys. Lett. 53: 995 (1988).
[9]H.E.G. Arnot, M. Watt, C.M. Sotomayor-Torres, R. Glew, R. Cusco, J. Bates, and S.P. Beaumont, Superlattices and Microstr. 5: 459 (1989)
[10]M. Kohl, D. Heitmann, P. Grambow, and K. Ploog, Phys. Rev. B37: 10927 (1988).
[11]M. Kohl, D. Heitmann, P. Grambow, and K. Ploog, Superlattices and Microstr. 5: 235 (1989).
[12]A. Forchel, H. Leier, B.E. Maile, and R. German, Advances in Solid State Physics, ed. U. Rössler, Vieweg, Braunschweig (1988).

[13]M. Tsuchiya, J.M. Gaines, R.H. Yan, R.J. Simes, P.O. Holtz, L.A. Coldren, and P.M. Petroff, Phys. Rev. Lett. 62: 466 (1989).
[14]For monographs on surface excitations and coupling processes, see e.g., "Electromagnetic Surface Modes", edited by A.D. Broadman (Wiley, New York, 1982); "Surface Polaritons", edited by V.M. Agranovich and D.L. Mills (North-Holland, Amsterdam, 1982).
[15]D. Heitmann and H. Raether, Surf. Sci. 59: 17 (1976).
[16]N. Marshall and B. Fischer, Phys. Rev. Lett. 28: 811 (1972).
[17]D.D. Sell, S.E. Stokowski, R. Dingle, and J.V. DiLorenzo, Phys. Rev. B7: 4568 (1973).
[18]J. Lagois and B. Fischer, Phys. Rev. Lett. 36: 680 (1976).
[19]D. Heitmann, Surf. Sci. 170: 332 (1986).
[20]V.M. Agranovich and T.A. Leskova, Pis'ma Zh. Eksp. Teor. Fiz. 29: 151 (1979); JETP Lett. 30: 538 (1979).
[21]C. Weisbuch, R.C. Miller, R. Dingle, A.C. Gossard, and W. Wiegmann, Solid State Commun. 37: 219 (1981).
[22]C. Zhang, M. Kohl, and D. Heitmann, Superlattices and Microstr. 5: 65 (1989).
[23]M. Nakayama, Solid State Commun. 55: 1053 (1985).
[24]M. Nakayama and M. Matsuura, Surf. Sci. 170: 641 (1986).
[25]K. Ogawa, T. Katsumura, and H. Nakamura, Appl. Phys. Lett. 53: 1077 (1988).
[26]C.Jagannath, E.S. Koteles, J. Lee, Y.J. Chen, B.S. Elman, and J.Y. Chi, Phys. Rev. B34: 7027 (1986).
[27]M. Kohl, D. Heitmann, P. Grambow, and K. Ploog, to be published.
[28]U. Mackens, D. Heitmann, L. Prager, J.P. Kotthaus, and W. Beinvogl, Phys. Rev. Lett. 53: 1485 (1984).
[29]H. Tuffigo, R.T. Cox, F. Dal'Bo, G. Lentz, N. Magnea, H. Mariette, and C. Grattepain, Superlattices and Microstr. 5: 83 (1989).
[30]M.H. Degani and O. Hipolito, Phys. Rev. B35: 9345 (1987).
[31]I. Suemune and L.A. Coldren, IEEE QE 24: 1778 (1988).
[32]G. Bastard in "Physics and Application of Quantum Wells and Superlattices", eds. E.E. Mendez and K. von Klitzing, Plenum Press, New York, p. 21 (1987).

RADIATIVE RECOMBINATION IN FREE STANDING QUANTUM BOXES

S. R. Andrews*, H. Arnot+, T. M. Kerr*, P. K. Rees*, and S. P. Beaumont+

* GEC Hirst Research Centre, East Lane, Wembley, HA9 7PP, UK
+ Dept of Electronic and Electrical Engineering, University of Glasgow, Glasgow, UK

ABSTRACT

We report photoluminescence measurements made on free standing lattice matched GaAs/AlGaAs and pseudomorphic InGaAs/GaAs quantum boxes fabricated by laterally patterning quantum wells using electron beam lithography and either reactive ion etching or ion beam milling. At temperatures below 20K the luminescence efficiency of the GaAs quantum box arrays tends to scale with the volume of quantum well material remaining after processing even for the smallest boxes which have lateral dimensions of only 40-50 nm. These observations suggest that the surface recombination rate in GaAs sub-micron structures is not necessarily large relative to the radiative recombination rate at low temperatures. In contrast radiative recombination in the InGaAs/GaAs quantum boxes is strongly quenched for lateral dimensions less than 500 nm. We suggest that this is because photoexcited carriers rapidly thermalise and become laterally localised in the GaAs boxes by potentials of order a few meV, perhaps arising from the effects of interface fluctuations or strain relaxation, and that such localisation effects are smaller in the InGaAs boxes.

INTRODUCTION

The novel optical properties of quasi-one-dimensional semiconductor quantum well structures are interesting both from the point of view of fundamental physics and also because of their applications in opto-electronic and also non-linear all-optical devices. Examples include lasers and optical switching and logic devices. Further quantum confinement in one or two dimensions can be achieved by laterally patterning quantum wells to form quantum wires or boxes (the latter also called dots or discs). In the case of quantum boxes the effect of carrier confinement in all three dimensions is to modify the density of states and increase the Coulomb interaction between electron and hole giving rise to an absorption spectrum consisting of a set of discrete lines with enhanced oscillator strength and saturation governed by energy level occupation. Compared with quantum wells, this leads to a prediction of enhanced electro

Science and Engineering of One- and Zero-Dimensional Semiconductors
Edited by S.P. Beaumont and C.M. Sotomajor Torres
Plenum Press, New York, 1990

absorption and refraction (Miller et al 1988, Schmitt-Rink et al 1987) and non-linear optical effects (Chemla and Miller 1986). The non-linear optical effects in quantum boxes should be similar to those observed in colloidal preparations of II-VI semiconductor microcrystallites (Brus 1986) but with the advantages of spatially selected positioning and more tailorable size and possibly dielectric environment of the microcrystallites.

In the III-V semiconductor systems lateral dimensions smaller than 20-30 nm will be necessary before enhanced confinement effects become significant. In practice the technology for patterning structures on such length scales is relatively immature compared with that involved in the growth of quasi-2D quantum wells and only subtle modifications to the optical properties of the quantum well starting material, such as small blue shifts (Temkin et al 1987, Gershoni et al 1988) and possibly spectral splitting of the ground state (Gershoni et al 1988), have been observed in the relatively coarse featured (lateral dimensions greater than 30 nm) structures produced so far. As in the literature, we shall refer to quantum wells with sub-micron 2D patterning as quantum boxes even if lateral confinement effects are very small. Apart from the size of quantum box necessary to observe enhanced confinement effects, other considerations are inhomogeneous broadening of spectral features due to lateral size variations (Wu et al 1987), crystal damage incurred by patterning and also surface related effects, especially in free standing structures.

In this paper we address the subject of recombination of photoexcited carriers in submicron free standing lattice matched $GaAs/Al_{0.3}Ga_{0.7}As$ and strained layer $In_{0.11}Ga_{0.89}As/GaAs$ quantum boxes with relatively large free surface to volume ratios. Oxygen exposed GaAs surfaces (and presumably InGaAs surfaces for small In composition) possess a large density of extrinsic states near the middle of the forbidden energy gap which effectively pin the surface Fermi level at that position. This midgap pinning is associated with trapping centres which can give rise to a large non-radiative surface recombination velocity at room temperature (Aspnes 1983) and can limit the performance of minority carrier devices such as bipolar transistors and light emitting diodes. Heterojunction interface recombination velocities are generally small by comparison and can be neglected (Bimberg 1986). There have been several studies of the low temperature light emitting properties of both free standing and buried quantum boxes and wires. Buried sub-micron structures fabricated in lattice matched GaAs/AlGaAs and InGaAs/InP either using ion implantation induced disordering and annealing techniques (Cibert et al 1986, Hirayama et al 1988) or by etching followed by regrowth (Petroff et al 1982, Miyamoto et al 1988) show little evidence of non-radiative interface recombination in the measured low temperature luminescence efficiencies. Free standing structures fabricated in InGaAs/InP similarly seem to show little evidence of surface recombination (Gershoni et al 1988, Forchel et al 1988). The situation in the GaAs/AlGaAs system is less clear. Forchel et al (1988) have interpreted their measurements of low luminescence efficiency in sub-micron GaAs/AlGaAs structures in terms of large surface recombination rates. This result disagrees with the work of Kash et al (1986) who observed enhanced luminescence efficiencies in their GaAs structures indicating that the exposed surfaces play little role. Comparison of the different reports is not straightforward since different authors use different fabrication and measurement techniques and give little indication of reproducibility. We describe measurements on a number of samples fabricated in different ways (so as to compare the effects of surface preparation and give some indication of reproducibility) which show that surface recombination can be a small effect at low temperatures in GaAs quantum boxes prepared by both reactive ion etching (RIE) and ion beam milling. In contrast we find that surface recombination is apparently important in InGaAs quantum boxes in the InGaAs/GaAs strained layer system and discuss possible differences between the two systems.

EXPERIMENTAL DETAILS

The GaAs/AlGaAs quantum box structures were made from material grown by molecular beam epitaxy on a n^+ GaAs substrate and consisting of the following undoped GaAs and $Al_{0.3}Ga_{0.7}As$ layers in order of growth: 500 nm GaAs, 80 nm AlGaAs, 80 nm GaAs, 34 nm AlGaAs, 20 nm GaAs, 34 nm AlGaAs, 10 nm GaAs, 34 nm AlGaAs, 5 nm GaAs, 34 nm AlGaAs. The InGaAs/GaAs quantum boxes were fabricated from the following undoped GaAs and $In_{0.11}Ga_{0.89}As$ layers grown in order on a n^+ GaAs substrate: 500 nm GaAs, 20 nm InGaAs, 30 nm GaAs, 9.5 nm InGaAs, 30 nm GaAs, 6 nm, InGaAs, 30 nm GaAs, 2.5 nm InGaAs, 30 nm GaAs. The well and barrier thicknesses were verified to 10% accuracy by transmission electron microscopy. Patterning was performed by electron beam lithography using a modified Phillips PSEM 500 scanning electron microscope with an 8 nm focus at 50 keV. Both positive and negative resist systems and various mask materials were used. A metal etch mask 20-25 nm thick was fabricated in either Nichrome or Titanium using the standard 'lift-off' procedure with positive resist. An organic etch mask was fabricated using a 0.12 μm or 0.28 μm thick high resolution negative resist (HRN). Free standing pillars containing quantum boxes were defined by RIE or argon ion milling. These techniques are more controllable and give better anisotropy than wet chemical etching as used by some authors (Petroff et al 1982, Miyamoto et al 1987, Miller et al 1989). Two different gases were used for RIE: $SiCl_4$ and a 1:5 CH_4/H_2 mixture (Cheung et al 1987). By optimising conditions it was possible to produce smooth, side walls and low damage (as evidenced by the photoluminescence measurements described below). The masks were not generally removed after fabrication.

Boxes with various nominal diameters between 40 nm and 500 nm were patterned in the form of arrays of free standing pillars in adjacent squares 100 μm on edge with typical pillar spacings of between 3 and 5 times the pillar diameter, corresponding to filling factors of 4-9%. The pillar diameters could be measured to an accuracy of ±5 nm in the SEM. Typically the pillar diameter variations within an array are of the same order. The uppermost two quantum wells were etched through in all cases. Large 100 x 100 μm^2 control mesas were also exposed in areas adjacent to the box arrays to assess etching damage and act as standards of luminescence efficiency.

Photoluminescence was excited above the AlGaAs band gap using the 632.8 nm (1.96 eV) line of a He-Ne laser with the sample suspended in a variable temperature helium atmosphere. Below band gap spectra were obtained using a pyridine 2 or styryl 9 dye laser. A 90° scattering geometry was used with the exciting incident light focused down to 30 μm and typically masking an angle of either 22° or 67° to the sample surface normal. Larger signals were obtained with the more grazing incident light, especially with metal masked samples, presumably because of improved coupling of light into the boxes by side wall illumination and most measurements were made in this geometry.

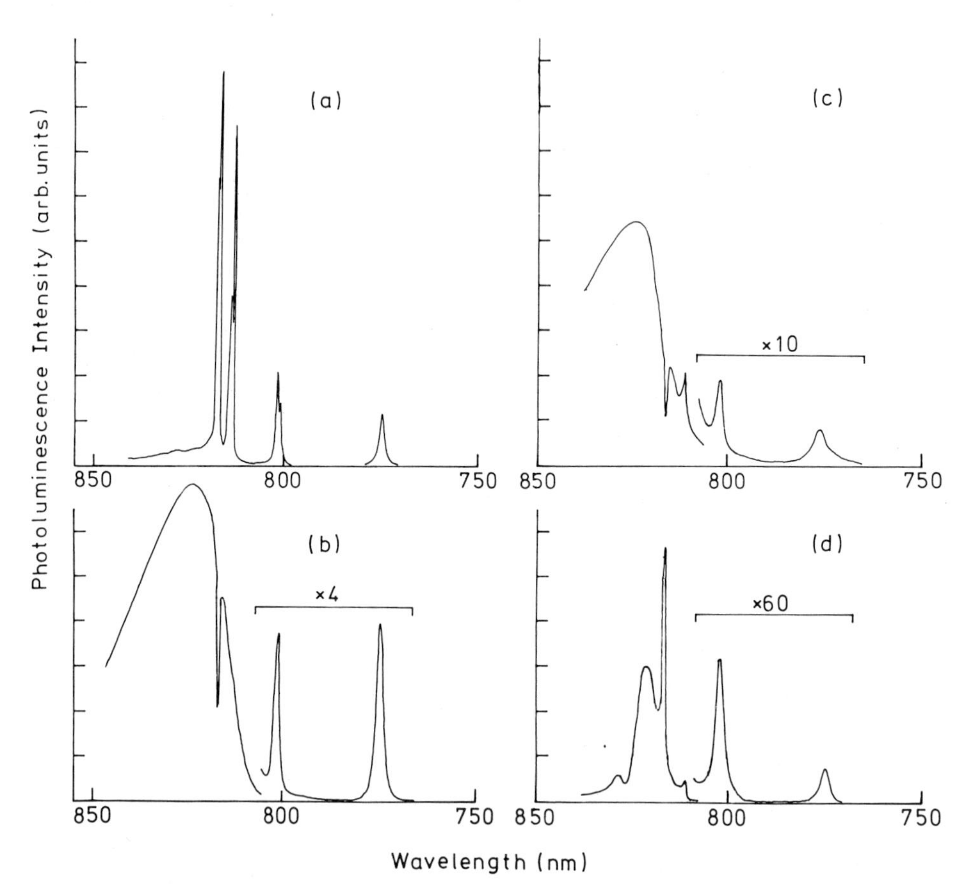

Fig 1 Representative photoluminescence spectra obtained at 4K with excitation at 1.65 eV and 1 Wcm^{-2} for the GaAs/AlGaAs sample before patterning (a) and after patterning into (b) 60 nm pillars (250 nm pitch) etched to depth of 0.2 μm using $SiCl_4$ RIE, (c) 60 nm pillars (200 nm pitch) etched to depth of 0.12 μm using Ar ion milling, (d) 60 nm pillars (180 nm pitch) etched to depth of 0.1 μm using CH_4/H_2 RIE.

Photoluminescence was collected at 90° to the incident beam and spectrally dispersed using a 0.85 m focal length double grating spectrometer with a typical resolution of 0.2 meV (full width at half maximum, FWHM).

EXPERIMENTAL RESULTS

GaAs/AlGaAs Quantum Boxes

Representative photoluminescence spectra of the GaAs/AlGaAs quantum well sample before and after patterning into pillars of 60 nm diameter are shown in Figure 1. The peaks near 775, 802, 814 and 818 nm in the spectra of the starting material, Figure 1a, arise from ground state free exciton recombination in the 5 nm, 10 nm, 20 nm and 80 nm (and thicker) layers respectively. The other important feature in the spectrum of Figure 1a is the broad, relatively weak background between 800 nm and 850 nm. This appears to be made up of a superposition of two broad peaks centred at 830 nm and 822 nm which we tentatively attribute to transitions involving an electron bound to a neutral carbon acceptor $(e\text{-}A^{\circ})_C$, and an exciton bound to a defect (d,X) such as a Gallium vacancy or vacancy complex (Briones and Collins 1982). The 100 x 100 μm^2 control mesas on the patterned samples showed similar spectra to the starting material (Figure 1a) but with the exciton peaks typically broadened by about 1 meV, possibly because of strain associated with differential thermal contraction between mask and sample on cooling or because of residual radiation damage. The spectra of the patterned material in Figures 1b,c,d clearly show peaks associated with recombination in the 5 nm and 10 nm wells. Linewidths (FWHM) in the starting material were 4 meV and 1.2 meV for the transitions in the 5 nm and 10 nm wells respectively. In the smallest patterned structures there was typically further broadening of 2-4 meV which we attribute mainly to radiation damage. For wavelengths longer than 810 nm the spectrum from the patterned areas is very similar to that of the surrounding unpatterned but etched regions of the samples. This will be discussed in more detail elsewhere.

Figure 2 shows the relative intensity of luminescence from the 5 nm and 10 nm quantum wells after patterning into boxes of lateral extent 40 nm - 500 nm using $SiCl_4$ RIE. The relative intensity is the ratio of the integrated intensity of the quantum well luminescence from the box array (divided by the box filling factor) to that from an adjacent 100 x 100 μm^2 mesa. The luminescence efficiency of the mesas was found to be the same as that of large, similarly masked pieces of the starting material. A relative intensity of 1 therefore indicates a luminescence efficiency which scales with the volume of excited quantum well material. Although some data show a lower efficiency, possibly because of processing problems, the majority of points do approximately support such a scaling. Possible processing problems include poor mask quality resulting in radiation damage and poor anisotropy in the etching resulting in over or under cutting leading to a reduction in surface coverage because of pillars falling over. The results in Figure 2 were obtained at 5K with excitation at 1.96 eV which is above the AlGaAs band gap. Similar results were obtained with excitation below the AlGaAs band gap at 1.65 eV and also with CH_4/H_2 RIE and argon ion milling. At first sight this result is surprising as we might expect non-radiative surface recombination at the air exposed free surfaces created in patterning to quench the luminescence. In the case of RIE chemical passivation was originally thought to offer a possible explanation but the results using argon ion milling, a simple physical sputtering process, seem to rule this out. The observation of an apparently low surface recombination rate is discussed at length below.

We have measured the energy dependence of the photoluminescence intensity (i.e. the photoluminescence excitation spectrum) in many of the box arrays and find that the luminescence efficiency has a similar variation with energy in both mesas and boxes as illustrated in Figure 3. We do not find any enhancement in the photoluminescence efficiency at high excitation energy relative to low energy as reported by Kash et al (1986). The shape and position of the n=1 heavy hole exciton resonance in the excitation spectrum was not found to vary with detection energy suggesting that the dominant line broadening mechanism

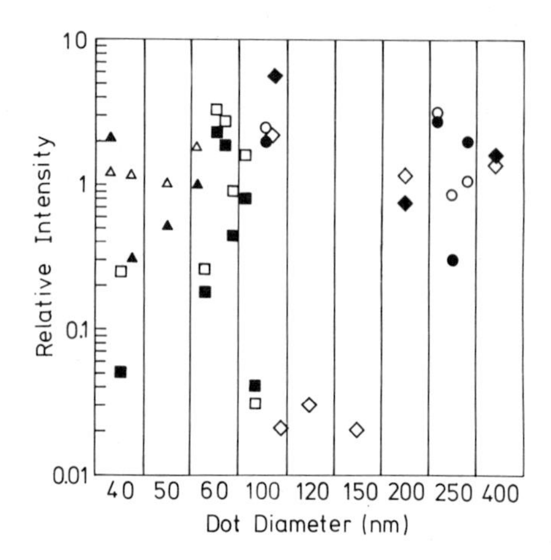

Fig 2 Integrated intensity of photoluminescence excited at 1.96 eV in four $SiCl_4$ etched samples (as shown by the key) relative to that in the control mesas on the same samples. The PL intensities are scaled by the volume of quantum well material excited as discussed in the text. Data from the 5 nm quantum well is shown by open symbols and the 10 nm well by filled symbols.

Symbol	Mask	Etch depth
triangle	0.12 μm HRN	0.60 μm
square	0.12 μm HRN	0.20 μm
circle	25 nm NiCr	0.20 μm
diamond	0.28 μm HRN	0.30 μm

is intrinsic to each quantum box (radiation damage) rather than reflecting size variations between boxes. No significant variation in ground state exciton energy with box size was found within the experimental uncertainty of 2 meV. These observations are consistent with calculations on 5 nm wells in 50 nm boxes with ±5 nm lateral size variations which give (taking into account the electron-hole interaction, Bryant 1988) a blue shift in the exciton ground state energy of 3 meV and an inhomogeneous broadening of 1 meV.

We have also measured the power dependence of the luminescence efficiency. Typical results are shown in Figure 4. We find a linear dependence, characteristic of radiative recombination, down to the lowest powers investigated. We conclude that centres for non-radiative recombination are saturated at all powers investigated. If we assume

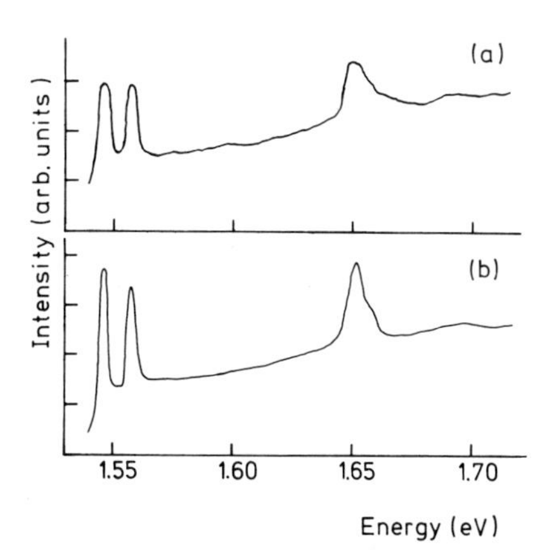

Fig 3 Photoluminescence excitation spectra of 10 nm quantum well at 5K in (a) array of 60 nm pillars and (b) 100 μm^2 mesa fabricated using CH_4/H_2 RIE with 0.28 μm HRN mask.

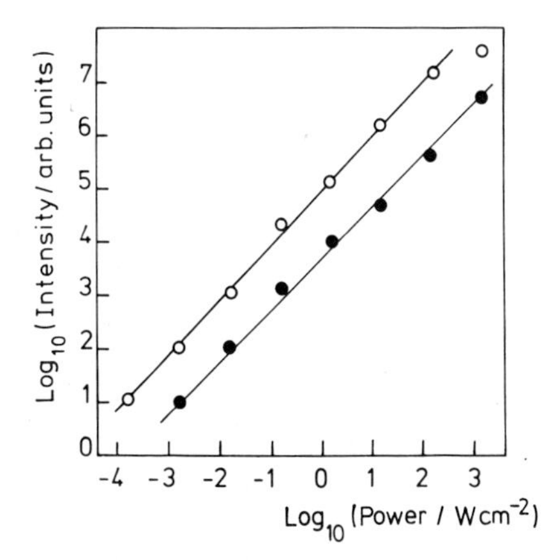

Fig 4 Excitation power dependence at 1.65 eV of integrated PL intensity from 10 nm well in an array of 60 nm dots (filled circles) and a control mesa (open circles) in a sample etched using CH_4/H_2 with a 0.28 μm HRN resist mask.

that the radiative recombination rate in the boxes is not enhanced and non-radiative recombination not related to the free surfaces is undiminished then the results in Figures 2,4 imply that non-radiative surface recombination in GaAs quantum boxes plays little role in limiting the luminescence efficiency. The first assumption that the radiative recombination rate is not enhanced can be made because the boxes are too large to show clear signs of lateral confinement which would enhance the oscillator strength in recombination and because the pillar diameters are smaller than those required for absorption enhancement by waveguiding action. The second assumption of unchanged 'bulk' non-radiative recombination needs more careful consideration. Kash et al (1986) observed an enhancement of the luminescence

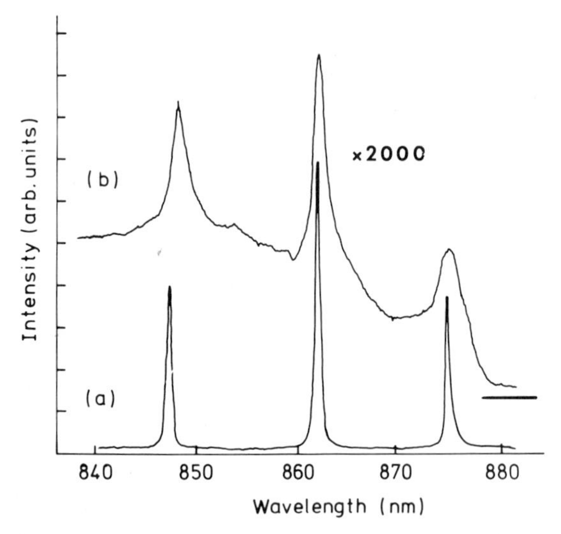

Fig 5 Photoluminescence spectra of InGaAs/GaAs sample at 5K excited at 1.49 eV (a) before and (b) after processing into 500 nm pillars (pitch 2500 nm, etch depth 0.3 μm), using $SiCl_4$ RIE and a 0.28 μm thick HRN etch mask.

efficiency of quantum wires and discs with dimensions of 40-45 nm by an excitation energy dependent factor of up to 50-100 compared with the unpatterned material. These authors suggested that patterning might eliminate many non-radiative recombination centres in the quantum wells thus increasing the luminescence efficiency. A similar argument has previously been used to explain the high luminescence efficiencies of quantum wells compared with bulk semiconductors (Ren and Dow 1985). In the latter case exciton confinement in the growth direction impedes transport to non-radiative centres. If an enhancement mechanism is operating in our samples then it must be balanced by increased surface non-radiative recombination. Since such a fortuitous balance seems unlikely we suggest that there is no enhancement mechanism in our samples and that the surface recombination is not significant in the GaAs/AlGaAs boxes.

InGaAs/GaAs Quantum Boxes

Typical photoluminescence spectra of the InGaAs/GaAs sample after processing into a 100 x 100 μm^2 mesa and an array of 500 nm diameter pillars are shown in Figure 6. The peaks at 847,861.5 and 874.5 nm arise from ground state recombination in the 2.5, 6 and 9.5 nm thick $In_{0.11}Ga_{0.89}As$ quantum wells respectively. In contrast to the GaAs/AlGaAs boxes the relative luminescence intensities in the InGaAs/GaAs boxes fabricated using both $SiCl_4$ and CH_4/H_2 RIE were found to be dramatically less than in the starting material or large mesas for excitation both above and below the GaAs bandgap. For 500 nm boxes the relative intensities were found to be between 0.1 and 0.01 of that in 100 x 100 μm^2 mesas. The luminescence from boxes with diameters of 300 nm and below was too weak to discriminate against the background GaAs luminescence.

DISCUSSION

Our overall conclusion is that the non-radiative surface recombination rate in our structures remains small compared with the radiative recombination rate in the GaAs boxes but not in the InGaAs boxes. Taking a typical exciton radiative lifetime of 300 ps for a 5 nm GaAs quantum well at 5K then a negligible non-radiative recombination rate in the smallest pillars suggests that the surface recombination velocity is less than 10^4 cms^{-1}. This result is different to that obtained by Forchel et al (1988) and Maile et al (1987) for their CCl_2F_2/Ar (1:4) RIE and argon ion milled GaAs/AlGaAs quantum wires. They observed strong luminescence quenching consistent with a surface recombination velocity of 5.10^5 cms^{-1}. This is similar to that in our InGaAs/GaAs boxes where we estimate a surface recombination rate of 10^6-10^7 cms^{-1}. Aside from the possibility of large surface recombination rates being attributable to excessive lattice damage during etching, two points need to be considered. Firstly, are small surface recombination velocities really unexpected for broad area, undoped, etched, GaAs surfaces at low temperatures and secondly, if not, is the recombination kinetics in a quantum box the same as in a large crystal. We discuss each point in turn.

To check the first point we performed measurements on samples containing undoped 0.5 μm thick GaAs epilayers bounded by AlGaAs barriers. These samples were etched at the same time as some of the quantum box arrays in order to expose a free GaAs surface. From photoluminescence measurements we have estimated surface recombination velocities of around 3.10^6 cms^{-1} at 5K and 10^4 cms^{-1} at room temperature. The temperature dependence is in qualitative agreement with phenomenological models of surface recombination (Stevenson and Keyes 1954) and the low value at room temperature is quantitatively consistent with the doping dependence described by Aspnes (1983). This then suggests that our observations of low surface recombination velocities are somehow specific to quantum boxes.

Carrier transport and recombination kinetics in quantum wells is very different to that in the bulk because the lateral potential fluctuations associated with local variations in well width which inhomogeneously broaden the exciton transitions can also laterally localise excitons (Hegarty et al 1984). In addition

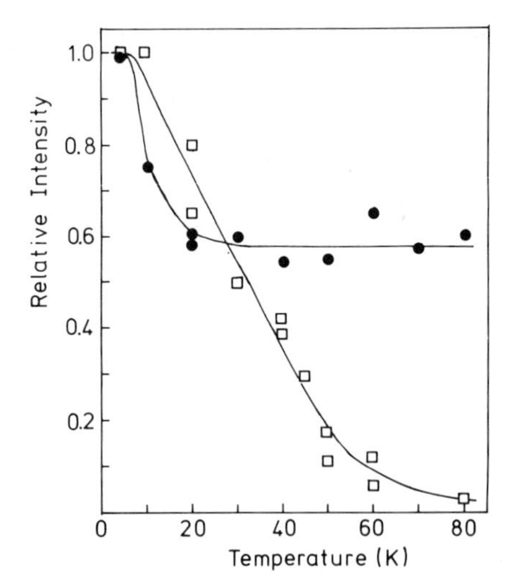

Fig 6 Temperature dependence of integrated luminescence intensity of the 5 nm quantum well (relative to that at 5K) in a GaAs/AlGaAs sample before processing (circles) and after patterning into an array of 60 nm pillars (squares).

the radiative recombination rate is enhanced by localisation perpendicular to the layers which increases the electron-hole overlap and radiative recombination rate compared with the bulk (Göbel et al 1983). In unpatterned structures at low temperatures hot photoexcited carriers rapidly (a few ps) form excitons and loose energy by LO phonon emission until within one LO phonon energy of the bottom of the band. Thermalisation is then by the slower process of acoustic phonon emission (a few tens of ps) until excitons relax into localised states where further thermalisation within the exciton lifetime is inhibited (Masumoto et al 1984). In quantum boxes localisation competes with further thermalisation to and trapping at mid-gap surface states. We suggest that localisation may successfully compete in our GaAs/AlGaAs boxes where there are fairly large potential fluctuations (or order 4 meV in the 5 nm well) as inferred from the Stokes shift between emission and absorption. In the InGaAs/GaAs boxes the interfaces appear to be much smoother with Stokes shifts of less than 0.2 meV in the 2.5 nm well so that localisation would compete less successfully.

A further possibility is that excitons are laterally localised in the GaAs/AlGaAs boxes as a result of relief of the mismatch strain in the AlGaAs barrier layers on removing the surrounding material (Kash et al 1988). Assuming total relaxation in the AlGaAs layers and a uniform strain in the GaAs layers after processing we estimate an increase in band gap in the quantum wells due to biaxial tension of 2.4 meV. In arriving at this estimate we have used the conventional strained layer model treatment with deformation potentials appropriate to GaAs (Huang et al 1987). If the strain is inhomogenous in the pillars, in particular, if it is greatest at the side walls then a strain induced lateral confining potential of order a few meV seems possible. Another possibility is that differential contraction between mask and semiconductor might impose significant strain on cooling. We checked that this was not important by measuring the luminescence efficiency of samples etched using CH_4/H_2 in which the HRN mask was subsequently removed in an oxygen plasma. No significant difference was found between the luminescence efficiency of samples cooled with and without masks in place. We might expect strain relaxation to be a smaller effect in the InGaAs/GaAs boxes, since although the as-grown layers are under biaxial compression the relatively thick barriers are lattice matched to the substrate.

If surface recombination in the patterned GaAs/AlGaAs structures is inhibited by lateral exciton localisation due either to fluctuations in well width or strain then we might expect to observe a change in behaviour with increasing temperature as carriers thermally activate into delocalised states and fall into surface traps. We have compared the temperature dependence of the photoluminescence efficiency of the 5 nm well in the patterned areas with that of the starting material as shown in Figure 6. The luminescence efficiency in the quantum boxes falls relative to that in the starting material for temperatures above about 20K, apparently confirming our conjecture. However, the exciton lifetime increases by a factor of 4 or 5 between 5K and 50K because of the effects of acoustic phonon scattering on the coherence volume of the wavefunction (Feldman et al 1987), which complicates the interpretation. Nevertheless, this data is consistent with our model.

To summarise, the dominant recombination mechanism of photoexcited carriers in laterally patterned lattice matched GaAs/AlGaAs quantum wells is found to be radiative at temperatures below 20K for lateral dimensions as small as 40-50 nm. In contrast, the dominant recombination mechanism in the strained layer InGaAs/GaAs system at low temperatures is found to be non-radiative, presumably involving surface traps, for lateral dimensions of order 500 nm or less. We have suggested two possible mechanisms whereby non-radiative surface recombination is inhibited by the existence of lateral potential barriers. One mechanism involves strain and the other disorder in the plane of the well. Both models are consistent with surface recombination being more important in InGaAs/GaAs boxes. On the basis of the disorder model it is possible in principle to understand why measurements of luminescence efficiencies in the same material system but on samples grown with different interface morphology might show different results. If strain relaxation is important then it is conceivable that different results might be obtained on boxes and wires.

REFERENCES

D E Aspnes 'Recombination at semiconductor surfaces and interfaces' Surf. Science 132, 406 (1983)

D Bimberg in "Solid State Devices" ed P Balk and O G Folberth (Elsevier, Amsterdam, 19860, p.101

F Briones and D M Collins 'Low temperature photoluminescence of lightly Si-doped and undoped MBE GaAs' J. El. Mats. 11, 847 (1982)

L Brus 'Zero-dimensional 'excitons' in semiconductor clusters' IEEE J. Quantum Electronics 22, 1909 (1986)

G W Bryant 'Excitons in quantum boxes: Correlation effects and quantum confinement' Phys. Rev. B37, 2635 (1988)

J Cibert, P M Petroff, G J Dolan, S J Pearton, A C Gossard and J H English 'Optically detected carrier confinement to one and zero dimension in GaAs quantum well wires and boxes' Appl. Phys. Lett. 49, 175 (1986)

D S Chemla and D A B Miller 'Mechanism for enhanced optical nonlinearities and bistability by combined dielectric-electronic confinement in semiconductor microcrystals' Optics Lett. 11, 522 (1986)

R Cheung, S Thoms, S P Beaumont, G Doughy, V Law and C D W Wilkinson 'Reactive ion etching of GaAs using a mixture of methane and hydrogen' El. Lett. 23, 857 (1987)

J Feldman, G Peter, E O Göbel, P Dawson, K Moore, C Foxon and R J Elliott 'Linewidth dependence of radiative exciton lifetimes in quantum wells' Phys. Rev. Lett. 59, 2337 (1987)

A Forchel, H Leier, B E Maile and R Germann 'Fabrication and optical spectroscopy of ultra small III-V compound semiconductor structures' Festkörperprobleme (Advances in solid state physics) 28, 99 (1988)

D Gershoni, H Temkin, G J Dolan, J Dunsmuir, S N Chu and M B Pannish 'Effects of two-dimensional confinement on the optical properties of InGaAs/InP wire structures' Appl. Phys. Lett. 53, 995 (1988)

E O Göbel, H Jung, J Kuhl and K Ploog 'Recombination enhancement due to carrier localisation in quantum well structures' Phys. Rev. Lett. 51, 1588 (1983)

J Hegarty, L Goldner and M D Sturge 'Localised and delocalised two-dimensional excitons in a GaAs-AlGaAs multiple-quantum-well structure' Phys. Rev. B30, 7346 (1984)

Y Hirayama, S Tarucha, Y Suzuki and H Okamoto 'Fabrication of a GaAs quantum-well-wire structure by Ga focused-ion-beam implantation and its optical properties' Phys. Rev. B37, 2774 (1988)

J D Huang, U K Reddy, T S Henderson, R Houdre and H Morkoc 'Optical investigation of highly strained InGaAs-GaAs multiple quantum wells' J. Appl. Phys. 62, 3366 (1987)

K Kash, A Scherer, J M Worlock, H G Craighead and M C Tamargo 'Optical spectroscopy of ultra small structures etched from quantum wells' Appl. Phys. Lett. 49, 1043 (1986)

K Kash, J M Worlock, M D Sturge, P Grabbe, J P Harbison, A Scherer and P S D Lin 'Strain-induced lateral confinement of excitons in GaAs-AlGaAs quantum well microstructures' Appl. Phys. Lett. 53, 782 (1988)

B E Maile, A Forchel, R Germann, K Streubel, F Scholz, G Weimann and W Schlapp 'Fabrication of nanometer width GaAs/AlGaAs and InGaAs/InP quantum wires' Microcircuit Engineering 6, 163 (1987)

Y Masumoto, S Shionoya and H Kawaguichi 'Picosecond time-resolved study of excitons in GaAs-AlAs multi-quantum-well structures' Phys. Rev. B29, 2324 (1984)

D A B Miller, D S Chemla and S Schmitt-Rink 'Electroabsorption of highly confined systems: Theory of the quantum-confined Franz-Keldysh effect in semiconductor quantum wires and dots' Appl. Phys. Lett. 52, 2154 (1988)

B I Miller, A Sharar, U Koren and P J Corvini 'Quantum wires in InGaAs/InP fabricated by holographic photolithography' Appl. Phys. Lett. 54, 188 (1989)

Y Miyamoto, M Cao, Y Shingai, K Furuya, Y Suematsu, K G Ravikumar and S Arai 'Light emission from quantum-box structures by current injection' Jap. J. Appl. Phys. 26, L225 (1987)

P M Petroff, A C Gossard, R A Logan and W Wiegmann 'Towards quantum well wires: Fabrication and optical properties' Appl. Phys. Lett. 42, 635 (1982)

S F Ren and J D Dow 'A mechanism of luminescence enhancement by classical well structures of superlattices' J. Luminescence 33, 103 (1985)

S Schmitt-Rink, D A B Miller and D S Chemla 'Theory of linear and nonlinear optical properties of semiconductor microcrystallites' Phys. Rev. B35, 8113 (1987)

D T Stevenson and R J Keyes 'Measurements of the recombination velocity at germanium surfaces' Physica 20, 1041 (1954)

H Temkin, G J Dolan, M B Panish and S N G Chu 'Low temperature photoluminescence from InGaAs/InP wires and boxes' Appl. Phys. Lett. 50, 413 (1987)

W-Y Wu, J N Schulman, T Y Hsu and U Effron 'Effect of size non-uniformity on the absorption spectrum of a semiconductor quantum dot system' Appl. Phys. Lett. 51, 710 (1987)

OPTICAL EMISSION FROM QUANTUM WIRES

A. Forchel, B. E. Maile, H. Leier, G. Mayer, and R. Germann
4. Physikalisches Institut, Universität Stuttgart
Pfaffenwaldring 57, D 7000 Stuttgart 80, FR-Germany

INTRODUCTION

The physical properties of semiconductors with dimensions of the order of the de Broglie wavelength of electrons depend strongly on the device dimensions.[1] In thin semiconductor heterostructures (quantum wells) the effective band gap is determined by the quantum well thickness in addition to the bulk properties of the quantum well material.[2] This allows to increase the band gap in thin quantum wells by hundreds of meV if a suitable confinement material is employed. Furthermore the energy dependence of the density of states changes from a proportionality to $E^{1/2}$ to a step function.

Two-dimensional semiconductor-heterostructures can be fabricated by different modern epitaxy-techniques like e.g. molecular beam epitaxy (MBE) or metal organic vapor phase epitaxy (MOVPE).[3,4] These technologies allow to control the growth of semiconductor materials down into the monolayer range. Due to the high potential for electronic and optoelectronic applications the research activities have been focussed on GaAs/GaAlAs and InGaAs/InP quantum wells.

Further significant changes of the electronic properties are expected if one reduces the dimensions of semiconductor structures in two or three directions down to values on the order of the de Broglie wavelength.[5] The fabrication of effectively one- and zero-dimensional semiconductor structures generally starts from epitaxially grown two-dimensional layers. By high resolution lateral patterning - usually based on electron beam lithography - lateral sections with extensions below 100 nm are

Science and Engineering of One- and Zero-Dimensional Semiconductors
Edited by S.P. Beaumont and C.M. Sotomajor Torres
Plenum Press, New York, 1990

defined. Current lithographical techniques are able to provide lateral structures on commercial III-V-substrates with dimensions down to less than 20 nm. For these dimensions significant changes of the electronic properties may be expected.

The most advanced technology for the definition of lateral patterns in the nanometer range is electron-beam lithography.[6] In special configurations structures with typical dimensions in the 1 nm-range have been defined.[7] The narrowest structures realized on thick semiconductor substrates reach well below 10 nm.[8] Using focussed ion-beams lateral patterns in the 50 nm range have been demonstrated.[9] In addition to the exposure of narrow linewidth patterns in ion-sensitive resists, focussed ion-beams can be used to modify the physical properties of the substrates directly by maskless implantation or etching on a sub-100 nm scale. X-ray lithography is capable of nm-resolution as well.[10] Due to the

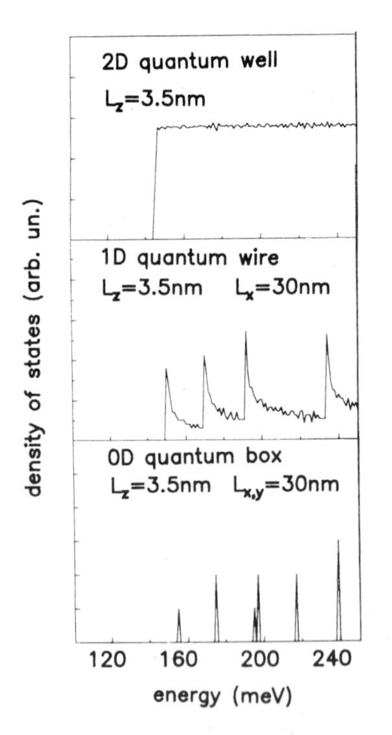

Fig. 1 displays the changes which occur for the energy dependence of the density of states in going from a two dimensional to a one- or a zero-dimensional stucture. For the calculation of the density of states of the 3.5 nm wide 2 D quantum well (top) the electron mass of GaAs and an AlGaAs barrier of 250 meV have been used. For the lateral confinement a width of 30 nm and a barrier height of 50 meV have been used in one (quantum wire, center part of fig 1) and two directions (quantum box, bottom part of fig. 1)

The lateral confinement of the quantum wire leads to significant changes in the energy variation of the density of states of the quantum wire and the dot. For the quantum wire a number of lateral subbands are formed. Their density of states decays proportional

difficult mask technology and the limited availability of suitable synchroton sources X-ray lithography up to now plays no important role for the definition of one- and zero-dimensional structures.
to $E^{-1/2}$. In the quantum dot the lateral subbands lead to a series of delta function shaped peaks in the density of states. In both cases the lateral quantization yields a significant shift of the subband edge to higher energy compared to the band edge in the quantum well.

The significant changes which arise for the density of states in energy space in one- and zero dimensional structures are expected to induce strong changes in the optical and transport properties of the semiconductor materials. The high potential of these structures for applications and the new physical phenomena expected in 1D and 0D-semiconductor systems has stimulated intense research efforts regarding the fabrication of low dimensional devices.

Currently most investigations on properties of 1D- and 0D-devices are related to structures for transport experiments. As these structures consider unipolar electron transport, confinement in the conduction band only is generally required. This confinement is most efficiently obtained by electrostatic depletion with suitably shaped gate electrodes.[11,12] The observation of 1D- or 0D quantization in optical experiments in contrast requires the simultaneous confinement of electrons and holes. Therefore electrostatic techniques can not be used. Different approaches are currently studied to obtain low dimensional structures for optical investigations. They include quantum wire and dot fabrication by dry etching,[13] implantation induced interdiffusion[14] and epitaxy on patterned substrates.[15]

The present paper describes some aspects of the fabrication and electronic properties of quantum wires developed for optical spectroscopy. The next section briefly reviews the high resolution electron beam lithography used for the definition of these structures with an emphasize on the special requirements of quantum wires fabricated for optical investigations. We then discuss results obtained for dry etched quantum wires in GaAs/GaAlAs and InGaAs/InP. Finally the technology and optical properties of newly developed quantum wires defined by implantation induced interdiffusion are described.

HIGH RESOLUTION ELECTRON BEAM LITHOGRAPHY FOR LOW DIMENSIONAL STRUCTURES FOR OPTICAL SPECTROSCOPY

The requirements of laterally patterned semiconductor structures for optical investigations of dimensionality dependent phenomena differ to a certain degree from those of nanometer patterns for transport studies. The spatial resolution in optical experiments is on the order of 1 μm. Individual quantized objects can only be addressed if the distance between them exceeds the resolution. The number of photons which are absorbed in a nanometer structure is quite small and limits the emission intensity: For a rather strong laser excitation with 10 ns pulses with an excitation intensity of 1 kW/cm^2 only about 10^3 photons/pulse will be absorbed in an area of 100 nm x 100 nm. The number of detected photons strongly depends on the external quantum efficiency and the detection efficiency and quickly approaches the noise level as the dimensions are reduced. In order to increase the sensitivity of optical experiments on nanometer wires one generally defines arrays in which the same structure is repeated over an area larger than the excitation spot. This implies a very good control of the dimensions of the structure.

We have fabricated wire structures from GaAs/GaAlAs- and InGaAs/InP-quantum well wafers by using high resolution electron beam lithography. Our commercial high resolution system provides a minimum beam diameter of 8 nm. The beam can be scanned over a field of 80 μm x 80 μm size using an address grid with 2.5 nm increments. With coarser address structures field sizes up to 1.6 mm x 1.6 mm can be used.

We mostly use the negative resist chloromethylated poly-α-methylstyrene (αM-CMS) for high resolution lithography.[16] Using proper development conditions in combination with a suitable descumming step to remove resist fragments between the exposed structures 20 nm wide lines can reproducibly be defined.

Negative resists are especially attractive for the fabrication of structures for optical studies. The resists are transparent at the excitation and the emission wavelengths. Therefore even resist covered structures may be characterized with high sensitivity. Furthermore the resist mask can be removed very easily in an O_2-plasma. In contrast, the metal masks used for the lift-off required after the exposure of positive tone resists often remain on top of the structure. These masks strongly attenuate the excitation and emission intensity from the wire or dot structures. Due to the different thermal expansion coefficients of the metal and the semiconductor the mask may induce significant stress in the structures, particularly in low temperature experiments. This can lead to shifts of the emission energy which are very difficult to distinguish from quantization effects.

In order to transfer the mask pattern into the semiconductor dry etching processes are generally used. In these processes the unprotected parts of the semiconductor are removed by an ion beam or in a plasma. If chemically reactive etch gases (e.g. Cl_2, CCl_2F_2) are used the semiconductor and the mask are attacked by physical and chemical effects. Reactive dry etching is expected to reduce surface damage and redeposition as well as to optimize the etch rate ratio between the mask and the semiconductor.

Fig.2 a) Dot pattern etched into InP by Ar/O_2 ion milling
b) GaAs wire defined by reactive ion etching with CCl_2F_2/Ar

Figure 2a shows a dot pattern etched into InP using ion milling with an Ar/O_2 mixture. The InP columns have a separation of 200 nm and a top diameter of about 50 nm.

Figure 2b shows etched GaAs wires defined by an αM-CMS etch mask. The width in the top section is about 30 nm. Together with the high reproducibility this feature size demonstrates the suitability of negative resists for the fabrication of quantum wire and dot arrays for optical studies.

LUMINESCENCE STUDIES OF ETCHED III-V SEMICONDUCTOR WIRES

We have fabricated wire structures from GaAs/GaAlAs- and InGaAs/InP quantum wells. For the optical investigation the samples contained a number of fields, in which the distance between the wires and the wire width were kept constant. Typically the fields extend over 150 μm x 2000 μm. Wire widths between 5 μm and 0.04 μm were studied. In addition to the patterned areas the samples contain reference fields which were completely etched or completely masked.

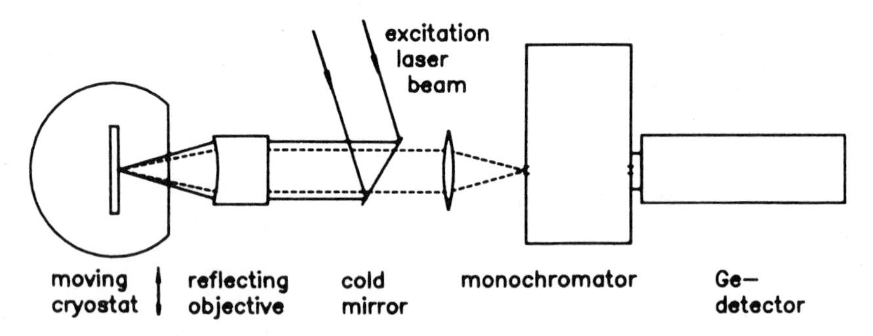

Fig.3 Schematic design of the set-up used to characterize the patterned substrates by photoluminescence spectroscopy with high spatial resolution.

For the luminescence measurements a special set-up was developed which permits a low temperature characterization of the patterns with approximately 5 μm spatial resolution.(Fig. 3) The high spatial resolution is obtained by using a specially designed cryostat and a reflective objective with a working distance of 24 mm in order to focus the laser on the sample.The achromatic reflection optics allows to use the same objective for excitation and detection. The samples are excited by an Ar-Laser. The excitation is coupled into the optical path by a cold mirror, which reflects light with wavelengths below about 560 nm and which is transparent for longer wavelengths. For spatial scans the cryostat can be moved by a stepping motor driven stage. The emission is spectrally dispersed in a 0.3 m monochromator and detected by a GaAs photomultiplier tube (GaAs samples) or a LN_2 cooled Ge-detector (InGaAs samples).

The samples used for the wire fabrication contain a single quantum well approximately 40 nm below the sample surface. The etch depth amounts typically to 60 nm. This implies that we etch completely through the top barrier layers and the quantum well in order to define the wire structures.

The lateral sidewalls of the wire structures significantly influence the quantum efficiency of the semiconductor wires. Fig. 4 shows the dependence of the emission intensity of etched InGaAs/InP quantum wires on the wire width L_x ($5\mu m \geq L_x \geq 0.5\mu m$). The detection wavelength corresponds to the maximum of the exciton emission intensity in the quantum well structures. The inset displays the orientation and the movement of the sample in the experiment.

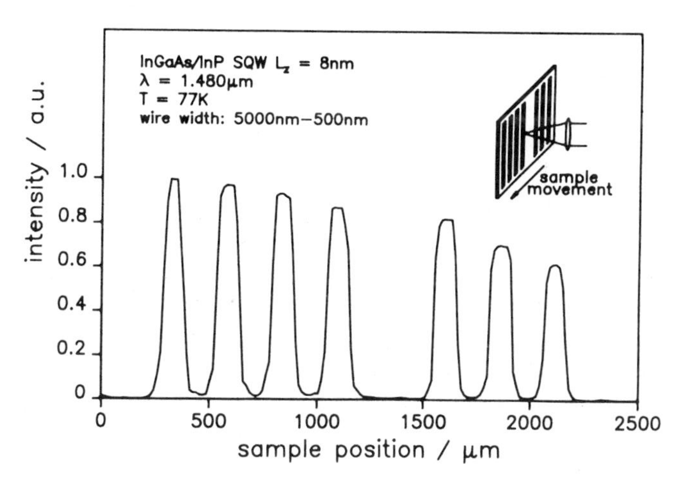

Fig. 4 Variation of the exciton emission intensity from $In_{.53}Ga_{.47}As/InP$ wires as a function of the wire width L_x ($5\ \mu m \leq L_x \leq 0.5\ \mu m$)

The lateral scan over the patterned sample shows the emission of two sets of grid patterns separated by a larger spacing which corresponds to a completely etched part of the sample. Fig. 4 demonstrates the high spatial resolution of our set up. Between the fields the emission intensity drops close to zero. This indicates that we have a negligible cross talk between fields with wires with different dimensions. In particular the completely etched reference has zero emission intensity. This field is very useful to determine the dynamics of the experiment.

Within the seven wire fields defined in the sample shown in Fig. 4 a constant area filling factor of 0.25 is used. Therefore the intensity variation versus wire width corresponds directly to the variation of the quantum efficiency. As shown in Fig. 4 the quantum efficiency drops approximately by 30 % if the widths of the InGaAs/InP wires is reduced from 5 microns to 0.5 microns.

The variation of the quantum efficiency as a function of wire width is due to non-radiative transitions at the etched sidewalls. With decreasing wire width the influence of non-radiative surface recombination increases. Furthermore dry etching damage may lead to the formation of optically inactive layers at the sidewalls ("dead layers").

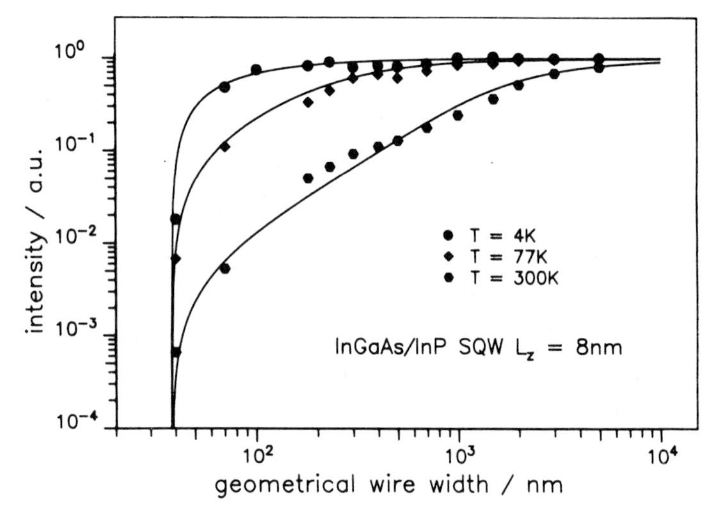

Fig.5 Lateral width variation of the quantum efficiency of the excitonic recombination in $In_{.53}Ga_{.47}As/InP$ for different temperatures. Points: experimental data; Lines: calculation.

We have investigated the influence of the side wall recombination and of the dead layer formation using InGaAs/InP wires with widths between 40 nm and 5 μm .[17] Fig. 5 displays the variation of the quantum efficiency versus wire width for three different temperatures. At liquid helium temperature the quantum efficiency is almost independent of the wire width above 100 nm. In going from a wire width of 70 nm to 40 nm a sharp drop of the emission intensity is observed. For the higher temperatures the side wall recombination effects become much more pronounced. In particular the decrease of the quantum efficiency starts at much larger wire width than at 2 K. A particularly strong temperature dependence of the quantum efficiency is observed for wire widths between 70 nm and 200 nm.

The solid lines in fig. 5 correspond to model calculations for the variation of the quantum efficiency due to the nonradiative recombination at the side walls. The model determines the electron hole pair density in a semiconductor slab taking into account the surface recombination velocity. Assuming excitonic recombination the change of the quantum efficiency due to side wall recombination can be obtained as a function of the geometrical wire width W, the bulk diffusion length L and the bulk diffusion coefficient D:

$$I(W)/I(W \rightarrow \infty) = 1 - \frac{\frac{SL^2}{AD} \sinh(\frac{A}{L})}{\frac{SL}{D} \cosh(\frac{A}{L}) + \sinh(\frac{A}{L})} \qquad (1)$$

with:

$$A = \frac{1}{2} W - W_d$$

W_d denotes the width of "dead layer" which may be present on both sides of the wire.

Using experimental values for the carrier life time in unprocessed quantum wells of 950 ps and 2100 ps at 4 K and 77 K, respectively, we can determine the surface recombination velocity and the width of the depleted layer very sensitively from fits to the experimental data. We obtain values of 1.6×10^3 cm/s and 5.9×10^3 cm/s for the surface recombination velocity at 4 K and 77 K. The observed temperature dependence of the surface recombination velocity is consistent with the usual assumption of a proportionality between S and the thermal velocity of the electron hole pairs. The width of the dead layer amounts to about 19 nm and is independent of temperature.

The variation of the quantum efficiency of etched wires versus the wire width depends sensitively on the patterned material system. Fig. 6 compares our results for the variation of the quantum efficiency with varying wire width for GaAs/GaAlAs and $In_{.30}Ga_{.70}As$/GaAlAs to the results of the previously discussed $In_{.53}Ga_{.47}As$/InP system (T=4 K).

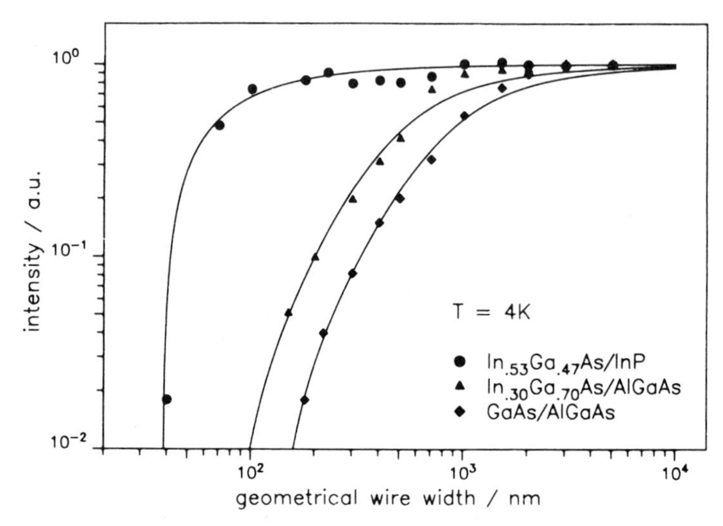

Fig.6 Comparison of the variation of the excitonic quantum efficiency (T=4K) versus well width for $In_{.53}Ga_{.47}As$/InP (circles), $In_{.3}Ga_{.7}As$/GaAlAs (triangles) and GaAs/GaAlAs (diamonds). Lines: Calculations for the variation of the quantum efficiency due to the side wall recombination and dead layer effects (equ. 1).

The wires have been etched using reactive ion milling with Ar/O_2 (InGaAs/InP) and reactive ion etching with CCl_2F_2Ar (GaAs, $In_{.3}Ga_{.7}As$).

We observe an increasing influence of the side wall recombination as the In content of the quantum wells is decreased. The width of the dead layer and the effective side wall recombination velocity depend strongly on the semiconductor material. We determine values of about 40 nm and 65 nm for the width of the dead layers and of $5x10^5$ cm/sec and 6.5×10^5 cm/sec for the side wall recombination velocity in $In_{.30}Ga_{.70}As$ and GaAs wires respectively.

The physical origin of the different influence of the dry etching on the quantum efficiency of the various semiconductor materials investigated here is not clear. Among the different factors which might play a role for the wire width dependence of the quantum efficiency passivation effects in the sidewall may be important. We would like to point out here that the sidewall recombination velocity closely aggrees with the surface recombination velocity in the different unprocessed material systems.

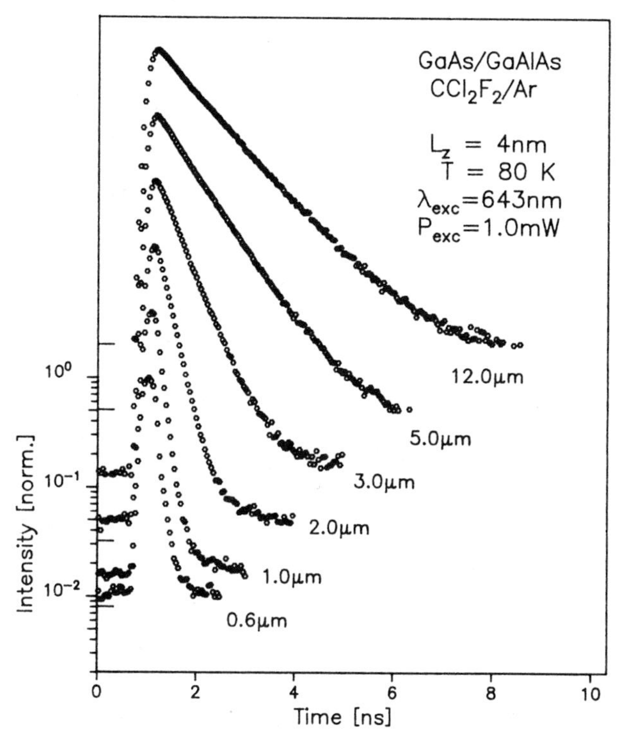

Fig.7 Time resolved luminescence intensity in GaAs-wires of different widths ((T=80K).

Time resolved measurements of the decay of the emission intensity in quantum wires constitute an independent access to the side wall recombination velocity. Fig. 7 displays time resolved measurements of the exciton emission intensity for a series of GaAs wires with widths between 12 μm and 0.6 μm at a temperature of 80 K. We observe a strong decrease of the recombination time constant as the wire width is reduced. In the case of a 12 μm wide wire the excitonic life time at 80 K amounts to about 1 ns, whereas a lifetime of only 30 ps is observed for 600 nm wide wires. From an extension of equ. 1 the influence of the side wall recombination on the total decay time can be used to evaluate the side wall recombination velocity. Consistent with our evaluation of the quantum efficiency for 80 K we determine values on the order of 1×10^7 cm/s for the side wall recombination velocity in etched GaAs wires.

INVESTIGATION ON INTERDIFFUSED GaAs/GaAlAs QUANTUM WIRES

As indicated by the discussion in the previous chapter side wall effects are a severe limitation for the use of etched quantum wires in optical experiments. Surface effects can be completely avoided in buried semiconductor structures. We have investigated the ion implantation induced interdiffusion technique for the fabrication of narrow buried GaAs quantum wires.

The quantum wire fabrication by implantation induced interdiffusion involves the following major steps. By electron beam lithography and lift off a gold wire mask is defined on top of a single quantum well substrate with a shallow top barrier (about 50 nm barrier thickness). The sample is then homogeneously implanted using e.g. Ga ions at an energy which in the unmasked areas allows the ions to penetrate through the quantum well whereas in the masked areas most ions are stopped in the Au layer. In our case we use a mask thickness of about 40 nm and a Ga energy of 100 keV. In the unmasked areas of the samples the Ga implantation creates a high density of defects (e.g. vacancies) whereas only a small defect concentration is created under the masks.

After the removal of the Au masks the sample is covered by Si_3N_4 and annealed for 30 min at 850 °C. For these annealing conditions Al atoms from the barrier layers may diffuse into the quantum well. The diffusion length and the activation energy of the diffusion process depend strongly on the defect concentration created during the implantation. Therefore strong interdiffusion effects occur in the unmasked areas whereas comparatively small effects are observed for the masked regions.

The different degrees of interdiffusion for masked and open sections of the sample lead to a spatially varying energy gap in the structure. For the unmasked areas a relatively strong increase of the energy gap occurs due to the large Al/Ga exchange. Basically this corresponds to an effective narrowing of the quantum well or - for diffusion lengths comparable with the quantum well thickness - to an admixture of Al at the center of the well. For the masked areas the energy gap may be kept close to that of the unprocessed samples by selecting suitable process conditions. A lateral confinement potential for the electrons and holes is then obtained from the difference in the conduction and valence band edges in the unmasked and masked regions of the sample.

The shape of the lateral potential defined by implantation induced interdiffusion on masked substrates depends on the lateral diffusion length of the defects prior to an Al/Ga exchange. In previous experimental studies on the quantum wire definition by interdiffusion typical values of the defect diffusion length in the range between 20 and 30 nm have been observed.[18,19] These large values of the lateral defect diffusion length severely limit the use of the structures for the observation of lateral quantization effects. A large lateral diffusion length corresponds to a very shallow quantization potential with small energetic differences between the 1-D-subbands. Furthermore the shallow lateral potential unavoidably leads to large changes of the quantum well width at the center of the mask.

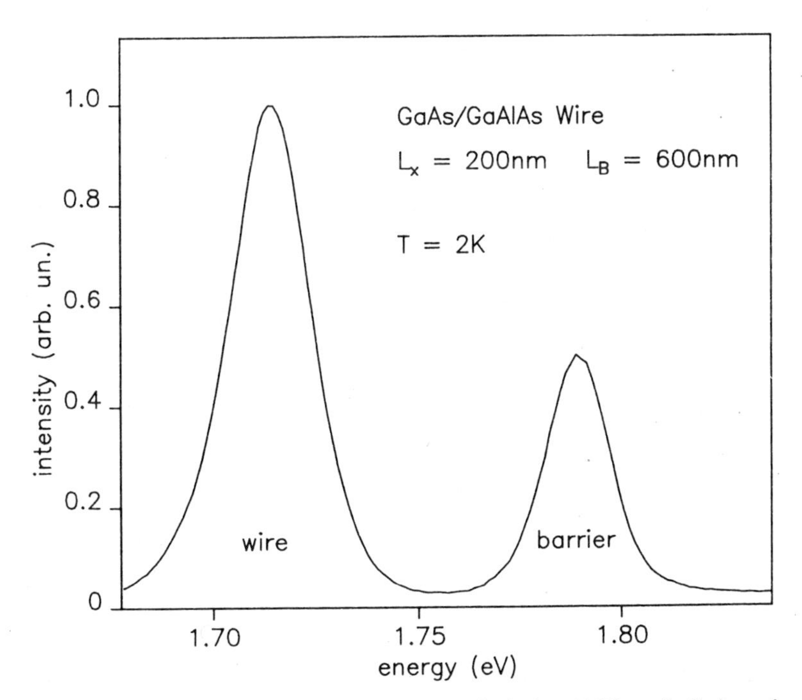

Fig. 8 Luminescence spectrum of an array of interdiffused GaAs wires and barriers. Wire width: 200 nm, barrier width: 600 nm.

We have optimized the implantation conditions with respect to a steep lateral potential. For Ga-doses on the order of $2 \times 10^{14} cm^{-2}$ we observe significantly smaller defect diffusion lengths than previously [14,18] reported. This allows us to observe significant lateral confinement effects in interdiffused wires defined with implantation masks with widths down to 40 nm.

Fig. 8 shows the optical emission spectrum of GaAs quantum wires with a width of 200 nm separated by 600 nm wide interdiffused barriers. The spectrum consists of 2 emission bands. The high energy band centered at about 1.79 eV corresponds to the lateral barrier region. The low energy emission band with a maximum at about 1.72 eV is due to recombination in the wire sections of the sample. It is noteworthy here that the emission intensity from the quantum wire is significantly larger than the intensity from the barrier region although the barriers are 3 times as wide as the wires. This indicates a transfer of electron hole pairs from the barrier into the wires. The energetic difference between wire and barrier emission corresponds to the sum of the lateral barrier height in the conduction and valence band. Under the present implantation and annealing conditions a total barrier height of 70 meV is obtained.

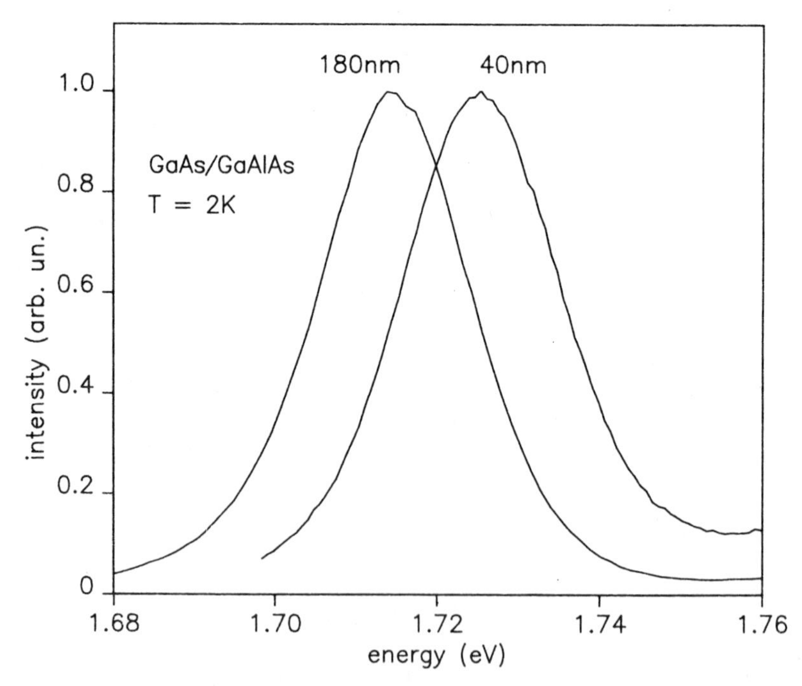

Fig. 9: Emission spectra of interdiffused GaAs wires with different widths

If we reduce the mask width down to 40 nm significant lateral quantization effects are observed. Fig. 9 displays the emission of wires defined by masks with 180 nm and 40 nm. The intensity maximum of the 40 nm wire is shifted by approximately 10 meV to higher energy compared to the case of the wider wire. The half width of the emission amounts to approximately 20 meV, independent of the wire width. This indicates the high quality of the wire masks used for the implantation.

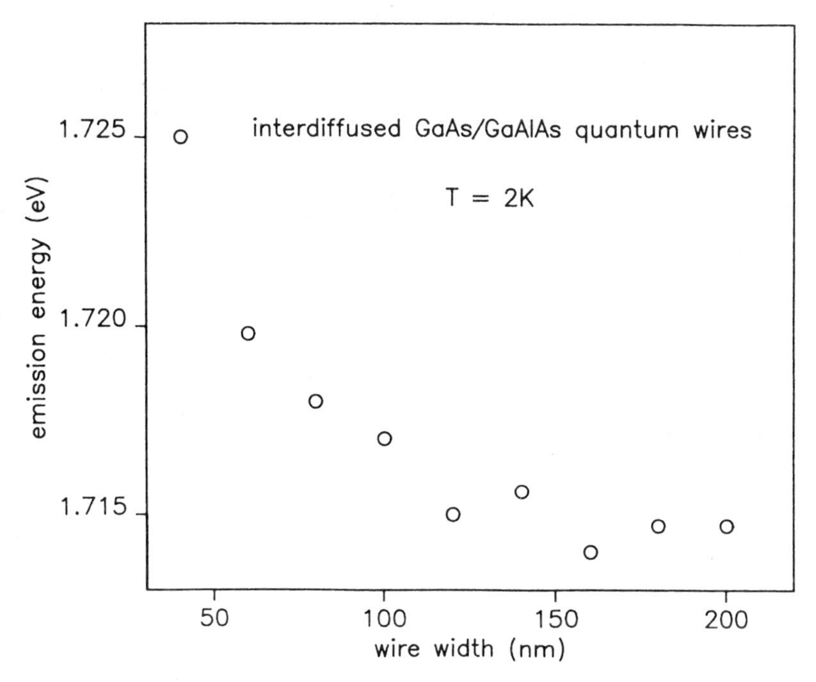

Fig. 10 Width dependence of the emission energy for interdiffused GaAs quantum wires.

We have investigated the energy shifts due to lateral quantization and interdiffusion using a set of quantum wires with widths between 200 nm and 40 nm. Fig 10 displays the wire width dependence of the energetic position of the emission peak at 2 K for the different samples. For wires with widths larger than 100 nm practically no lateral quantization effect is obtained. Below 100nm the emission starts to shift to higher energy. Particularly strong energetic shifts are observed for the two narrowest mask widths studied (60 nm and 40 nm).

The variation of the emission energy with wire width allows us to determine the key-parameter of the lateral potential shapes, i.e. the effective lateral diffusion length. Assuming an error function profile we determine an effective lateral diffusion length of 9 nm for our process conditions. This value is much smaller than previously reported.[18,19] In particular, this value is comparable to the value of the lateral straggling length which occurs in the implantation process. This implies that our high dose definition of quantum wires has largely suppressed any defect diffusion under the mask.

Using the results of our photoluminescence investigation of interdiffused wires with different widths and assuming an error function for the lateral potential variation we can calculate the one dimensional subbands in our structures. Using a mask widths of 40 nm, the combined depth of the potential wells in the conduction and valence band of 70 meV and an effective diffusion length of 9 nm the 1 D subband ladders in the conduction and valence band in the 40 nm structure can be calculated.

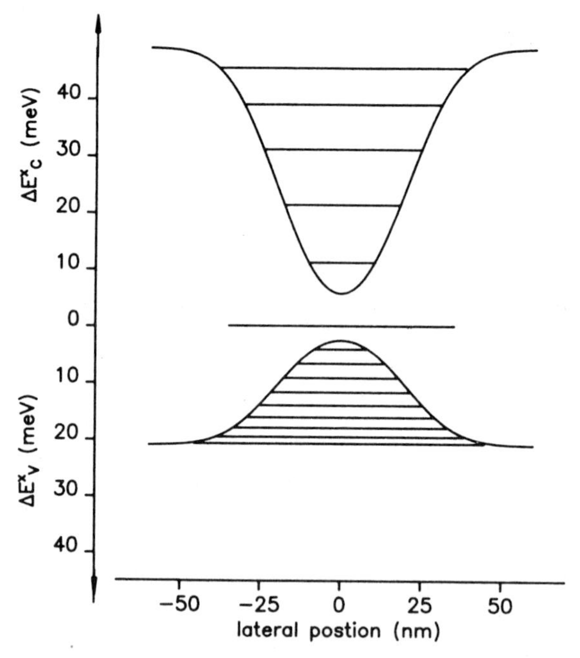

Fig. 11 Calculated 1D subband edges and lateral potential of an interdiffused GaAs wire (mask width 40 nm)

As shown in fig. 11 strong lateral quantization effects should arise in the 40 nm wire. In the approximately parabolic potential well of the conduction band a typical subband spacing of 10 meV is obtained. Using the bulk heavy hole mass the valence band splitting is much smaller (on the order of 2.5 meV). The zero of energy in fig. 10 corresponds to the energetic position of the emission in unprocessed 2 D-layers. As can be seen from the figure, the bottom of the lateral potential well is shifted to higher energy for the conduction and valence band. This is a consequence of changes induced at the mask center by the lateral straggling of the ions.

The optical transitions shown in fig. 8 and 9 correspond to transitions between the lowest 1D subbands and the conduction and valence band. As shown by the detailed evaluation in fig. 11, the shift to higher energy is due to about 50% to lateral quantization effects and 50% due to changes of the quantum well width at the wire center. These are the strongest lateral quantization effects observed in interdiffused quantum wires up to now.

SUMMARY

We have described two approaches for the fabrication of quantum wires for optical studies. By electron beam lithography and dry etching "open" GaAs, $In_{.3}Ga_{.7}As$ and $In_{.53}Ga_{.47}As$ wires have been defined. Luminescence spectroscopy on open quantum wires shows the dominant influence of nonradiative side wall recombination on the optical properties. The side wall recombination effects are strongly temperature dependent and are used here to extract values for the side wall recombination velocity and the width of an optically inactive dead layer for the different material systems investigated. We observe a strong material dependence of the side wall effects. The influence of the side walls on the optical properties decreases with increasing In-content of the active layers.

We have optimized the implantation induced interdiffusion process for wire fabrication. By this technique buried wires can be defined by a local variation of the quantum well thickness or the Al-content of the quantum well. We have determined implantation conditions leading to relatively sharp lateral potential profiles. By optical spectroscopy the basic properties of interdiffused wires with widths between 200 nm and 40 nm were investigated. The wires give rise to a strong emission. With decreasing wire width a significant shift of the emission of the wires to higher energy is observed. From the width dependence of the energetic positions and the emission of the wire and barrier layers the energetic structure of the wires has been obtained. For a 40 nm wide wire the typical conduction subband spacing amounts to 10 meV whereas the valence band spacing amounts to about 2.5 meV.

The present results on interdiffused quantum wires clearly indicate that semiconductor structures exhibiting one- and zero-dimensional features in optical experiments can be reproducibly fabricated. Due to the detrimental influence of side wall effects in etched structures, buried structures are generally preferential for the successful definition of quantum wires and dots. Another very promising approach is the direct growth of wires by molecular beam epitaxy on suitably oriented substrates.[20]

ACKNOWLEDGEMENTS

We would like to thank M.Pilkuhn for stimulating discussions. We are grateful to H.P.Meier, IBM Zürich (GaAs/GaAlAs) to J. P. Reithmaier ($In_{.3}Ga_{.7}As$/GaAlAs) and to D. Grützmacher ($In_{.53}Ga_{.45}As$/InP) for the high quality samples used in the study of the properties of etched semiconductor wires. We would like to thank G. Weimann and W. Schlapp, FTZ Darmstadt/Walter Schottky Institut, München, for the high quality GaAs/GaAlAs samples used for the interdiffusion experiments. The financial support of this work by the Stiftung Volkswagenwerk and the Deutsche Forschungsgemeinschaft is gratefully acknowledged.

REFERENCES

[1] L. Esaki and R. Tsu, IBM Research Note RC-2418 (1969; L. Esaki and R. Tsu, IBM J. Res. Develop. **14**,61 (1970)

[2] R. Dingle, Festkörperprobleme/Advances in Solid State Physics, Vol. XV, ed. by H.J. Queisser (Vieweg, Braunschweig 1975), p.21

[3] G. Weimann, W. Schlapp, in Springer Series in Solid State Physics **53** S. 88, herausgegeben von G. Bauer, F. Kuchar, H. Heinrich, Springer Verlag, Berlin, Heidelberg, New York, Tokyo, 1984

[4] H. Kawai, K. Kaneko, N. Watanabe, J. Appl. Phys. **56**, 463 (1984)

[5] Y. Arakawa and H. Sakaki, Appl. Phys. Lett. **40**, 939 (1982)

[6] R. E. Howard, L.D. Jackel and W.J.Skocpol, Microelectronic Engineering **3**,3 (1985)

[7] M. Isaacson and A. Muray, J.Vac. Sci. Technol. **19** 1117 (1981)

[8] F. Emoto, K. Gamo, S.Namba, N. Samoto and R. Shimizu, Jap.J. Appl. Phys. **24**

[9] M.Komuro, H. Hiroshima, H. Tanoue and T. Kanayama, J. Vac. Sci. Technol.**B4**, 985 (1983)

[10] A.C. Warren, I.Plotnik, E.H. Anderson, M.L. Schattenburg, D.A.Antoniadis and H.I. Smith, J. Vac.Sci. Technol. **B4**, 3655 (1986)

[11] D.A. Wharam, T.J. Thornton, R. Newbury, M. Pepper, H.Ahmed, J.E.F. Frost, D.G. Hasko, D.C. Peacock, D.A. Ritchie, G.A.C. Jones, J.Phys. C. **Vol 21**, L209 (1988)

[12] B.J. van Wees, H. van Houten, C.W.J. Beenakker, J.G. Williamson, L.P. Kouvenhoven, D. van der Marel, C.T. Foxon, Phys. Rev. Lett., **Vol 60**, No 9, 848 (1988)

[13] H. Temkin, G.J. Dolan, M.B. Panish, and S.N.G. Chu, Appl Phys. Lett **50**, 413 (1987)

[14] J. Cibert, P.M. Petroff, G.J. Dolan, S.J. Pearton, A.C. Gossard, and J. H. English, Appl. Phys. Lett. **49**, 1275 (1986)

[15] M. Tsuchiya, J.M. Gaines. R.H. Yan, R.J. Simes, P.O. Holtz, L.A. Coldren, P.M. Petroff, Phys. Rev. Lett. **62**, 466 (1989)

[16] B.E. Maile, A. Forchel, R. Germann, A. Menschig, H.P. Meier, D. Grützmacher, J. Vac. Sci. Technol. B **6**, 2308 (1988)

[17] B.E. Maile, A. Forchel, R. Germann, D. Grützmacher, Appl. Phys. Lett., 17. April 1989

[18] Y. Hirayama, S. Tarucha, Y. Suzuki, and H. Okamoto, Phys. Rev. **B37**, 2274 (1988)

RAMAN SCATTERING AND PHOTOLUMINESCENCE OF GaAs-BASED NANOSTRUCTURES

C. M. Sotomayor Torres, M Watt, H E G Arnot, R Glew*, W E Leitch, A H Kean, R Cusco Cornet, T M Kerr**+, S Thoms, S P Beaumont, N P Johnson and C R Stanley

Nanoelectronics Research Centre, Dept Electronics and Electrical Engineering, Glasgow University, Glasgow G12 8QQ, GB
* STC Technology Ltd., London Road, Harlow, CM17 9AN, GB
** GEC Hirst Research Centre, East Lane, Wembley, HA9 7PP, GB

INTRODUCTION

Recent developments in Electron Beam Lithography (EBL) and Reactive Ion Etching (RIE), among other semiconductor fabrication techniques, have enabled semiconductor material to be patterned into arrays of quantum well wires (QWW) and quantum dots (QD). The regime of 1- and 0-D quantization is achieved by reducing the dimensions of the semiconductor to lengths comparable to the de Broglie wavelength. Optical spectroscopic evidence of 1-D quantization has been reported by the group at A T & T Bell Laboratories[1], at Stuttgart University[2] and at the NTT laboratories in Japan[3]. 0-D quantization has been harder to confirm in optical spectroscopy. So far, significant work has taken place in assessing the interaction of photons with QD with particular emphasis placed on radiative recombination mechanisms[4].

In this paper we report phonon Raman Scattering of cylinders of GaAs and the first observation of surface phonons in nanolithographic structures. The geometry of the cylinders allows both TO and LO phonon scattering and for dimensions approximating an infinite cylinder, surface phonons appear in the spectrum with frequencies between those of the TO and LO modes. Complementary photoluminescence studies of free standing and overgrown quantum dots of diameter between 75 and 350nm were carried out. The main result is that the integrated emission from free-standing and overgrown QD scales with the emitting volume. Overgrowing a semiconductor layer on QD arrays made in non-optimised samples can enable the photoluminescence emission to be recovered.

EXPERIMENTAL DETAILS

GaAs cylinders were fabricated by EBL and $SiCl_4$ RIE in undoped GaAs in arrays of 200x200 μm^2. Details of the fabrication process are given elsewhere[4]. Several samples were prepared with cylinder diameters ranging from 60 to 200nm and heights from 140 to 570nm. Raman scattering spectra were recorded at room temperature in the nearly backscattering geometry on and off the arrays. Experiments varying the angle of incidence of the light were also carried out. The 488nm line of an Argon laser and a JY U1000 double spectrometer with standard photon counting electronics were used.

GaAs quantum dots were fabricated in a MOCVD layer (sample A), grown at 750°C with a single GaAs 10nm quantum well 50nm below the surface cladded by $Al_{0.3}Ga_{0.7}As$ layers; in one MBE layer (sample B) consisting of 1.4, 2.8, 5.7 and 11.5nm GaAs wells cladded by 20nm of $Al_{0.3}Ga_{0.7}As$, except for the uppermost barrier

Science and Engineering of One- and Zero-Dimensional Semiconductors
Edited by S.P. Beaumont and C.M. Sotomajor Torres
Plenum Press, New York, 1990

layer of 100nm thickness; and in a second MBE layer (sample C) of 4, 7 and 10nm GaAs wells separated by 20nm $Al_{0.3}Ga_{0.7}As$ barriers. Sample C was grown on an $n^{+}GaAs$ susbstrate.

Samples A and B have a thin GaAs capping layer. All three samples were subjected to the same fabrication procedure outlined above, except that the high resolution negative (HRN) resist masks used were removed in an O_2 plasma etch in the case of samples A and B. The QD diameters ranged from 75 to 500nm separated by approximately five times their diameter from each other in arrays of $100x100\mu m^2$ and

Figure 1 (a) SEM micrographs of 110nm QD in sample A:

$200 \times 200\mu m^2$. A mesa adjacent to each dot array provided a control sample area. SEM micrographs of QD in the three samples are shown in Figs. 1(a), (b) and (c). MOCVD overgrowth of the QD in samples A and B was carried out at 750°C with $Al_{0.4}Ga_{0.6}As$ with a thickness of 0.2 and 0.6μm for samples A and B, respectively. Both overgrown layers were capped with 40nm of GaAs. The overgrown layer thickness was estimated from growth rates. SEM micrographs of overgrown QD arrays are shown in Figs. 1(d) and (e). TEM analysis of the overgrown layers is in progress. Emission spectra were recorded at 60 and 2K for sample A and sample B, respectively. The 633nm line of a He-Ne laser focused to 50μm was used. Sample C was excited at 2K with the 488nm line of an Argon Laser. The typical spectral resolution was 1nm.

Figure 1 (b) SEM micrographs of 110nm dots in sample B:

Figure 1 (c) SEM micrographs of 100nm dots in sample C:

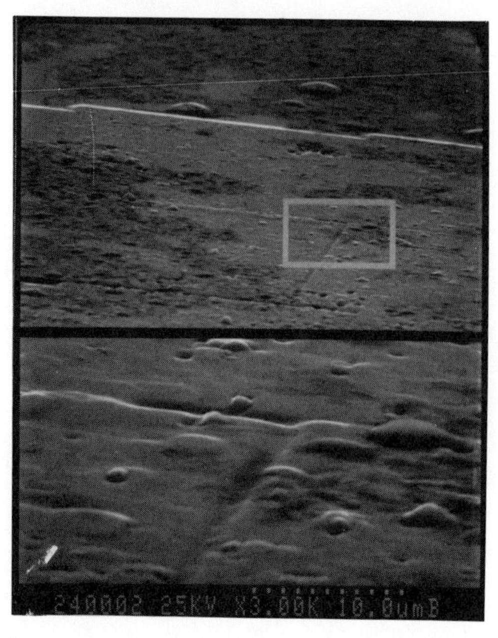

Figure 1 (d) SEM micrographs of MOCVD overgrowth on a QD array of sample A:

Figure 1 (e) SEM micrographs of MOCVD overgrowth on a QD array of sample B.

RESULTS AND DISCUSSION

Raman scattering of GaAs cylinders.

In the backscattering configuration from (100) GaAs surfaces the LO phonon is allowed and the TO is forbidden. In a previous study of RIE GaAs[6], we showed that if the RIE process is not optimised, the surface damage induced by the impinging ions extends to the first few tens of nm. The net effect was to alter the surface symmetry and consequently the electric field at the semiconductor surface thus allowing the otherwise forbidden TO phonon. In the Raman spectra of our cylinders, strong TO

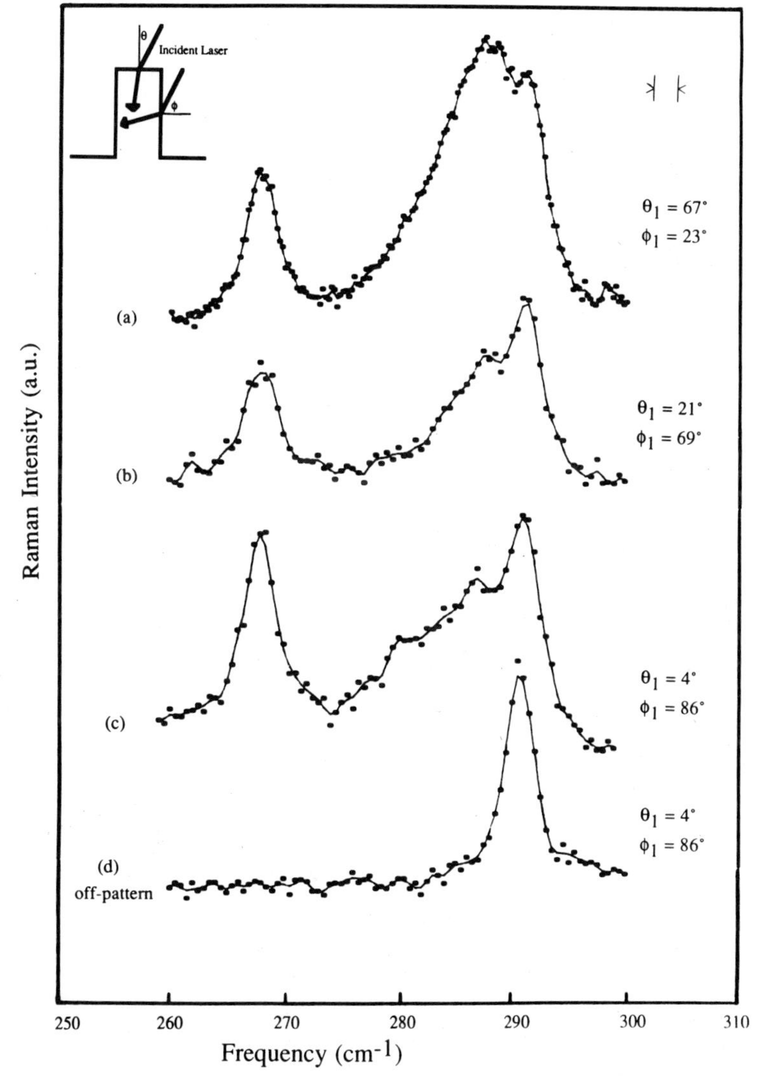

Figure 2 Raman scattering of GaAs cylinders (80nm diameter, 310nm height) from various scattering geometries.

and LO phonons are observed. The TO mode arises from the coupling to other surface planes allowed by the patterned surface. The emergence of surface phonons is unambiguous for cylinders of diameter equal or less than about 80nm, length of 200nm or longer and straight vertical sidewalls[7]. As expected, the surface mode has a frequency lying between the TO and LO phonons and its intensity is strongly dependent upon the angle of incidence of the laser beam (Fig. 2). The intensity increases as the angle between the cylinder axis and the laser beam is increased. The surface phonon

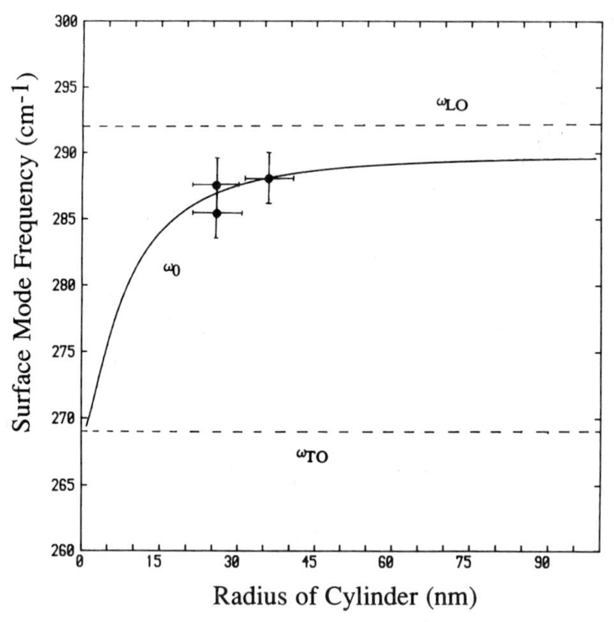

Figure 3 Calculated lowest-order suface phonon frequencies and experimental points from cylinder samples which must resemble an infinite cylinder.

frequency is also seen to depend upon cylinder diameter, decreasing for decreasing diameter. In Fig. 3, the solid lines correspond to the model developed by Englman and Ruppin[8] based on an electrostatic continuum of an infinitely long cylindrical crystal neglecting retardation effects. As a test of our assignment, the cylinders were covered with a layer of Si_3N_4, which at around $33\mu m$ has a dielectric constant of 2.2. The model of Englman and Ruppin predicts that for an increasing dielectric constant, the surface phonon frequency decreases. We observed a decrease from 288 to $285 cm^{-1}$ which agrees with the theoretical predictions. A full discussion on these observations is given in Watt et al[9].

Emission from GaAs Quantum Dots

Emission was obtained from all free-standing QD (75, 100, 110, 250, 300 and 500nm diameter) fabricated in samples B and C. In Fig. 4 emission from the top quantum well (1.4nm) of sample B for the three dot diameters is shown and compared to that of the same quantum well in the adjacent mesas. The QD emission lies within 5meV of the peak position of the same well in the mesas and is slightly broadened by

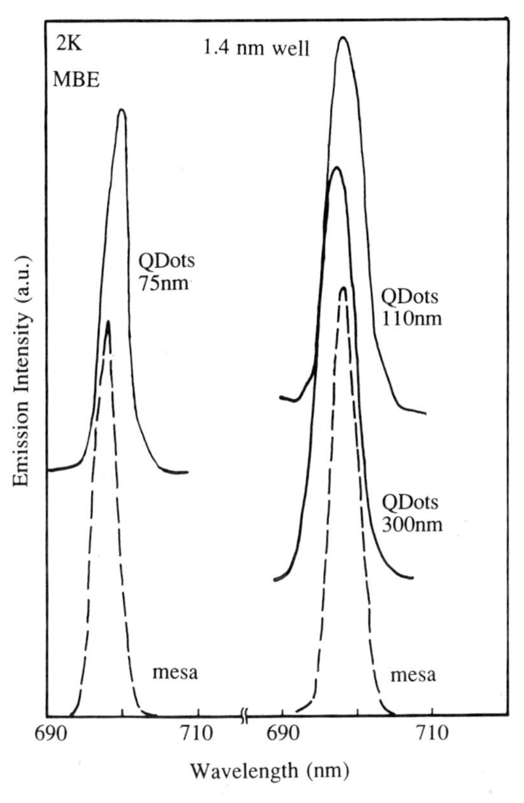

Figure 4 Emission from the 1.4nm quantum well in free standing QD of various diameters in sample B. The emission from adjacent mesas is also shown.

at most 5meV compared to that of the mesas: FWHM = 9 to 14 meV, cf FWHM = 8.6 meV in the as-grown sample for this quantum well. Emission from the 2.8nm quantum well from all these QD was also observed, although no significant emission signal from the two wells furthest away from the top surface was obtained cf before etching. Emission was also obtained from free-standing QD (100, 250 and 500 nm diameter) fabricated in sample C.

In Fig. 5 the 4nm well emission from three dot arrays and one mesa is shown. No significant shifts or broadening are observed. In all cases the QD integrated emission intensity scales with the emitting volume. In contrast, no significant emission signal was obtained from the QD fabricated in the MOCVD material (sample A) although emission from the 10nm well in the control mesas was readily observable. However, in recent experiments emission from 100nm QD on MOCVD material has been observed[10]. After overgrowth, the MOCVD QD of 35nm diameter emitted. Their

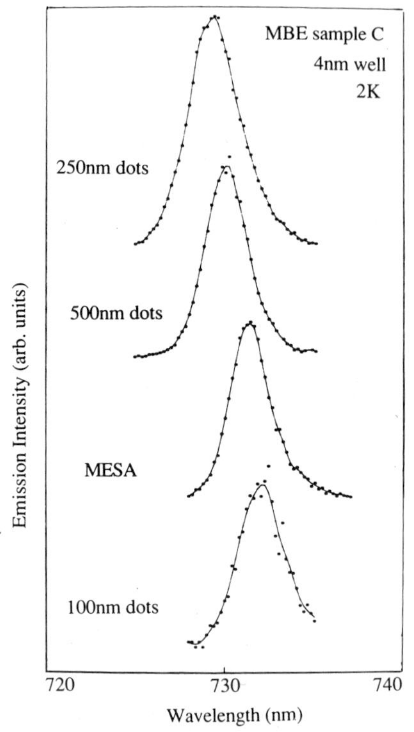

Figure 5 Emission from the 4nm well in 4 areas of sample C: 3 dot arrays and one mesa. The spectra are placed in order of their relative positions on the wafer. The shift of the peak is due to a change in the QW width across the wafer. The points are experimental data and the lines are a guide to the eye. The excitation wavelength was 488nm and the spectral resolution was 0.5nm.

spectrum is shown in Fig. 6 including those of the adjacent as-etched and overgrown mesas. There are no significant energy shifts between the peak energy of the overgrown mesas and QD. However, the emission peak from the overgrown mesa is shifted by 5meV to higher energies compared to that of the as-etched mesa emission. The emission intensity of the overgrown mesa is about seven times weaker than for the as-etched mesa. After overgrowth only the larger, 300nm, QD fabricated in sample B luminesced detectably.

In Fig. 7, the emission from the 1.4 and 2.8 quantum wells in the 300nm QD and mesas is shown before and after overgrowth. There is no significant broadening in the emission spectra shown. However, emission peaks from the overgrown QD and mesas have shifted to higher energies compared to the as–etched QD by as much as 26 and 52 meV, for the 1.4 and 2.8nm quantum wells, respectively. The integrated emission intensity of the overgrown mesas and dots scaled with the emitting volumes, and a comparison between that before and after overgrowth is within a factor of ten. Although 633nm optical excitation created electron–hole pairs not only in the quantum well layers but also in the $Al_{0.3}Ga_{0.7}As$ barrier layers, the overgrown AlGaAs layer with 40% Al was transparent to this wavelength, thus enabling comparisons of integrated intensity between mesas and QD at any stage.

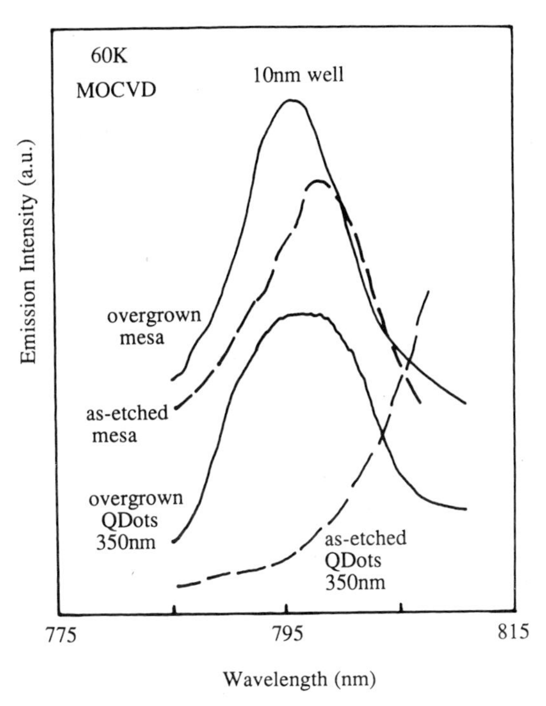

Figure 6 Emission from the 10nm quantum well in the 350nm dots in sample A, as etched and after overgrowth.

From our work in assessment of RIE[6], it is known that RIE induces various changes in the semiconductor surface including loss of As and possible passivation of shallow levels. These two effects may cancel because passivation of impurities enhances light emission whereas As vacancies encourage quasi–deep (non–radiative) level formation in GaAs. Furthermore, removal of the HRN mask in an O_2 plasma etch is likely to drive oxygen into the As vacancies in the mesas and dots, thus impairing radiative recombination. In addition, it is known that the density of surface states in GaAs is sufficiently high to inhibit radiative processes in nanostructures[11]. Rougher quantum well interfaces, such as those in (MOCVD) sample A compared to (MBE) samples B and C, are likely to be more susceptible to these chemical changes. The thermal cycling involved in overgrowth may anneal some of the surface damage, depassivate shallow impurities in an As–rich environment and, by coverage of the QD and mesa surfaces, minimise the impact of surface states (and surface recombination velocity). Considering diffusion of photoexcited carriers to be greater in sample A (emission recorded at 60K) which would allow them to reach the QD surface faster and recombine nonradiatively there, then overgrowth would passivate the surface states and possibly reconstruct the crystal surface between the QD and the overgrown layer, thus allowing radiative recombination. Furthermore, the temperature dependence of the emission intensity, of the as–grown material and overgrown quantum dots is similar in the range of 2 to 77K remaining almost constant. By contrast the emission of the as–etched dots has recently

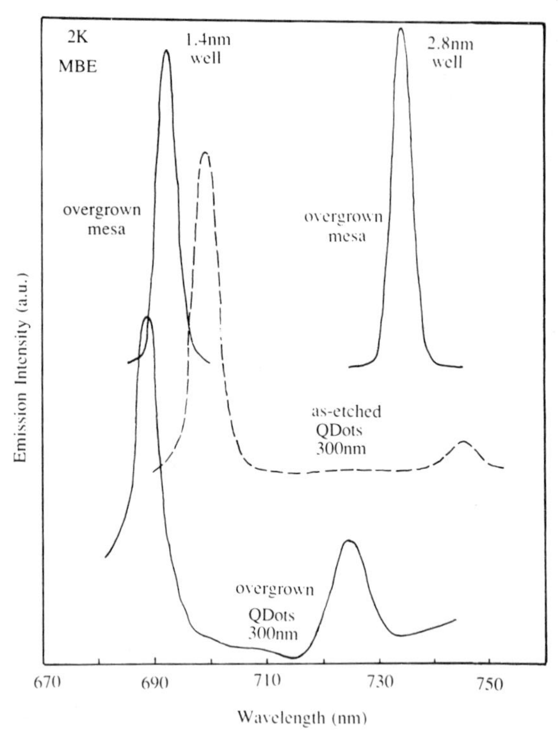

Figure 7 Emission from the 1.4 and 2.8nm quantum wells in overgrown 300nm QD and mesas in sample B. The dashed line spectrum corresponds to the as–etched QD.

been measured to be strongly temperature dependent, virtually disappearing by 30–40K[10]. This would explain the absence of free–standing QD emission in sample A at 60K. The small energy shift in the as–etched (MBE) samples (QD and mesas) suggests that this MBE layer was stable to RIE processing. However, it is not clear why no emission was observed from the lower lying quantum wells even in the larger QD in sample B. The net effect of overgrowth was the blue shift in mesas and QD. Since the overgrown layer was thicker for the (MBE) sample B, it is likely that a longer exposure to a temperature of 750°C may have enhanced lateral Al diffusion, which would explain why the QD emission shifts more than that of the mesas although this argument does not explain fully why the 2.8nm well peak shifts almost twice as much as the 1.4nm one closer to the top. If strain were to play a role, blue shifts of the reverse magnitude would be expected since the thinner well would experience the stronger strain. Further experimental tests are in progress to clarify these results.

CONCLUSIONS

Novel phonon features are observed in GaAs cylinders of nanometer dimensions which are successfully interpreted as surface phonons. Emission from free-standing and overgrown quantum dots has been observed. The emission intensity scales with the emitting volume in all cases. In general, the FWHM does not deteriorate significantly but blue shifts, which are not due to quantisation, are observed. Further work on overgrowth of MBE and MOCVD QD, study of the role of interface roughness on QD emission intensity and photoluminescence excitation to investigate strain effects are in progress.

ACKNOWLEDGEMENTS

This work is supported by the UK Science and Engineering Research Council. HEGA has an SERC CASE studentship with GEC and RCC is supported by the Spanish Government. The continuous support and encouragement of Professor J Lamb and the technical assistance of Mr D Irons are gratefully acknowledged.

+Present address: Northeast Semiconductor Inc., 134 Lexington Drive, Ithaca, NY 14850, USA

REFERENCES

1. J Cibert, P M Petroff, G J Dolan, D J Werder, S J Pearton, A C Gossard and J H English, Superlattices and Microstructures, 3 35 (1987) and H Temkin, G J Dolan, M B Panish and S N G Chu, Appl Phys Lett, 50, 413 (1987) and references therein.
2. A Forchel, H Leier, B E Maile and R German, Festkörperprobleme, 28, 99 (1988) and references therein
3. Y Hirayama and H Okamoto in: "Physics and Technology of Submicron Structures", Springer Series in Solid State Sciences, 83, p.45 Eds. H Heinrich, G Bauer and F Kuchar (Springer, Berlin, 1988).
4. see, for example, S R Andrews et al in this volume
5. S P Beaumont in: "Nanostructure Physics and Fabrication", Eds M A Reed and W P Kirk, Academic Press (in press)
6. M Watt, C M Sotomayor Torres, R Cheung, C D W Wilkinson, H E G Arnot and S P Beaumont, J Mod Opt, 35, 365 (1988)
7. M Watt, H E G Arnot, C M Sotomayor Torres and S P Beaumont in: "Nanostructure Physics and Fabrication", Eds. M A Reed and W P Kirk, Academic Press (in press)
8. R Englman and R Ruppin, J Phys C, 1, 614 (1968)
9. M Watt, C M Sotomayor Torres, H E G Arnot and S P Beaumont (to be published)
10. H E G Arnot et al (to be published)
11. see, for example, L Lassabatere in: "Semiconductor Interfaces: Formation and Properties", Springer Proc. in Physics, 22, p.239, Eds. G Lelay, J Derrien and N Boccara (Springer, Berlin, 1987).

ON THE IMPACT OF LOW-DIMENSIONALITY IN QUANTUM-WELL, WIRE, DOT-SEMICONDUCTOR LASERS

Claude Weisbuch and Julien Nagle

Laboratoire Central de Recherches, Thomson-CSF
Domaine de Corbeville, 91404 Orsay cedex, France

I INTRODUCTION

The impact of semiconductor quantum wells on lasing, electrooptic and non-linear properties is by now well established[1]. The improved material parameters originate in such various physical phenomena as reduced Density-of-States (DOS), quantum-confined wavefunctions, increased light-matter interaction through room-temperature excitons, square two-dimensional DOS, etc. It is therefore natural to expect that using lower dimensionality structures such as quantum wires or quantum dots one should obtain even better properties. We will restrict our discussion here to quantum wire and quantum dot lasers. Some recent papers have discussed the impact of low dimensional structures in the electro-optic[2,3] and non-linear fields[4,5].

Before examining the 1D and 0D cases, we will first review the 2D quantum well laser[6] , in which all relevant concepts have been evidenced. The excellent properties stem from two main positive effects and two additional cancelling properties :

(i) In quantum wells, the number of quantum states of the system is reduced due to the wavefunction quantization perpendicular to the layer plane. This reduced number of quantum states directly implies that the number of states to be inverted to reach net gain, i.e. the transparency population, will be reduced as compared to 3D Double-Heterostructure (DH) laser (by an amount which is the number of populated quantum states of the DH in the direction perpendicular to the layer).

(ii) The square DOS in 2D leads to a more efficient use of injected carriers to create gain than in 3D : the maximum of the gain curve always lies at the bottom of the quantized 2D band, instead of shifting towards higher energies due to the ever-increasing 3D-DOS. Therefore, a larger fraction of the injected carriers participate in the gain (in 3D those below the maximum of the gain curve are unuseful and the tail of the Fermi-Dirac distribution function above the maximum of the gain populates an increasing DOS instead of a constant DOS). 2D gain builds up more rapidly with increased carrier density, which translates into a non-linear gain curve with a larger starting differential gain. The price to pay here is that the gain will saturate at some value (typically 10.000 cm^{-1} for a 50 Å GaAs quantum well) when the square DOS' s of electrons and holes are completely filled (see figure 1).

Science and Engineering of One- and Zero-Dimensional Semiconductors
Edited by S.P. Beaumont and C.M. Sotomajor Torres
Plenum Press, New York, 1990

(iii) As the active volume is smaller in a quantum well laser, the volume gain g_{vol} obtained for a given injected carrier number will be correspondingly larger than in a 3D laser, varying approximately as d^{-1}, where d is the active layer thickness.

(iv) The confinement factor (Γ), i.e. the fraction of the optical wave contained in the active volume, diminishes with the layer thickness. Correspondingly, to obtain the same amplification of the optical wave, i.e. the same modal gain Γg_{vol}, one will require a larger volume gain g_{vol} in a quantum well laser to compensate the decreased Γ. In single quantum well lasers, Γ decreases as d^2 which cannot be compensated by the improved volume gain of QWL's. One therefore either uses

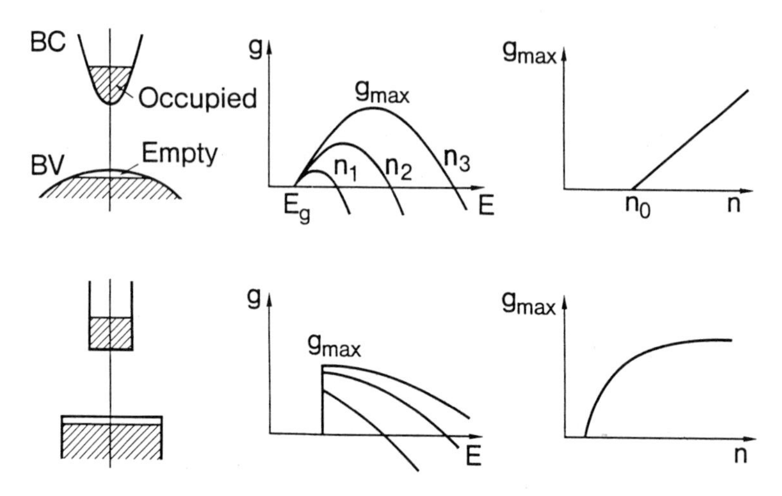

Fig.1. Schematics of the gain formation in 3D and 2D lasers

multiple quantum wells (however at the expense of increased number of states to be inverted to reach transparency) or separate-confinement heterostructures in which the confinement of the guided optical wave by a separate heterostructure cavity allows a smoother variation of Γ as d. In that case, one can expect that the increase in volume gain will exactly compensate the decrease in confinement factor for quantum well lasers, and the advantages of the Separate-Confinement Heterostructure (SCH) laser therefore only originate from the first two points. A typical value for Γ is 4.10^{-4} .d in optimized GaAs/GaAlAs SCH structures, where d is expressed in Å.

The above short discussion gives a feeling of the important factors in low-dimensionality lasers. We discuss them now in more details.

II MODELLING THE QUANTUM WELL LASER

The gain of a quantum-well laser can be written (when lasing occurs on the first allowed transition between subbands) as

$$g_{vol} = K\ M^2\ \rho_{joint}\,(f_c - f_v) \qquad (1)$$

Where K is a coefficient involving fundamental and material constants, M^2 is the interband matrix element, ρ_{joint} is the joint DOS of the conduction and valence bands corresponding to the laser transition, f_c and f_v are the Fermi-Dirac occupancy factors for electrons in the conduction and valence bands respectively. From the square DOS, it is obvious that when $f_c = 1$, $f_v = 0$ (complete inversion), g_{vol} saturates.

The usual way to calculate g as a function of injected carriers density is to inject a value for electronic density n_e, then calculate the electron quasi-Fermi level entering f_c, then deduce the hole quasi-Fermi level assuming electrical neutrality $n_e = n_h$, then calculate g. The complicacy comes from the fact that

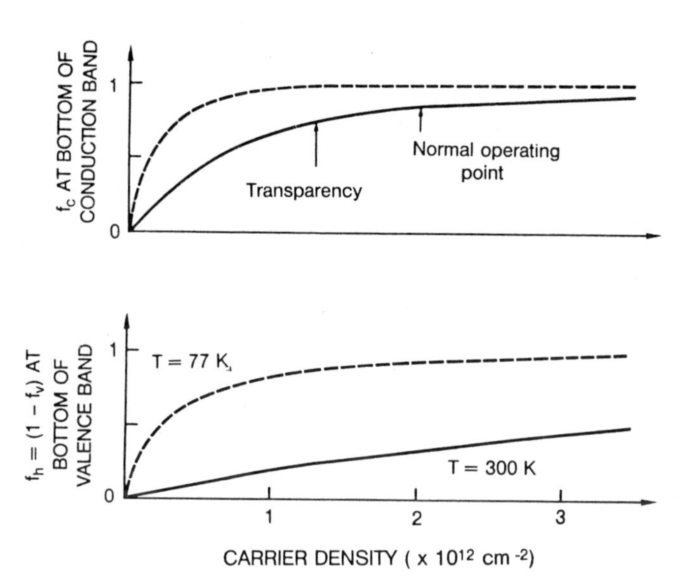

Fig.2. Inversion factor for electrons and for holes as a function of carrier density in a 60 Å GRINSCH GaAs/AlGaAs QW laser

excited bands must be considered, both for conduction electrons and holes. The latter case is of paramount importance, as several hole bands are populated under usual conditions as energy separations between confined hole levels are small due to the heavy hole mass. Therefore, for electron densities yielding quasi-fully inverted electron bands, hole levels are only slightly populated (see figure 2). (To simplify the problem of the valence bands mixing, we only consider parabolic hole masses with k = 0 transverse masses).

This is the more directly evidenced when temperature is reduced : in that case, thermal population of excited hole states becomes negligible, and similar inversion levels are reached for electrons and holes, yielding a gain value ~4-5 times larger for an electron occupancy factor in the range 0.6 - 0.9.

Knowing these gain curves, it is straightforward to extract the threshold current. The carrier density is related to injected current through n = I/Red where R is the carrier recombination rate, which includes both radiative recombination rate and non-radiative recombination rates such as surface recombination, Auger effect etc...

The threshold condition can be written

$$\Gamma g_{th} = \Gamma \alpha_a + (1 - \Gamma)\alpha_c + \frac{1}{2L} \mathrm{Log} \frac{1}{R_1 R_2} \qquad (2)$$

where α_a and α_c are the optical loss coefficients in the active and confining layers, and the last term represents the loss through the output mirrors of the laser cavity of length L with reflexion coefficients R_1 and R_2. The total losses are typically 40 cm^{-1} for a 300 μm long laser, which decompose in 10 cm^{-1} for the two first terms, and 30 cm^{-1} assuming 30 % reflecting cleaved facets for the last term.

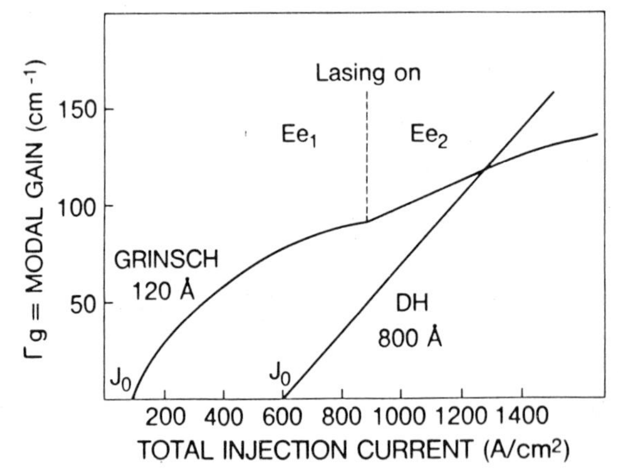

Fig.3. Volume gain (maximum TE gain) as a function of current density for a 120 Å GRINSCH QW laser and a 800 Å DH laser

If linearization is an acceptable procedure (very good in 3D, not so good in 2D), (1) can be transformed into a more practical shape

$$g_{vol} \sim B(n - n_o) \quad (3a) \qquad\qquad g_{vol} = A\,(J - Jo) \quad (3b)$$

where A and B are differential gains, and n_o and J_o are the transparency electron density and current density respectively. As mentioned above, the main difference between the threshold currents in 3 and 2D will stem from J_o.

This property is evidenced in figure 3. One sees that the main effect in QW lasers is the reduction in transparency current J_o (ref.7). Depending on losses, threshold condition will be fulfilled either by lasing on the n = 1 transition or on the n = 2 quantum well transition[8]. Also, the differential gain will be very depending on the threshold condition, which is of great importance for high-speed, and narrow-linewidth lasers.

Many additional features of QW lasers have been explained, such as the superiority of Graded-Index SCH structures over straight SCH's[8], anomalous temperature dependence when operating on the n = 2 transition etc[6]... Very low threshold stripe lasers (I_{th} < 1mA) have been obtained, the record being 0.55 mA, by reducing the optical loss term by antireflective coating deposition on the mirror facets[7]. It is clear that in that case the improvement in laser threshold will be much greater for QW lasers than in DH lasers, because of the dominant effect of J_o in the latter case in determining the threshold current.

III EXTENSION TO 1D AND 0D LASERS[9-12]

From the preceding discussion, one expects improvements and deteriorations when further decreasing the material dimensionality

(i) diminishing the volume of the active medium will diminish the transparency current

(ii) the density of states tends to more and more concentrate carriers in quantum states which are useful for the lasing transitions. In the limit of quantum dots, every injected electron is active at the lasing wavelength

(iii) the confinement factor diminishes as barrier material is required to separate the quantum-confinement regions which are the only ones to interact with the optical wave

(iv) care must be exerted to check that the saturated gain is large enough to overcome losses to reach threshold. This clearly implies that numerous low-D systems are used in parallel to obtain satisfactory performance.

A major unknown is the level-broadening parameter γ : in the case it were small, very large monochromatic gain coefficients could be obtained, as instead of being distributed over ~ kT like in 3 and 2D, carriers would be concentrated in much narrower energy ranges. One often assumes $\gamma = h / \tau = 6$ mev ($\tau \sim 10^{-13}$s), from a typical 3D scattering time. It is however not sure that scattering events would be conserved at such high rates in low dimensions. In any case, dimensional fluctuations of the numerous required low-D systems will create an inhomogeneous broadening of the gain curve. One should remark that a 6 mev inhomogeneous broadening of 100 Å quantum dots will require ~± 3 Å precision on the quantum dot dimension.

The situation is then represented schematically on figure 4. Focussing on the quantum well and quantum dot cases, one easily deduces that injecting equal numbers of electrons in a quantum well layer or in quantum boxes, one obtains a gain about 10 times larger for quantum boxes. (A typical quasi Fermi-level of 40 mev for laser operation with uncoated facet mirrors has been chosen). This originates in the fact that the optical matrix element is independent of the dimensionality of the system for a k-allowed transition and therefore the integrated oscillator strength over allowed k transitions at a given energy only reflects the joint density-of-states. The advantage of lower dimensionality systems is to concentrate carriers into k-states which are more and more monoenergetic as the dimensionality decreases.

This improvement is compounded with a much better hole inversion at room temperature (near 80 %), provided the hole levels are better separated in the quantum box case than in the QW case, bringing in an additional improvement factor of ~4.

How many quantum boxes are required ? The quasi-Fermi level in the QW case corresponds to ~2.10^{12}cm^{-2} occupied states. One needs 40 times less quantum boxes in the system to reach the same volume gain, but as the confinement factor is decreased (typically by a factor of 4 if one uses box-center separations of twice the box lateral size), one will need 2.10^{11} electrons in boxes per cm^2 to reach threshold. The calculation by Takahashi and Arakawa (fig.5) shows a saturated modal gain value of ~100 cm^{-1} (complete inversion) for an equivalent quantum box density of ~5.10^{11}cm^{-2} (i.e. 10^{12} electrons per cm^2 when including spin).

It should be remarked that this huge improvement predicted in threshold current (~ 20 μA in fig.5) is however not due to low-D effects. As the total number of quantum states is in a 300μm-long Q-box laser is 2 x 9 x 300μm x 10^6cm^{-1} = 5.4 10^5, as compared to 5μ x 300μ x 10^{12}cm^{-2} = 1.5 10^7, (for a 5μm-wide stripe QW laser), one indeed expects of factor of ~30 improvement in threshold current,

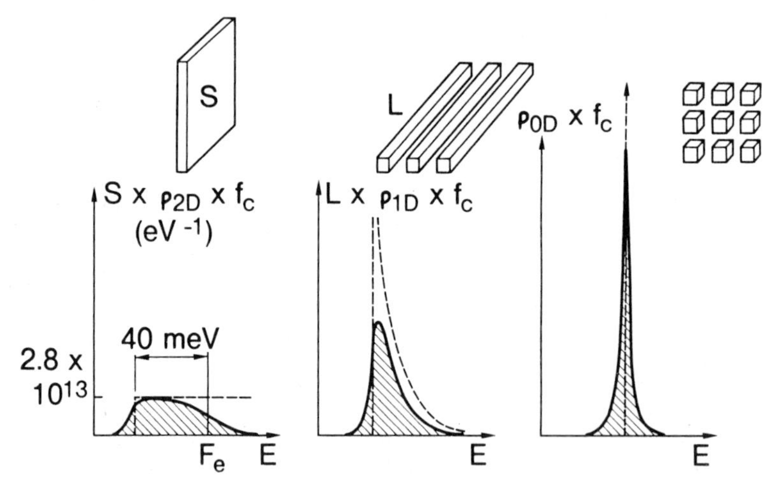

Fig.4. Broadened electronic density of states times electron occupancy factor for a quasi-Fermi level 40 meV above the first quantized level. Broadening determines the gain maximum in 1D and 0D case (schematics)

assuming equal recombination time. (which is certainly not true, confinement leading to shortened lifetimes). It should however be emphasized that the improvement comes from the narrowed active region. If one were able to fabricate a 2000 Å QW narrow-stripe laser, one would theoretically require a threshold current of ~ 10 μA assuming a threshold current density of 150 A cm^{-2}.

IV WHY OD OR 1D LASERS ?

In regards of preceding short discussion, one may wonder why one still wants to pursue the difficult path to quantum wire and dot laser fabrication. It seems that the optimization of increased volume gain and decreased confinement factor, has been reached with the quantum well laser. There are however some additional features to be discussed :

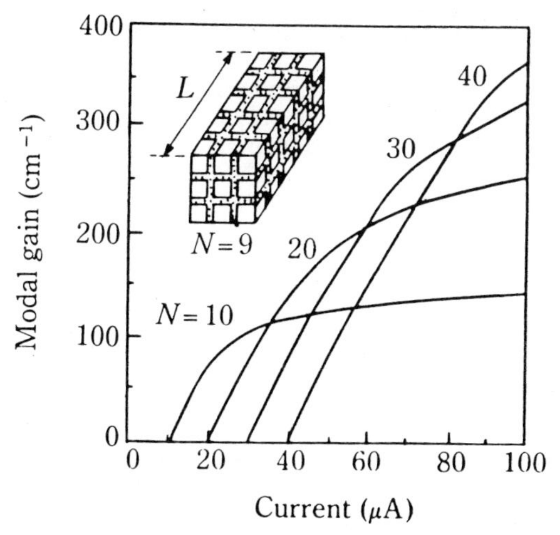

Fig.5. Modal gain as function of injected current for various number of 50 Å quantum boxes (from ref.14)

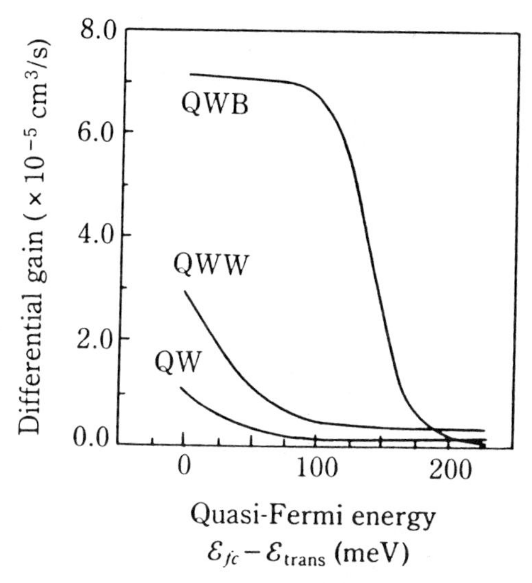

Fig.6. Differential gain for QW, QW wire and QW box lasers as function of the Fermi-energy level of the conduction band (from ref.14)

- the number of quantum boxes required is large because a large modal gain is required due to the large laser losses, mainly due to uncoated facets. This can be reduced by a factor ~5, using reflecting facets, leading to limit threshold currents below 10 μA. One should however recall that such lasers would most certainly be limited to extremely low powers, as lasers are usually limited by thermal considerations to operating currents not exceeding 10 times the threshold current.
- other features of quantum wire and quantum dot lasers are highly desirable, such as high-differential gain in quantum wire and box lasers is much higher than in QW lasers[13,14] (figure 6). This is a main driving force for making lower dimensionality lasers

- so far, we discussed a modelization which is the extrapolation of our current knowledge. New phenomena could bring positive or negative effects : electron-hole correlation could enhance the optical matrix element. Auger non-radiative recombination could be badly enhanced, as evidenced in microcrystallite semiconductor-doped glasses[15]. The quantum semiconductor field has been in the past so full of surprises that one should be prepared to the unexpected and not only rely on seemingly-wise extrapolation.

REFERENCES

1. See e.g. the contributions in "Semiconductors and Semimetals" vol.24, "Applications ofMultiquantum Wells, Selective Doping and Superlattices", volume editor R. Dingle, Academic, New-York, 1987
2. I. Suemune and L.A. Coldren, Band-Mixing Effects and Excitonic Optical Properties in GaAs Quantum Wire Structures-Comparison with the Quantum Wells, IEEE J.Quantum Electronics QE-24 : 1778 (1988)
3. Y. Chiba and S. Ohnishi, Quantum-Confined Stark Effects on a GaAs Cluster Embedded in $Al_xGa_{1-x}As$, Phys.Rev. B38 : 12988 (1988)
4. E. Hanamura, Very Large Optical Nonlinearity of Semiconductor Microcrystallites, Phys.Rev. B37 : 1273 (1988)
5. T. Takagahara, Excitonic Optical Nonlinearity and Exciton Dynamics in Semiconductor Quantum Dots, Phys.Rev. B36 : 9293 (1987)
6. J. Nagle and C. Weisbuch, The Physics of Low-Dimensional Effects in Quantum-Well Lasers, in "Quantum Wells and Superlattices Physics II", SPIE Proceedings 943, F.Capasso, G. Döhler and J.N. Schulman eds, p.76 (1988)
7. P. Derry, A. Yariv, K.Y. Lau, N. Bar Chaim, K. Lee and J. Rosenberg, Ultra Low-Threshold Graded-Index Separate Confinement Single-Quantum Well Buried Heterostructure (Al,Ga)As Lasers with High-Reflectivity Coatings, Appl.Phys. Lett. 50 : 1773 (1987)
8. J. Nagle, S.D. Hersee, M. Krakowski, T. Weil and C. Weisbuch, Threshold Current of Single Quantum Well Lasers : The role of the confining Layers, Appl.Phys. Lett. 49 : 1325 (1986)
9. Y. Arakawa and H. Sakaki, Multidimensional Quantum Well laser and Temperature Dependance of its Threshold Current, Appl.Phys.Lett. 24 : 195 (1982)
10. M. Asada, Y. Miyamoto and Y. Suematsu, Gain and the Threshold of Three-Dimensional Quantum Box Lasers, IEEE J.Quantum Electronics QE-22 : 1915 (1986)
11. Y. Miyamoto, M.Cao, Y. Shingai, K. Furuya, Y. Suematsu, K.G. Ravikumar and S.Arai, Light Emission from Quantum Box Structure by Current Injection, Japan.J.Appl. Phys. 26 : L225 (1987)
12. K.J. Vahala, Quantum Box Fabrication Tolerance and Size Limits in Semiconductors and Their Effect on Optical Gain, IEEE J.Quantum Electronics QE-24 :523 (1988)
13. Y. Arakawa, K. Vahala and A. Yariv, Dynamic and Spectral Properties of Semiconductor Lasers with Quantum Well and Quantum Wire Effects, Surf.Sci. 174 : 155 (1986)
14. T. Takahashi and Y. Arakawa, Theoretical Analysis of Gain and Dynamic Properties of Quantum Well Box Lasers, Optoelectronics 3 : 155 (1988)
15. P. Roussignol, M. Kull, D. Ricard, F. de Rougemont, R. Frey and C.Flytzannis, Time-Resolved Direct Observation of Auger Recombination in Semiconductor-Doped Glasses, Appl.Phys.Lett. 51 : 1882 (1987)

EXCITONS IN LOW DIMENSIONAL SEMICONDUCTORS

L. Viña, E. E. Mendez*, W. I. Wang+, J. C. Maan**, M. Potemski** and G. E. W. Bauer++

Instituto de Ciencia de Materiales–C.S.I.C., and Departamento de Fisica Aplicada. Universidad Autonoma, E-28049 Madrid, Spain
*IBM T J Watson Research Center, P O Box 218, Yorktown Heights, NY 10598, USA
+Electrical Engineering Department, Columbia University, New York, NY 10027, USA
**Max-Planck-Institut, Hochfeld-Magnetlabor, BP 166X, F-38042 Grenoble, France
++Philips Research Laboratories, P O B 80000, 5600 JA Eindhoven, The Netherlands

We present high resolution pseudo-absorption spectra of GaAs/GaAlAs quantum wells. Information on the energy spectrum of excitons is obtained from low temperature photoluminescence excitation spectroscopy. The application of an external electric field tunes the energy of the excitons, and fine structure is observed as a result of the interaction of high-angular momentum states and the ground state of the first light-hole exciton. Their oscillator strength is enhanced by means of an external magnetic field. The observed structures are assigned to excited states of excitons by comparison with calculations of excitonic mixing in the presence of external electric and magnetic fields.

INTRODUCTION

The study of the absorption spectra of excitons in the presence of electric and magnetic fields is closely related to other important problems in spectroscopy such as the structure of atoms in very strong magnetic fields on the surface of neutron stars, the splitting and broadening of atomic spectral lines by electric and magnetic fields in a plasma, the Stark and Zeeman effects in the hydrogen atom,[1] etc. In spite of the numerous studies dedicated to excitons in semiconductors, fine structure in bulk materials has been resolved and identified only up to the first excited state (2s).[2,3] Two dimensional semiconductors provide an excellent scenery to study the energy spectrum of excitons: their binding energy is enhanced with respect to bulk crystals, they exist even at room temperature,[4] and the electric and magnetic fields necessary to attain similar conditions to those typical of atomic physics can be easily obtained in the laboratory. The high quality of $GaAs/Ga_{1-x}Al_xAs$ quantum wells (QWs), grown by epitaxial techniques, has allowed the routine observation of the first excited state of the heavy-hole exciton [$h_1(2s)$] as a well defined peak.[5-7]

The application of external electric and magnetic fields to semiconductor QWs and superlattices has provided important information about excitonic properties and band structure parameters.[8-11] The tuning of the states by the external fields yields to level crossing, a method widely used in atomic spectroscopy to reveal fine and hyperfine structure in the spectra. We review in this paper the effects of combined electric and magnetic fields on the first heavy-hole and light-hole excitons in $GaAs/Ga_{1-x}Al_xAs$ quantum wells.

Science and Engineering of One- and Zero-Dimensional Semiconductors
Edited by S.P. Beaumont and C.M. Sotomajor Torres
Plenum Press, New York, 1990

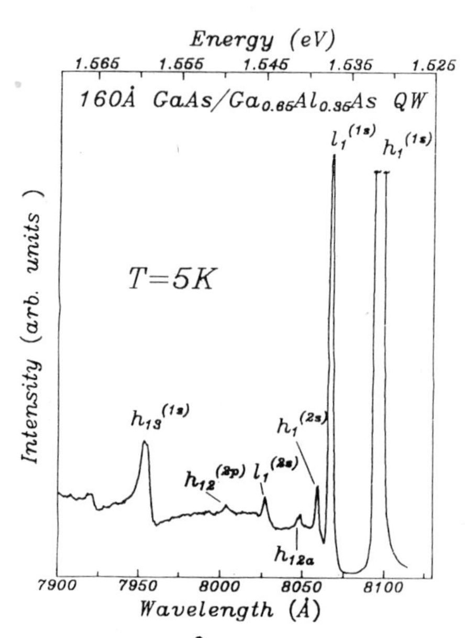

Fig.1 PLE spectra at 5K of a 160Å GaAs/GaAlAs quantum well. A small electric field of the order of 2kV/cm is present in the sample.

RESULTS AND DISCUSSION

The sample used in our studies was a high-quality p-i-n $GaAs/Ga_{0.65}Al_{0.35}As$ heterostructure, with the intrinsic region consisting of five isolated wells (160Å) with barriers of 250Å. Photoluminescence excitation (PLE) spectra were obtained at low temperature exciting the sample with the light from an LD700 dye, pumped by a Kr^+-ion laser. Electric and magnetic fields were applied in the direction perpendicular to the layers. To record the spectra, the spectrometer was set at the wavelength of the ground state of the heavy-hole exciton, which changes with the magnitude of the applied fields.

The richness of the absorption of our sample can be seen in Fig. 1, which shows a PLE spectrum at an electric field of ~2kV/cm. The main structures correspond to the ground state of the first heavy-hole and light-hole excitons, $h_1(1s)$ and $l_1(1s)$, respectively. The slightly forbidden exciton, $h_{13}(1s)$, corresponding to the first electron and the third heavy-hole subbands is also clearly seen in the figure. The forbidden transition h_{12a} appears because of the presence of the small electric field. This structure disappears when the voltage across the sample matches exactly the flat-band condition.[12] Besides the ground-state excitons, the high-quality of our sample allows the observation of excited states of excitons as well defined peaks. The 2s states of the heavy-hole and light-hole exciton, as well as the 2p state of the h_{12} exciton are clearly resolved in the spectrum. The two-dimensional character of the QW is reflected in the step-like shape of the joint density of states.

A striking feature of excitons in two-dimensional systems is their existence at conditions where they are absent in bulk materials: excitons in $GaAs/Ga_{1-x}Al_xAs$ QWs can be observed, even at room temperature, in the presence of high density, light-induced plasma, which does not seem to destroy the excitonic character of interband transitions.[14] In spite of the much smaller binding energy of the excited states compared with that of the ground states they also exhibit unusual properties. Figure 2 shows PLE spectra at different temperatures in the spectra region of $l_1(1s)$ and $h_1(2s)$. With increasing temperature, a red shift and a small broadening of both structures can be observed. The binding energy of $h_1(2s)$ amounts only to ~0.4meV and one would expect that the exciton becomes thermally ionized at temperatures higher than 5K, however, this exciton exists at temperatures as high as ~30K.

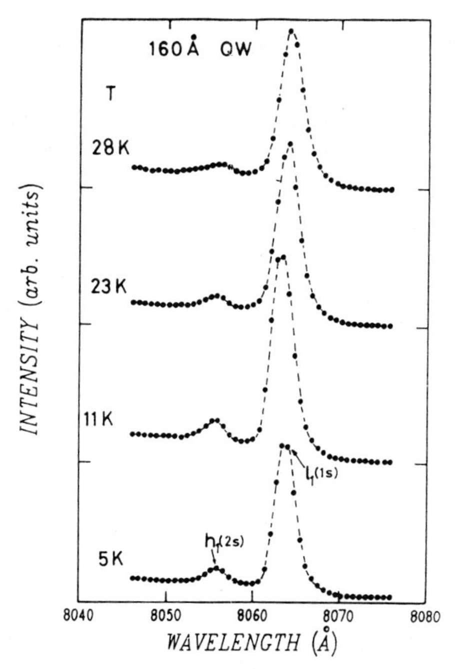

Fig. 2 PLE spectra in the region of the ground state of the light–hole exciton and the first excited state of the heavy–hole exciton at different temperatures.

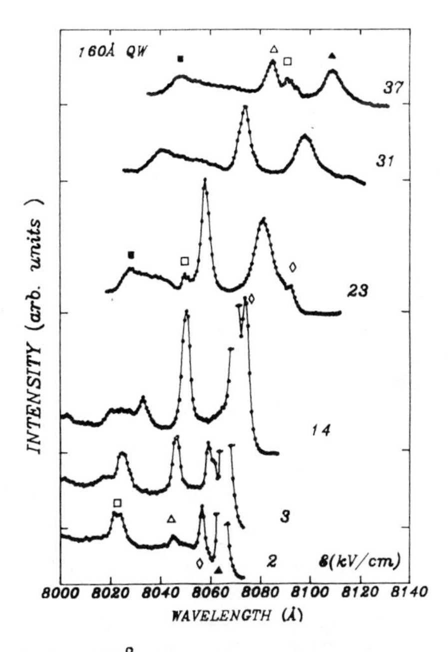

Fig. 3. PLE spectra of the 160Å GaAs/GaAlAs QW at different electric fields. The ground state of the heavy–hole exciton is not shown.

An electric-field induced coupling between the ground state of l_1 and the excited states of h_1 has been reported recently.[6] As a result of the interaction the $h_1(2p)$ state becomes observable. We shall concentrate here on the effects of an electric field on forbidden transitions and the excited states of the light-hole exciton. Excitation spectra are depicted in Fig. 3 for different electric fields. With increasing field the peaks shift to lower energies and forbidden transitions become allowed. Envelope-function calculations predict that only the ground state of h_{12} should appear in the energy range of the spectra.[15] However, two transitions, $h_{12a}(\Delta)$ and $h_{12b}(\blacksquare)$ are observed. Photocurrent measurements with light-polarized parallel and perpendicular to the plane of the layers and with uniaxial stress parallel to the [110] direction have shown a strong heavy-light mixed character of these excitons.[15] At high electric fields h_{12b} corresponds to the ground-state of the h_{12} exciton, and h_{12a} is believed to arise from an interaction between the excited states of l_1 and the ground state of h_{12}. Recently, h_{12a}, has been attributed to the 2p state of the light-hole exciton.[16]

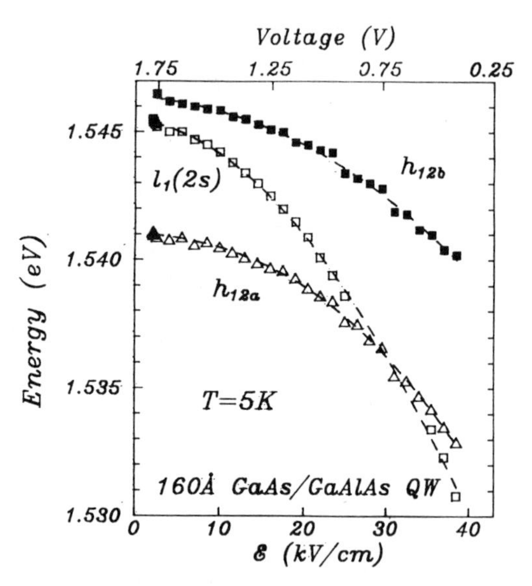

Fig. 4. Stark shift of forbidden excitons and the 2s state of the light-hole exciton. The dashed lines are a guide to the eye.

Figure 4 shows the energy of h_{12a}, h_{12b} and $l_1(2s)$ as a function of electric field. The excited state of l_1 shifts much more rapidly than h_{12a} and therefore they approach as the electric field is increased. These states cross at a field of ~30kV/cm without any appreciable interaction. $l_1(2s)$ is hidden below h_{12a} (see spectrum at 31kV/cm in Fig. 3) and appears again, at higher fields, at its low-energy side. However, the tunning of the excited states of the heavy-hole exciton induces a coupling with the ground state of l_1. This interaction can be already observed in the spectrum at 14kV/cm in Fig. 3, where $h_1(2x)$, with x standing for excited, has an oscillator strength comparable to that of $l_1(1s)$.

The application of an additional small magnetic field confines the excitons in the plane of the wells, enhancing their oscillator strength, and removes the Kramers degeneracy. Therefore, with the use of circularly polarized light, excitons associated to holes with spin-up and spin-down components can be selected, and fine structure in the excited states of h_1 is observed. The excitonic states are labeled according to the irreducible representations of the direct product of hole Bloch and exciton envelope functions[17], with an additional superscript indicating their ordering in energy. The main character of the excitonic states is compiled in Table I for the low and high electric field limits.

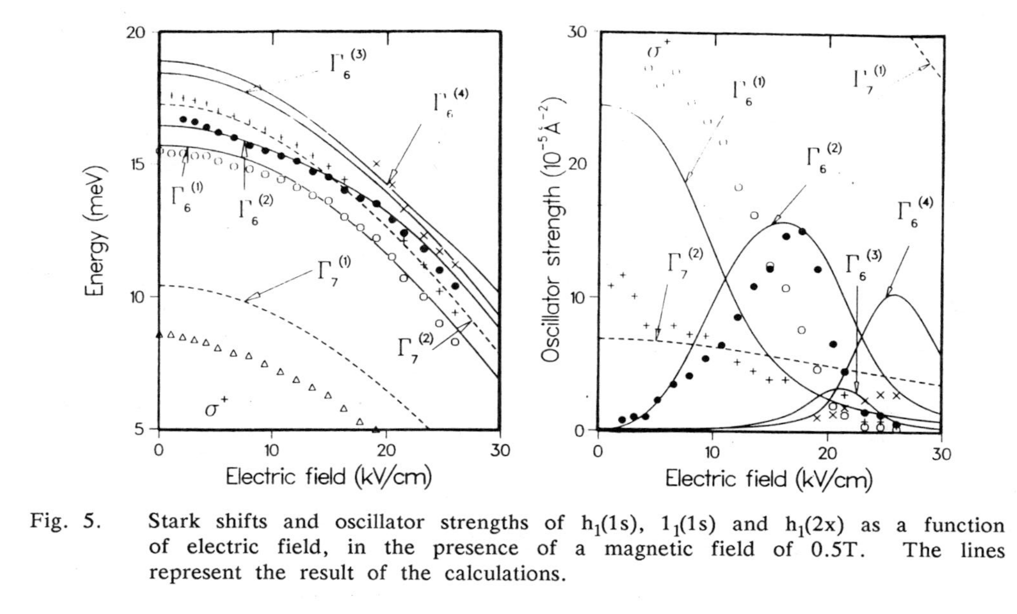

Fig. 5. Stark shifts and oscillator strengths of $h_1(1s)$, $1_1(1s)$ and $h_1(2x)$ as a function of electric field, in the presence of a magnetic field of 0.5T. The lines represent the result of the calculations.

Table 1. Main character of the excitonic states in the low and high electric field limits for σ^+ polarization. The symbols in parentheses correspond to those used in Fig. 5.

σ^+

	low field		high field	
$\Gamma_7(1)$	$h_1(1s)$	(Δ)	$h_1(1s)$	(Δ)
$\Gamma_6(1)$	$l_1(1s)$	(o)	$h_1(2p-)$	(o)
$\Gamma_7(2)$	$h_1(2s)$	(+)	$h_1(2s)$	(+)
$\Gamma_6(2)$	$h_1(2p-)$	(•)	$h_1(3p-)$	(•)
$\Gamma_6(3)$	$h_1(3p-)$	(x)	$h_1(3d+)$	(•)
$\Gamma_6(4)$	$h_1(3d+)$	(x)	$l_1(1s)$	(x)

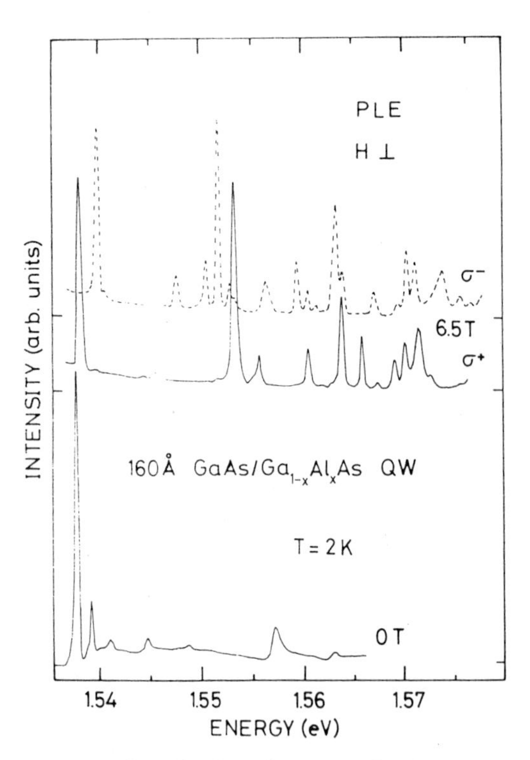

Fig. 6. PLE spectra at low light electric field at 0T and 6.5T for σ^+ (solid line) and σ^- (dashed line) polarized light.

As the electric field is increased, the separation between $h_1(x)$ and l_1 decreases, thereby enhancing the interaction from l_1 and become resolved in the spectra. The interaction of $l_1(1s)$ takes place first with $h_1(2p-)$ and subsequently with $h_1(3p-)$ or $h_1(3d+)$, as these states approach the ground state of l_1. The ambiguity in the latter assignment arises from the difficulty to resolve these two closely lying states. The theory predicts that the coupling of the d-states should be stronger, but the results are very sensitive to the choice of band structure parameters in the high-field limit, the excited states lose again their oscillator stregnth and are below the ground state of 1_1, reversing the original ordering. It is worthwhile to mention the excellent agreement between theory and experiment concerning both, energy positions and oscillator strengths. The larger Zeeman splitting for the light s-states compared to that of the heavy states can only be explained by a theory which includes exciton mixing.[8,18]

Finally, let us show how the complete energy spectrum of the excitons is revealed by means of an external magnetic field. Figure 6 shows PLE spectra at at 0T and 6.5T for σ^+ and $\sigma-$ polarized light. The main peaks in the spectra at finite field correspond to the *ns*-series of the heavy-hole and light-hole excitons. However *p* and *d* states are also resolved, specially when they interact with *s* excitonic states and share with them their oscillator strength. The Zeeman splitting of the structures is clearly seen in the figure, and the importance of recording the spectra with circularly polarized light is also evident. Due to the extremely complicated fan-diagrams and changes in oscillator strength of the excitons, as the magnetic field is changed, the states can be only identified by comparison with a theory which takes into account simultaneously the effect of the Coulomb interaction and that of the magnetic field, while considering the real band-structure of the quantum well.[17] the magnetic field imposes an additional constrain on the motion of the carriers across the field and, for large fields, there is a strong reduction in the transverse motion and consequent transformation of the dimensionality of the excitons in the well. This effect can be observed in the density of states, that changes from a step-like shape (0T) to an almost vanishing density in the regions between the excitonic transitions (6.5T).

ACKNOWLEDGEMENTS

We want to thank L.L.Chang, L.Esaki and M.F.H Schuurmans for helpful discussions. This work was sponsored in part by CICYT Grant No. MAT-88-0116-C02-02.

References

1. For an excellent review see: V S Lisitsa, Sov.Phy. Usp. 30, 927 (1987) [Usp.Fiz. Nauk 153, 379 (1987)]
2. D D Sell, Phys, Rev. B 6, 3750 (1972)
3. S Zemon, C Jagannath, S K Shastry, and G Lambert Solid State Commun. 65, 553 (1988)
4. X L Zehng, D Heiman, B Lax, F A Stair, Appl. Phys. Lett. 52, 984 (1988).
5. K J Moore, P Dawson, and C T Foxon. Phys Rev. B 34, 6022 (1986).
6. L Viña, R T Collins, E E Mendez, and W I Wang, Phys. Rev. Lett. 58, 832 (1987).
7. L W Molenkamp, G E W Bauer, R Eppenga, and C T Foxon, Phys, Rev. B 38, 6147.
8. L Viña, G E W Bauer, M Potemski, J C Maan, E E Mendez, and W I Wang, Phys. Rev B 38, 10154 (1988).
9. E E Mendez, F Agullo-Rueda and J M Hong, Phys. Rev. Lett. 23, 2426 (1988).
10. J C Maan, G Belle, A Fasolino, M Altarelli, and K Ploog, Phys. Rev. Lett B 30, 2253 (1984).
11. D C Rogers, J Singleton, R J Nicholas, C T Foxon, and K Woodbridge, Phys. Rev. B 34, 4002 (1986).
12. L Viña, Surf. Science 196, 569 (1988).
13. D A B Miller, D S Chemla, T C Damen, A C Gossard, W Wiegmann, T H Wood, and C A Burrus, Phys. Rev B 32, 1043 (1985)
14. J C Maan, private communication.
15. R T Collins, L Viña, W I Wang, L L Chang, L Esaki, K v Klitzing and K Ploog. Phys. Rev, B36, 1531(1987).
16. L C Andreani and A Pasquarello, Europhys. Lett. 6, 259 (1988).
17. G E W Bauer and T Ando, Phys. Rev B 38, 6015 (1988).
18. G E W Bauer and T Ando, Phys. Rev B 37, 3130 (1988).

RESONANT TRANSVERSE MAGNETOTUNNELING THROUGH DOUBLE BARRIER SYSTEMS

G. Platero, C. Tejedor*, L. Brey* and P. Schulz*

Inst. Ciencia de Materiales de Madrid, C.S.I.C. and
Dpto. de Física de la Materia Condensada, Univ. Autónoma
Facultad de Ciencias. Universidad Autónoma de Madrid
Cantoblanco, 28049 Madrid, Spain

Tunneling through semiconductor barriers has been the matter of many scientific works in the last years, not only because of the potential applications of these systems as electronic devices but also for their interesting physical properties.

The effect of a magnetic field on the electronic structure and therefore, on the transport properties of this kind of systems is very different depending on the field orientation with respect to the sample growth direction.

We have studied the tunneling current through a double barrier (DB) when a magnetic field is applied parallel to the interfaces (B⊥J). In this case, the carrier's motion in the growth direction is affected by the field because of the magnetic potential added to the barrier potential in this direction. The magnetic levels are not degenerated as in the bulk and the spectrum consists in dispersive bands instead of discret levels[1]. Before to start with the magnetotransport through a DB, we are going to describe briefly the main results for the case of a single barrier, obtained by means of the transfer hamiltonian formalism[2].

The Schrödinger equation for a barrier potential $V_b(z)$ separating two semifinite crystals in the presence of a transverse magnetic field, is given in the effective mass approximation by:

$$\left[-\frac{\hbar^2}{2m^*m_o}\frac{d^2}{dz^2}+\frac{e^2B^2}{2m^*m_o}(z+z_o)^2+V_b(z)-E_{n,ky}\right]\phi(z)=0 \qquad (1)$$

Where ky is related to the electronic orbit center through:

$$z_o=\frac{\hbar\, ky}{eB}=lm^2ky \qquad (2)$$

and lm is the magnetic length.

The electronic levels are then:

$$E=\frac{\hbar^2\, kx^2}{2m_o m^*}+E_{n,ky} \qquad (3)$$

Science and Engineering of One- and Zero-Dimensional Semiconductors
Edited by S.P. Beaumont and C.M. Sotomajor Torres
Plenum Press, New York, 1990

In order to calculate the tunneling current through a simple barier we follow ref[2] and define two aditional hamiltonians H_L and H_R such that:

$$H \equiv H_L + V_L \doteq H_R + V_R \qquad \text{(see Figure 1)}$$

The Figure(2) shows the electronic levels $E_{n,ky}$ for a single barier. The states at anticrossings between different branches are the only ones with a wave function shared between the right and left crystals, therefore they form the available channels for the magnetotunneling because they allow that electron wave-packets initially to the left side pass through the barrier to the right side. These anticrossing correspond to crossings between the left and right hamiltonians levels[2].

The transition probability for this case is given by:

$$P_{LR} = \frac{2\pi}{\hbar} \delta(E_L - E_R) \left|\langle L|V_L|R\rangle\right|^2 \tag{4}$$

The current density comes out summing up all the transition probabilities between occupied states to the left and empty states to the right[2].

The figure (3-a) shows the current density as a function of the magnetic field for an AlGaAs 100 Å barrier between two semiinfinite GaAs media doped with $n=10^{18}$ cm^{-3} and an external bias of 0.2 V. The current density shows oscillations which are due to two different phenomena: The bulk Fermi level oscillations with the magnetic field and the reduction of available tunnelling channels comming from the crossing between an occupied ϕ_L (left hamiltonian wave function at a crossing) and an unoccupied ϕ_R as the field increases.

For small fields there are many of these channels and only the bulk oscillations of E_F are reflected in the current density. On the contrary, for high fields very few channels exist and the disappearence of one of them as the field increases is strongly reflected in the current density. In figure (3-b) the I-V characteristic curve is represented for this single barrier sample. The small oscillations overimposed on the expected increase of j with the bias come also from the variation in the number of channels available for tunneling[2].

In a previous paper[3] we have extended the transfer Hamiltonian formalism to study resonant tunneling (The Generalized Transfer Hamiltonian Method, GTH), in which virtual transitions are involved and we applied it to study DB systems with no magnetic fields. Our purpose in this paper in to analyze, by means of this formalism, both contributions: coherent and sequential, to the magnetotunneling. The coherent processes take place for electrons tunneling between an initial state in the first electrode and a final one in the right electrode, through virtual transitions to the well levels, with energy an parallel momentum conservation. The sequential tunneling is due to the electronic current produced by electrons from the left to the well states and from them to the right[4-6]. As we will see below, the available pair of channels for each sequential process are not at the same energy and parallel momentum, therefore memory loss phenomena are involved.

We can calculate the coherent transmission probability by means of (3):

$$P_{LR}^{coh} = \frac{2\pi}{\hbar} \delta(E_L - E_R) . T_{LR} \tag{5}$$

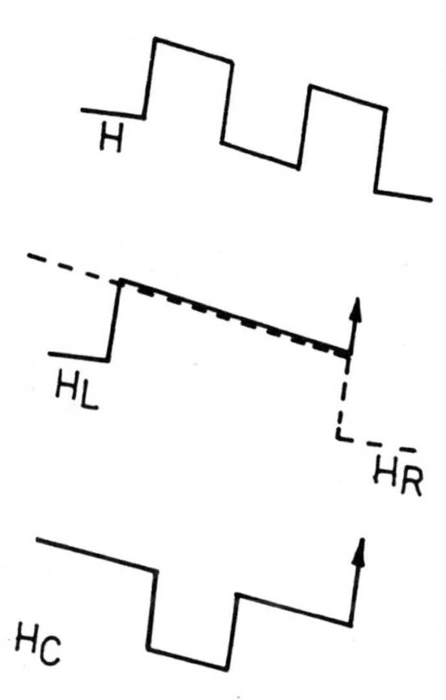

Fig. 1. Sketch of the different Hamiltonians used to study tunneling with the Transfer Hamiltonian method.

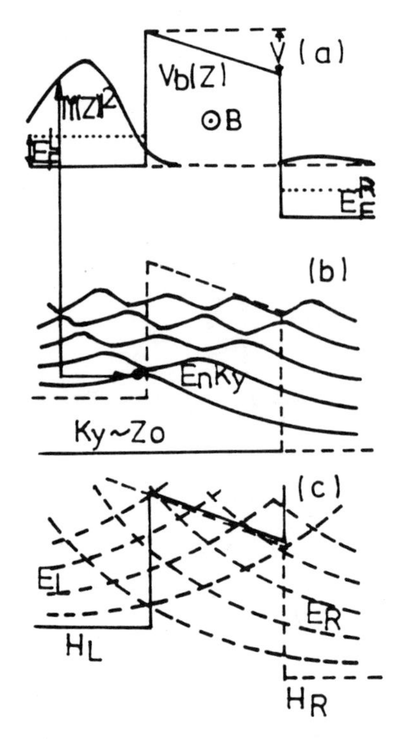

Fig. 2. a) Potential profile $V_b(7)$ of a barrier between two crystals with Fermi levels E_F^L and E_F^R and an applied bias. The probability density $|\phi(z)|^2$ of a magnetic level in the presence of a transverse field is also shown. (b) Magnetic levels $E_{n,ky}$ as a function of ky (or the orbit center Z_o) The dot labels the level for which $|\phi(z)|^2$ is given in (a). c) Model left and right hamiltonians used to compute the tunneling current densities in the transfer Hamiltonian method and their corresponding magnetic levels in the presence of a transverse magnetic field.

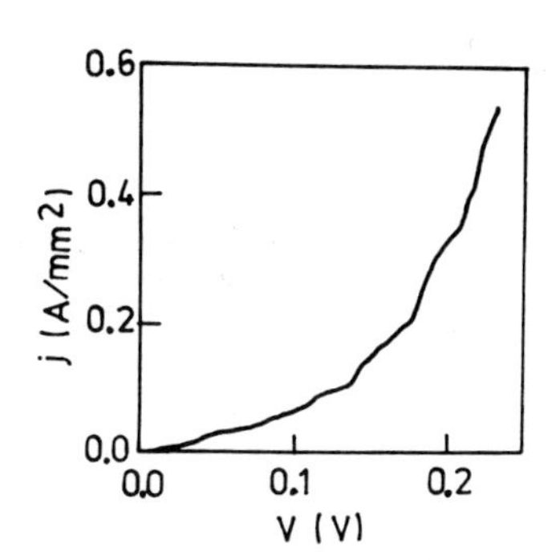

Fig. 3a. Current density j(A/mm^2) and bulk GaAs Fermi level (meV) as a function of an in-plane B (Tesla) for a GaAs-GaAlAs-GaAs barrir with lb=100 Å, V_b = 0.3 eV; V=0.2 V and n=10^{18}cm^{-3}.

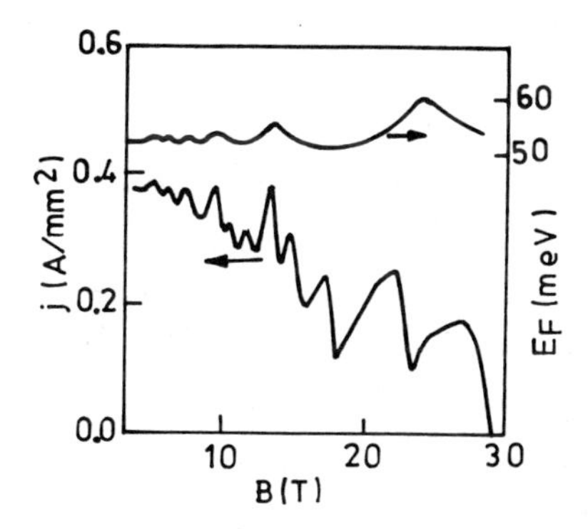

Fig. 3b. Current density j(A/mm^2) as a function of the bias voltage for the same sample as in figure 3-a and a fixed field of 10T (E_F=53.4 meV).

where

$T_{LR} = |t_{LR}|^2$

and

$$t_{LR} = \langle L|V_L + V_L\ (E_L - H + P_R V_R)^{-1} P V_R|R\rangle \tag{6}$$

P_R being the projection operator on the state $|R\rangle$ and $P=1-P_R$.

The term $\langle L|V\cdot|R\rangle$ in the matrix elements of (6) in the one corresponding to first order perturbation theory and is the term which controlls the tunneling through simple barrier systems. For resonant problems, however, higher order perturbation terms are also important and take into account the virtual transitions through the quantum well states. We have checked that a good approximation is to substitute $(E_L - H + P_R V_R)^{-1}$ by the Green's function G_C for a center hamiltonian H_C (Figure 1). This approximation, which simplifies considerably the numerical procedure is also justified by the fact that resonant states are mainly localized in the well.

In figure (4) we plot the characteristic I-V curve due to coherent tunneling for an AlGaAs double barrier and two different magnetic fields: B=6 and 10 Tesla. The sample consists in two barriers of 100 Å separated by a GaAs quantum well of 70 Å and a carrier density of $n=10^{18}$ cm^{-3}. We do not consider the band bending produced by the bias but the electric potential dropping enterely across the DB structure.

We obtain well resolved peaks for each magnetic field coresponding to the available tunneling channels (left and right levels crossings) with energy between the left and right Fermi levels. The main peaks are determined by the crossings of E_L and E_R which are close in energy to a resonant state (Fig. 5).

The available experiment information we know[7-10] gives no trace of this structure for the I-V curve but shows only one peak much broader for each magnetic field. The explanation for this discrepancy is that the experimentally observed features[7-10] correspond to sequential tunneling processes which are, for these cases, orders of magnitud larger than the coherent ones. The tunneling channels for sequential tunneling come from the left center (lc) and center-right (cr) crossings (figure 5). The electrons to the left side tunnel to the resonant well levels through the lc channel and to the right side through the cr one. These channels are not at the same energy and ky, therefore losss memory processes are included.

In figure (5) the sequential channels lc and cr are represented. As the energy range in which the tunneling processes giving contribution to the current is determined by the left Fermi level, only the pairs of channels lc and cr with energy very close, contribute to the sequential process . We have made the approximation shown in figure (5) that these channels are at the same energy and ky, the last one being the corresponding to the lr crossing, and the contribution of each sequential channel is weighted with an exponential function of the difference of ky momentun of the two lc and cr channels. By means of this function, the channels in which a small change in the quantum numbers takes place are selected giving the main contribution to the sequential transmission. We have considered the barriers as two resistors in serie, then the expression for the transmission coefficient is:

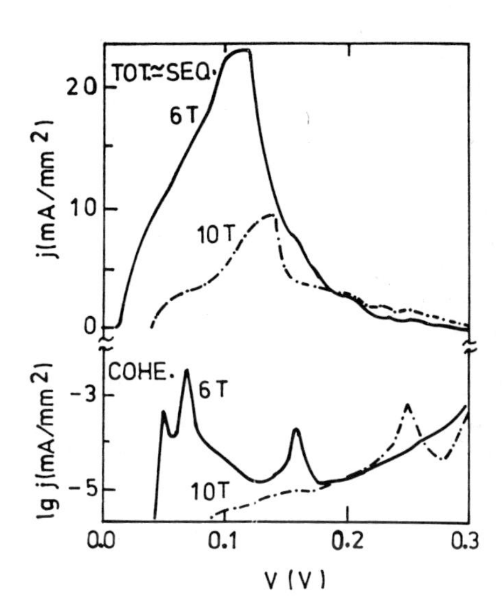

Fig. 4. a) Total current density as a function of the bias for a double barrier of AlGaAs between GaAs ($l_{barriers}$ = 100 Å, l_{well}=70 Å) with $n=10^{18}$ cm^{-3} for B = 6 and 10T. b) Logarithm of the coherent current density as a function of the bias for the same system as a).

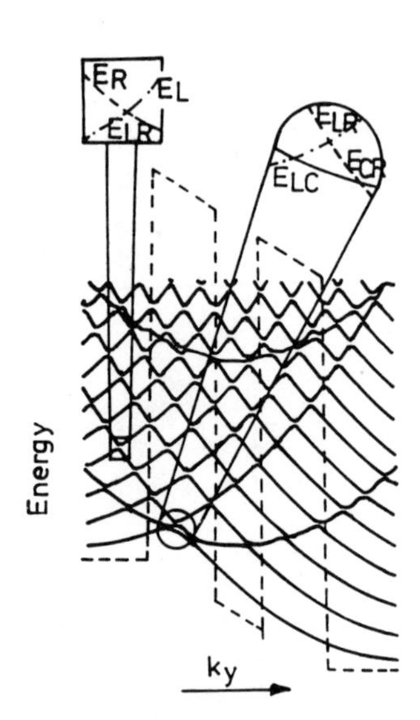

Fig. 5. Dispersion relation ot the total hamiltonian describing a DB with a transverse magnetic field. The square inset shows the lr tunneling channels and the circle inset the lc and cr ones.

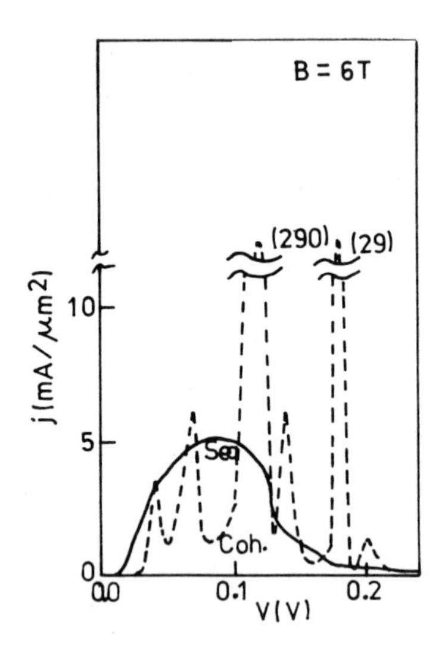

Fig. 6. Sequential (continuous line) and coherent (dashed line) current density as a function of the bias for a double barrier of AlGaAs between GaAs ($l_{barrier}$=20 Å l_{well}=70Å) with n=10^{18} cm^{-3} for B=6T. Numbers in parenthesis give the intensity of the peaks out of scale.

$$(T^{i}_{seq})^{-1} = ((T^{i}_{lc})^{-1} + (T^{i}_{cr})^{-1})\ e^{(lm\Delta ky)^2} \quad (7)$$

where the superindex i refers to the i channel, lm is the magnetic length and $\Delta ky = ky^{cr} - ky^{lc}$.

This approximation has been taken from the use of path integral methods to compute the impurity induced transitions (proportional to the exponential of the classical action) between edge states at the two surfaces of a narrow channel with a magnetic field [11,12].

The total transmission is evaluated adding both the coherent and the sequential contributions:

$$T_{tot} = \sum_{i} T^{i}_{coh} + T^{i}_{seq}$$

The main contribution to the current through samples with wide barriers, as in the case of figure 4, is the sequential one, in agreement with experimental results. As the magnetic field increases, the localization of the states increases too, the current diminishes and the peak moves to higher bias.

In figure (6) we represent both coherent and sequential currents for a sample with very thin GaAlAs barriers (20 Å) separated by a 70 Å GaAs well. In this case the coherent tunneling is more important than the sequential one and a structure of peaks appears in the I-V characteristic curve.

In conclusion, we have analyzed theoretically resonant transverse magnetotunneling through a DB within the framework of the Generalized Transfer Hamiltonian formalism (3). Both sequential and coherent contributions to the current are calculated and we discuss in which configurations sequential or coherent processes determine the current transmission.

ACKNOWLEDGMENTS

We are indebted to Dr. J.C. Maan for providing us with a program to solve differential equations by means of a finite-element method. We acknowledge Dr. C. Rössel, E. Méndez and S. Ben Amor for providing us experimental information before its publication. This work has been supported in part by the Comisión Interministerial de Ciencia y Tecnología of Spain under MAT88-0116-C02-01.

REFERENCES

1. J.C. Maan, Festkörper-probleme, 27 137 (1987).
2. L.Brey, G. Platero and C. Tejedor, Phys. Rev. B, 38, 9649 (1988).
3. L. Brey, G. Platero and C. Tejedor, Phys. Rev. B, 38, 10507 (1988).
4. M. Buttiker, IBM J. Res. Dev. 32, 63 (1988).
5. M. Buttiker, Phys. Rev. B, 38, 9375 (1988).
6. M.C. Payne, J. Phys. C, 19, 1145 (1986).
7. M.L. Leadbeater, L. Eaves, P.E. Simmonds, G.A. Toombs, F.W. Sheard, P.A. Claxton, G. Hill and M.A. Date, Solid State Electronics, Vol. 31 707 (1988).
8. S. Ben Amor, K.P. Martin, J.J.L. Rascol, R.J. Higgins, A. Torabi, H.H. Harris and C.J. Summers, Appl. Phys. Lett., 53, 2540 (1988).
9. P. Gueret, C. Rossel, E. Marcla and H. Meier, J. Appl. Phys. (to be published).
10. E.E. Méndez, L. Esaki and W.I. Wang, Phys. Rev. B, 33, 2893 (1986).
11. H.A. Fertig and B.I. Halperin, Phys. Rev. B., 36, 7969 (1987).
12. J.K. Kain and S.A. Kivelson, Phys. Rev. Lett., 60, 1542 (1988).

AUTHOR INDEX

SUBJECT INDEX